ALSO BY MARCIA BARTUSIAK

Thursday's Universe
Through a Universe Darkly
Einstein's Unfinished Symphony
Archives of the Universe

The Day We Found
the Universe

The Day We Found the Universe

MARCIA BARTUSIAK

Pantheon Books • New York

Library of Congress Cataloging-in-Publication Data

Bartusiak, Marcia, [1950–]
The day we found the universe / Marcia Bartusiak.
p. cm.
Includes bibliographical references and index.
ISBN 978-0-375-42429-8
1. Astronomy—History.   I. Title.
QB15.B37 2009
520.9—dc22   2008034377

www.pantheonbooks.com

Printed in the United States of America
First Edition

2 4 6 8 9 7 5 3 1

To Steve

The center of my universe, who shared
every light-year along the way

Contents

Contents

Discovery

Preface

January 1, 1925

The twenties were not just roaring, they were blazing.
Moviegoers were flocking to the cinema to watch in amazement
as Moses parted the Red Sea in Cecil B. DeMille's silent film epic
The Ten Commandments, Greece overthrew its monarchy and proclaimed itself a republic, the first dinosaur eggs were discovered in
Mongolia's Gobi Desert, and crossword puzzles became all the rage. It
was the height of the Jazz Age, when Victorian ideals came tumbling
down in a frenzy of flappers, Freudian analysis, and abstract art. While
majestic ocean liners crossed the Atlantic in under five days, Clarence
Birdseye introduced the public to the novelty of frozen food and a failed
artist named Adolf Hitler published *Mein Kampf*. It was a world, wrote
F. Scott Fitzgerald in his classic novel *The Great Gatsby*, "redolent of
orchids and pleasant, cheerful snobbery and orchestras which set the
rhythm of the year, summing up the sadness and suggestiveness of life
in new tunes."

It was also an era of immense scientific fervor. On December 30,
1924, some four thousand scientists descended upon Washington,
D.C., to attend the annual conference of the American Association for
the Advancement of Science. Taking advantage of the three-day gathering, the American Astronomical Society held its meeting in the capital
at the same time, with nearly eighty astronomers attending from across
the United States. They lodged at the Powhatan, a plush eight-story
hotel located on the corner of Eighteenth Street and Pennsylvania
Avenue, where a room with private bath cost $2.50 a night and weary
travelers could relax in its rooftop garden. Two blocks away Calvin
Coolidge opened the doors of the White House to the visiting AAAS

members. While notorious for being a man of few words, the thirtieth president of the United States was uncharacteristically chatty the day of the reception. "It has taken endless ages to create in men the courage that will accept the truth simply because it is the truth," Coolidge told his guests. "We have advanced so far that we do not fear the results of that process. We ask no recantations from honesty and candor.... Those of us who represent social organization and political institutions look upon you with a feeling that includes much of awe and something of fear as we ask ourselves to what revolution you will next require us to adapt our scheme of human relations." Six months later high school biology teacher John Scopes would go on trial in Tennessee for illegally teaching Charles Darwin's theory of evolution.

The astronomers, though, were scarcely aware that Washington was host to the largest number of scientists ever assembled for an AAAS meeting. Their interest was intently focused on the astronomy program, which included talks on the atmosphere of Mars, how fast celestial objects could move, the temperature of Mercury, and the latest computed orbit of the eclipsing double-star system Algol.

On Wednesday, the second day of the meeting, the astronomers were taken by glass-topped buses to the U.S. Naval Observatory, in the northwest sector of the town, for a tour of the facility and a buffet luncheon in its stately main hall. Later that evening, New Year's Eve, "occurred an event which was marked on the program and celebrated by a number of the faithful," *Popular Astronomy* recounted. As the clock struck twelve, astronomers happily changed to civil reckoning for determining the start of a day. No longer would the astronomical day begin at high noon, a tradition launched in the days of Ptolemy that often led to great bookkeeping confusion. Instead, it now began at midnight, just as it did for everyone else. "It will probably be remembered and noted long after other astronomical happenings of the current year are forgotten," stated the magazine.

But a presentation made on Thursday, New Year's Day, ultimately overshadowed all other events at the meeting. Looking out their hotel windows that inaugural morning of 1925, convention-goers discovered a blanket of snow covering the city, enough to give holiday sleds a good tryout, reported the *Washington Post*. Despite the ongoing snowstorm, however, the astronomers kept to their schedule and walked the short distance to the newly constructed Corcoran Hall, on the nearby campus of George Washington University, for a joint session with the mathematicians and physicists of the AAAS. They first heard a talk on stellar evolution, followed by a lecture posing the question "Is the Universe Infinite?" which led to a lively discussion among the conferees. Then right before the noon break, a paper modestly titled "Cepheids in Spiral Nebulae" was presented to the assembled audience. Those not familiar with astronomy likely imagined it was a minor technical work, of interest only to a specialist. But the astronomers in the room immediately grasped its significance. For them, it was electrifying news. Despite its lackluster title, this paper was no less than the culmination of a centuries-long quest to understand the true nature and extent of the cosmos. January 1, 1925, was the day that astronomers were officially informed of the universe's discovery.

The author of the paper was thirty-five-year-old Edwin Hubble, a staff astronomer at the Mount Wilson Observatory, in southern California. Hubble had aimed Mount Wilson's 100-inch reflector, the largest telescope in its day, toward a pair of celestial clouds known as Andromeda and Triangulum, the only spiral nebulae in the nighttime sky that can be seen with the naked eye. By having access to significant telescopic power, Hubble was at last able to resolve individual stars in the outer regions of the two mistlike clouds, and to his surprise and delight some turned out to be Cepheids, special stars that methodically dim and brighten as if they were slow-blinking cosmic stoplights.

The signals revealed that our galaxy, the Milky Way, was not alone. The Cepheids were telling Hubble that the Andromeda and Triangulum nebulae were very distant, situated far beyond our galactic borders. Our celestial home was suddenly humbled, becoming just one of a multitude of galaxies residing in the vast gulfs of space. In one fell swoop, the visible universe was enlarged by an inconceivable factor, eventually trillions of times over. In more familiar terms, it's as if we had been confined to one square yard of Earth's surface, only to suddenly realize that there were now vast oceans and continents, cities and villages, mountains and deserts, previously unexplored and unanticipated beyond that single plug of sod. Hubble directed our eyes to billions of other galaxies—other Milky Ways formerly unknown—scattered like separate atoms through space and time, as far outward as telescopes could peer. Indications of the Milky Way's true place in the universe had been cropping up for years, but the evidence was indirect, conflicting, and controversial. Hubble stepped into the fray and finally provided the decisive proof. He confirmed an idea to everyone's satisfaction that beforehand had been on far shakier ground.

It was the astronomical news of the century and yet Hubble, astonishingly, was not present—at this, his moment of triumph. Instead, the staid and respected Princeton University astronomer Henry Norris Russell stood in for Hubble that morning and relayed his findings to the conferees. From all accounts, Hubble was neither sick nor detained by family matters. He might have been put off by the long and wearying cross-country train ride, but the reason for his absence was possibly more idiosyncratic. Hubble, a former legal scholar trained in weighing evidence, was concerned that by the time of the astronomy meeting he hadn't countered every feasible argument against his finding. At his own observatory, in fact, a colleague had gathered the strongest ammunition against his conclusion, evidence Hubble couldn't yet refute. This loose end bothered him immensely. What Hubble craved was an airtight case—no stone unturned, no question left unanswered—before stepping up to the podium himself. Being caught in a scientific error was Hubble's greatest nightmare. Back in California the young astronomer was fretfully asking himself, Could I possibly be wrong?

With the stunning pictures of our resplendent cosmos now so widely circulated, such a part of the routine imagery that surrounds us daily, it's difficult to remember that less than a hundred years ago astronomers' conception of the universe was very different than it is today. There were no quasars, no distant galaxies, no exotic black holes or wildly spinning neutron stars. No one even knew for sure how the Sun could keep generating its tremendous energies over billions of years. What was called "the universe" consisted of a single, disk-shaped collection of stars that cuts a magnificent swath across the celestial sky. With Earth located within this great stellar assembly, we peer outward through the disk and perceive it as a band (much the way a plate looks viewed from its side). Known since ancient times as the Milky Way because of its ghostly white visage, our galaxy a century ago was not just the sole inhabitant of the cosmos. It *was* the cosmos—a lone, star-filled oasis surrounded by a darkness of unknown depth.

A few voices of dissent could be heard, arguing against this perspective. A growing number of small spiraling clouds were being sighted in

The Milky Way over the Kitt Peak National Observatory, Arizona
(*Photo by Michael R. Cole, UrbanImager*)

the heavens; these faint celestial objects were lurking wherever a telescope gazed away from the Milky Way into deep space. Were these spiral nebulae close to us or were they farther off? No one knew, because at
the turn of the twentieth century astronomers didn't yet have the means
to determine their distance with assured accuracy. The only thing they
could do was speculate. Some looked at these nebulae, shaped like
springs unwinding, and thought, "Ah, nearby solar systems in the making." Others observed the same tiny clouds and imagined them as a host

of sister Milky Ways so distant that their stars melded into faint and misty whiteness. That would mean the Milky Way was not special at all but merely one island of stars caught in the midst of a far larger archipelago. But the majority of astronomers rejected this strange—even frightening—concept. That other galaxies existed seemed inconceivable, and so they fiercely clung to what they perceived to be their pivotal place in the cosmos. Nicolaus Copernicus may have moved Earth and its inhabitants from the hub of the solar system in the sixteenth century, but humanity remained comforted by the notion that it retained a privileged position in the very heart of the Milky Way, the sole galaxy. They rested easy knowing they resided in the very center of the universe. There was no hard-and-fast evidence to suggest otherwise.

That contentment was shattered, though, as astronomy underwent a spectacular transformation, starting in the waning years of the nineteenth century. "This was an era of extraordinary change in every phase of human life on this planet," recalled Edwin Frost, an astronomer who had personally witnessed the transition at the Yerkes Observatory in Wisconsin. "[It] was truly a Victorian age drawn to a close with the end of the century." When Frost was growing up in the 1880s, Europe was the touchstone in matters of literature, painting, and science. "Even steel rails for the trunk-lines were imported from Britain as late as my college days," he said. "Then Andrew Carnegie and others found that rails could be made better and cheaper in America. . . . The child was rapidly getting out of its infancy." Discoveries and inventions were on the rise. Seemingly overnight, there were electric lights, heating by coal, hot-air furnaces, indoor bathrooms, and automobiles smoothly traveling down asphalt-paved roads.

Astronomy blossomed within this atmosphere of teeming innovation. Cameras became standard equipment on telescopes, enabling observers to gather light over an entire night and so generate images of faint stars and nebulae never before seen. And spectroscopes, devices that separate starlight into its component colors, allowed astronomers to figure out what the stars and other celestial objects were truly made of. Suddenly the very chemistry of the heavens was in their grasp. Meanwhile, prominent industrialists, enriched by the bounty of the Gilded Age, provided the money that allowed big dreamers to construct the large telescopes they had so long desired.

Given the swift emergence of these technological improvements, dry textbook accounts, reduced to a discovery's most essential elements, make it appear as if Hubble's historic achievement had taken place overnight. He goes to the world's largest and best-equipped telescope and, voilà, he reveals a cosmos populated with myriad galaxies spread over space as far as the telescopic eye could see. The Milky Way suddenly becomes a minor player in a much larger drama, and Hubble is anointed cosmology's "prime architect" for making this astounding breakthrough. But that is not the case at all. In reality, Hubble stood on the shoulders of a series of astronomers farsighted enough to tackle a problem others had been ignoring. Answers did not arrive in one eureka moment, but only after years of contentious debates over conjectures and measurements that were fiercely disputed. The avenue of science is more often filled with twists, turns, and detours than unobstructed straightaways.

Astronomers trained in the older, classical ways, who dwelled on calculating the motions of the planets and measuring the positions of stars to the third decimal place, had not been distressed at all by the mystery of the spiral nebulae. They figured that once the matter was resolved it would not greatly change their perception of the overall structure and contents of the heavens. Simon Newcomb, the dean of American astronomy in the late nineteenth century, remarked at an observatory dedication in 1887 that "so far as astronomy is concerned . . . we do appear to be fast approaching the limits of our knowledge. . . . The result is that the work which really occupies the attention of the astronomer is less the discovery of new things than the elaboration of those already known, and the entire systemization of our knowledge."

Within ten years James Keeler, director of the Lick Observatory, in California, proved Newcomb was exceedingly shortsighted. Against everyone's advice, Keeler got a troublesome reflecting telescope—the first of its kind at high elevation—back in working order and demonstrated its power with singular panache. Even though the telescope's mirror was relatively small, it allowed him to estimate that there were tens of thousands of faint nebulae arrayed over the celestial sky, ten times more than had been known before. In the 1910s Lick astronomer Heber Curtis followed up on Keeler's findings and gathered additional evidence to suggest that these many spiraling nebulae were nothing less than separate galaxies. At the same time, a few hundred miles south at Mount Wilson, near Los Angeles, Harlow Shapley resized the Milky

Way, measuring it as far larger than previously thought and shoving our Sun off to the side, away from the galaxy's hub. As Shapley liked to put it, "The solar system is off center and consequently man is too."

The story of our universe's discovery centers mightily on Shapley and Hubble, scientific knights who jousted with each other for years over the universe's true structure. These archrivals shared similar backgrounds and yet couldn't have been more different in temperament and tactics. Both were born in rural Missouri and both came to astronomy through unusual routes: Hubble as a discontented high school teacher, Shapley as a crime reporter. And each, after obtaining his doctoral degree, was selected by the visionary George Ellery Hale to work at the Mount Wilson Observatory, the greatest astronomical venue in its day. Each pursued a question that few others were asking. For Shapley, it was our precise location within the Milky Way; for Hubble, our place in the universe at large.

Their work took place during a crucial moment of transition. While European astronomers were diverted by World War I and its resulting turmoil, American astronomers were free to push forward on the question of the spiral nebulae. Figuring out the universe's exact configuration became an American obsession, its participants drawn from the Lick, Mount Wilson, and Lowell observatories, newly built in the western United States. The world's older observatories had no chance at all, for at the Lick and Mount Wilson observatories, in particular, astronomers had access to advanced telescopes situated on prime high-elevation sites, a combination essential to cracking the mystery.

Hubble gets deserved credit for providing the last, painstaking turn of the lock. "Hubble's drive, scientific ability, and communication skills enabled him to seize the problem of the whole universe, make it peculiarly his own, contribute more to it than anyone before or since, and become the recognized world expert of the field," wrote astronomer Donald Osterbrock, archivist Ronald Brashear, and physicist Joel Gwinn for a centennial celebration of Hubble's birth.

By 1929, just five years after his initial finding on the galaxies, Hubble made an even more astounding discovery. He and his colleague Milton Humason gathered the key evidence that opened the door to proving that the universe was expanding, with the galaxies continually riding the wave outward. Space-time was in motion! Half the work to reach this startling conclusion was actually performed on an Arizona mountaintop a decade earlier by Vesto Slipher, a Lowell Observatory astronomer whose vital role in arriving at this finding is now largely forgotten outside the halls of academia. Such is the power of Hubble's legend. It pushed the contributions of others into the shadows as the years progressed. This book intends to shine the spotlight once again on the entire cast of characters who contributed to revealing the true nature of the universe and laid the groundwork for Hubble's success.

Knowledge of the cosmic expansion was a transforming event. It allowed astronomers to escape the confines of their home galaxy, letting them explore a far larger cosmological vista. The Milky Way was now fleeing outward, giving theorists free rein to contemplate the universe's very origin. They mentally put the cosmic expansion into reverse and imagined the galaxies drawing closer and closer to one another, until they ultimately combined and formed a compact fireball of dazzling brilliance. In this way, they realized that the universe had emerged in the distant past from an enormous eruption—the Big Bang. No longer was our cosmic birth a matter of metaphysical speculation or a biased whim; it had become a scientific principle that could be tested and probed.

This new cosmic outlook came about through a unique convergence—the perfect storm—of sweeping developments. Not only did a burgeoning economy provide the money—and new technologies the instruments—to make these discoveries, but newly introduced ideas in theoretical physics supplied some answers. No less a scientific figure than Albert Einstein had arrived on the scene with a novel theory of gravity that provided a unique explanation for the universe's bewildering behavior.

A dynamism entered into the universe's workings. Einstein's equations introduced the idea that space and time are woven into a distinct object, whose shape and movement are determined by the matter within it. His general theory of relativity anticipated the universe's expansion and turned its study into an intellectual and theoretical adventure. Early globetrotters had crossed the oceans in search of terra firma—solid land, new continents—previously unknown to them and ready for exploration. With his relativistic vision of space-time as a pliable fabric that can bend and stretch, Einstein allowed astronomers to recast the ancient search into a quest for *cosmos firma*. Glued together by the genius physicist, space and time became cosmic real estate to be appraised, mapped, and scrutinized, with Hubble serving as its first surveyor.

Hubble eventually summarized his cosmological findings in a work titled *Realm of the Nebulae,* which is part history, part college textbook, and part professional memoir. This book was labeled a "classic" by his peers at the very time it was published in 1936. And Hubble's initial take still holds up in its broad outline. "[His] picture differs from today's only in details," Caltech astronomer James Gunn noted decades after its publication. "One looks through the pages almost in vain for things that are known to be wrong. One finds a few . . . [but] we still determine the distances of the nearest galaxies by methods described [by Hubble]. We still mostly use Hubble's classification scheme. We still pay a great deal of attention to the questions Hubble asks."

However, there is one glaring exception to Gunn's statement. Although Hubble's name is now strongly attached to the discovery of the expanding universe, he was never a vocal champion of that interpretation of his data. That was because there were other hypotheses in play in the 1930s and 1940s. Hubble was reluctant to choose sides, at a time when his newly mined data and Einstein's theory were so fresh. Hubble always coveted an unblemished record: the perfect wife, the perfect scientific findings, the perfect friends, the perfect life. His observations that the galaxies were fleeing outward were to him always *apparent* velocities. He wanted to protect his legacy in case a new law of physics sneaked in and changed the explanation. So far, it hasn't.

Hubble was lucky in a way. The Hubble Space Telescope could easily have been given another name had certain events turned out differently: if someone had not prematurely died (Keeler), if someone else had not taken a promotion (Curtis), or if another (Shapley) was not mulishly wedded to a flawed vision of the cosmos. The discovery of the modern universe is a story filled with trials, errors, serendipitous breaks, battles of wills, missed opportunities, herculean measurements, and brilliant insights. In other words, it is science writ large.

Setting Out

[1]

The Little Republic of Science

An immense continent of rock known as the North American plate slid inexorably over an oceanic slab of Earth's crust moving east- ward. At the tectonic juncture, where the two gargantuan plates smashed together, the ocean floor plunged downward, the tremendous compression forging massive blocks of shale and sandstone. In due course some of this material lifted upward from its depths, relentlessly rising toward the sky to form the Diablo Mountain Range—two hun- dred miles of peaks and vales stretching from the San Francisco Bay southward along California's coastline. As if readying for a perfor- mance, nature sculpted the landscape that, millions of years later, offered astronomers a unique observing platform for their studies of the cosmos. Situated on the eastern edge of the Pacific, this lofty terrain became the perfect vantage point from which to make the first great dis- coveries in twentieth-century astronomy.

One noticeable peak in the Diablo Range, some forty miles from the sea, was known to early settlers as La Sierra de Ysabel. The first to record an ascent to its uppermost reaches were William Brewer, a geologist who worked on California's first complete geological survey, and Charles Hoffman, a topographer. Laurentine Hamilton, then a Presby- terian minister from San Jose, tagged along for the 1861 summertime adventure. While journeying over the lower elevations the men used mules but struggled over the last three miles on foot. With the two sci- entists burdened down by their heavy equipment, the minister was able to sprint ahead, pushing through the chaparral, mesquite, and thick groves of scrub oak that filled the mountain's furrowed sides like well-sprinkled seasoning. Upon reaching the summit, Hamilton waved his hat in the air and exclaimed, "First on top, for this is the highest point." In honor of the achievement, Brewer graciously named the peak after his "noble and true" friend.

Within three decades Mount Hamilton was the site of a radical new endeavor in astronomy. Fueled by America's escalating wealth, "the public mind in this country is now directed to the importance of original scientific research," wrote Joseph Henry, head of the Smithsonian Institution, in 1874 to the noted English biologist Thomas Huxley, "and I think there is good reason to believe that some of the millionaires who have risen from poverty to wealth will in due time seek to perpetuate their names by founding establishments for the purpose in question." In the vanguard to answer that call was San Francisco entrepreneur James Lick, who funded the world's first astronomical observatory permanently established at high elevation. Before this, professional telescopes were routinely constructed in relatively low-lying areas, near major cities or on university campuses for easy access.

In 1888, from its commanding perch atop Mount Hamilton, the Lick Observatory began operating the largest telescope in its day, which featured a pair of imposing lenses a full yard wide to gather and magnify the celestial light. It was the same type of telescope through which Galileo first peered, one that directed the light through two aligned pieces of glass, but the diameter of the Lick instrument was a couple of dozen times larger. Its founder spared no expense to house this giant refractor. The massive building was designed in a classical style by Washington architect S. E. Todd. From afar, it appeared as if a European palace had been magically transported to the American West. Inside its dome, hand-carved moldings decorated the walls. The floors were curved wooden planks, polished to a sheen and stylishly following the shape of the circular dome. Tourists traveled on stagecoach for hours for a glimpse of this new wonder of the scientific world.

Unbeknownst to those visitors, though, the most innovative work at Lick was actually being done in more modest surroundings, about a quarter of a mile south of the showstopping telescope, at the end of a mountain spur known as Ptolemy Ridge. There, in a far smaller dome that resembles a quaint medieval chapel, James Keeler labored to put a reflecting telescope into operation, which used a silvered mirror instead of a lens to magnify its image. It was an instrument that everyone

Lick Observatory, c. 1910. The 36-inch telescope is housed
in the big dome; the smaller Crossley-telescope dome is at the far left.
(*Mary Lea Shane Archives of the Lick Observatory, University Library,
University of California–Santa Cruz*)

warned him would be nothing but trouble. Big-lensed refractors were
the telescopes of choice in the late 1800s, but Keeler bravely broke from
that tradition, establishing an approach in professional astronomy that
eventually spread to every major observatory throughout the world.

Though now reduced to a minor figure in many histories of astronomy, Keeler was actually a forerunner to the birth of modern cosmology, a crucial player in helping launch the new field. A man who could
manipulate a spectroscope like no other, he pioneered uses for the new
instrument at a time when astronomers were just beginning to apply the
methods of physics in their work. This specialty, newly tagged "astrophysics," enabled observers at last to discern the chemical and physical
natures of stars, planets, and nebulae.

In Keeler's time the universe was a far simpler place, at least to our
modern-day eyes. The cosmos consisted solely of a vast collection of
stars, a disk-shaped distribution somewhat flattened, with the Sun situated in an honored place near the center. Beyond that, said most
astronomers, was simply a void—possibly extending out to infinity.

But there were oddities in the celestial sky, difficult to explain. There were these mysterious nebulae that through a telescope looked like watery whirlpools, mistlike clouds exhibiting spiraling shapes. Astronomers were long acquainted with other types of nebulae, such as the vast and chaotic cloud in Orion and the ringlike nebulae, but these vaporous objects resided within the bounds of the Milky Way. The spiral nebulae, on the other hand, were solely found away from the milky band of our galaxy. Why did they prefer the more open cosmic spaces, as if they were avoiding the stars? Astronomers didn't have a good, rational explanation for this unique distribution. Keeler, to his credit, made these nebulae his prime subject for investigation, at a time when the study of stars and planets was far more attractive to astronomers. Before the introduction of photography, the number of nebulous clouds in the heavens was estimated to be in the thousands. Mounting a camera onto his reflecting telescope, Keeler began to realize there were likely tens of thousands. His discovery was a revelation, and in making this masterful leap in observing, Keeler opened up a vast new arena for astronomy.

Keeler's celestial curiosity may have initially been sparked by a dramatic solar eclipse he witnessed in 1869 when he was eleven—its narrow path of totality creating a sensation as the Moon's shadow stretched across the United States. A few months later, his family moved from Illinois to Mayport, Florida, where Keeler was homeschooled, surrounded by the stacks of *Scientific Americans* to which his father subscribed. Ordering some lenses from an optical dealer who advertised in the magazine, young Keeler built his first telescope—a 2½-inch refractor with a cedar tube. He was soon spending long nights at his scope drawing sketches of lunar craters and the planets. He was riding the wave of a new American fancy.

Earlier in the nineteenth century astronomical research in the United States had been a rather haphazard affair, until two key events dramatically altered the situation. In the autumn of 1833 people throughout the country witnessed a meteor storm, torrents of shooting stars, like no other. It was described as a "constant succession of fire balls, resembling sky rockets, radiating in all directions from a point in the heavens," which led to this spectacular celestial fireworks show being called the "Falling of the Stars." A decade later the public went agog once again over the Great Comet of 1843, proclaimed by Yale astronomer Denison Olmsted in that preelectric era as "the most remarkable in its appearance of all that have been seen in modern times." The comet was visible even in daylight, with a nucleus as bright as the full Moon and a tail that stretched for nearly two million miles. Together, the meteor blizzard and the comet sparked a huge surge of public interest in the heavens. It also made U.S. politicians woefully aware of their country's lack of first-rate scientific institutions to study such captivating phenomena. The English novelist Frances Trollope, who spent some time in America in the 1820s, had found it "extraordinary that a people who loudly declare their respect for science, should be entirely without observatories. Neither at their seats of learning, nor in their cities, does anything of the kind exist."

This deficiency was quickly remedied with the opening of observatories in both Cincinnati, Ohio, and Chapel Hill, North Carolina, and at colleges such as Yale, Harvard, and Williams. The U.S. Naval Observatory, the country's first national observatory, also obtained its first decent telescope. During this period before the Civil War, additional observatories quickly sprang up, becoming the scientific facility de rigueur for major American cities and colleges. These efforts at last fulfilled the vision of John Quincy Adams, America's sixth president, who had long pushed for the country to erect a "lighthouse in the sky." "Some Americans, haunted by a nagging sense of cultural inferiority and smarting from invidious comparisons with Europe, fostered astronomical research as a matter of national pride," writes historian Howard Miller. And once in place, these pioneering astronomical outposts stimulated a continuing interest, especially among young boys like Keeler who dreamed of one day taking part in this new American endeavor.

Described by acquaintances as a "lankey green country boy" with a backwoods "cracker drawl," Keeler came to develop special skills in building instruments, which enabled him to enter Johns Hopkins, America's first research university, just a year after the institution opened in Baltimore, Maryland. Upon graduation in 1881, Keeler began work near Pittsburgh at the Allegheny Observatory, headed up by Samuel P. Langley, the man who two decades later would almost beat the Wright brothers at getting a piloted, self-propelled aircraft flying. A year of graduate work in Germany in 1883–84 sharpened Keeler's expertise in spectral analysis. All this preparation turned out to be indispensable when he received an offer to join the staff of the Lick Observatory, the revolutionary new mecca for astronomy in central California.

. . .

James Lick initiated a remarkable trend. He stood at the head of a long line of prosperous benefactors in the late nineteenth and early twentieth centuries who used the business fortunes they accrued in the United States to construct some of the most productive observatories in astronomical history. Given his largesse, Lick raised the stakes in astronomy. Before this, the most acclaimed observatories were in Europe and sponsored by either universities or governments. With resources sparse, these institutions were often slow to adopt new techniques and instruments. At each observatory, surveys plugged along for decades, often using just one key telescope. But the Lick Observatory offered a new model for research, one that ran at a quicker pace enriched by private capital. Lick made big telescopes the commemorative monument of choice among the American nouveaux riches. Moreover, with these privately funded observatories being established from scratch, they were able to purchase the finest instruments and adopt the latest technologies. As a consequence, astronomy advanced in the United States at a faster pace than in any other country in the history of science. "Starting from essentially zero at the beginning of the nineteenth century," says historian Stephen Brush, "the Americans had overtaken the Germans to jump into second place by the end of that century and were already challenging the British for the top spot." Domination of the heavens appeared to go hand in hand with economic riches.

Lick earned his riches. Born in 1796 to a rural Pennsylvania Dutch family, just as the new republic of the United States was getting under way, he learned the trade of woodworking at the side of his father. After making a fairly comfortable living running his own shop in New York City, Lick abruptly decided in 1821 to move to South America, bent on amassing a fortune. There he became a master builder of fine-wood piano cases, a venture that proved highly lucrative in a culture where dancing and music were greatly valued. After twenty-seven years, though, living at first in Argentina, then Chile, and finally Peru, he decided to sell his varied business concerns and return to the United States. Arriving in San Francisco by ship in 1848, just as California was about to secede from Mexico, he came ashore with $30,000 in gold doubloons and six hundred pounds of Peruvian chocolate made by his friend Domingo Ghirardelli.

Wasting no time, Lick quickly put his incisive business acumen to

James Lick
(*Mary Lea Shane Archives of the Lick Observatory, University
Library, University of California–Santa Cruz*)

work. He shrewdly used his gold to purchase real estate in San Francisco, then just a scrubby town with scarcely a thousand inhabitants.
When residents started heading to the hills to make their fortune in the
California gold rush, Lick was there to provide them with a stake by
buying up their town land at bargain prices. He also bought a gristmill,
greatly expanding it, and built California's first great luxury hotel, the
opulent Lick House, which occupied an entire city block (and was later
destroyed in the fire that tore through San Francisco after its horrific
1906 earthquake).

Lick never married but still built a homestead at the south end of
San Jose, where he lovingly cultivated rare plants and shrubs from
around the world. The community considered him an eccentric miser;
he dressed like a tramp and at times slept on a bare mattress laid out
atop a piano crate. As a youth, he had gotten a girl pregnant, but her
father, a prosperous miller, refused his offer to marry her, judging Lick
too poor and socially inferior. The miller could hardly have imagined
how astronomy, decades later, would benefit from this snobbish rebuff.

Without a legitimate heir, Lick, in his old age, began to think of using some of his tremendous wealth (he had accumulated nearly $4 million, around $100 million in today's dollars) to erect a gargantuan monument to himself. For Lick it was a chance at immortality. He particularly favored the idea of constructing a giant marble pyramid on the corner of Fourth and Market streets in downtown San Francisco, a structure that would have surpassed Egypt's Great Pyramid of Giza in size.

But a few auspicious encounters revised this vainglorious plan. Lick had once spent a few days with a visiting amateur astronomer and lecturer, George Madeira, who captivated him with talks about astronomy's latest discoveries. They met again a few years later for some telescope viewing when Madeira allegedly asserted, "If I had your wealth, Mr. Lick, I would construct the largest telescope possible to construct." Around the same time Joseph Henry, then head of the National Academy of Sciences as well as the Smithsonian, was visiting San Francisco and arranged a meeting with Lick to discuss how wealthy men could use their money to cultivate science. The following year, 1872, the Harvard naturalist Louis Agassiz gave a widely reported lecture at the California Academy of Sciences, where Agassiz echoed Henry's refrain.

All these lessons struck a chord. Lick soon astonished the California Academy when he granted the institute, without prior notice, the gift of a downtown lot to build a museum and more expansive headquarters. Academy president George Davidson, a geodetic surveyor and astronomer, promptly called on Lick to thank him, initiating a friendship. When Lick was later felled by a stroke and confined to a two-room suite at his hotel for nearly a year, Davidson regularly visited, engaging Lick with chats about the rings of Saturn, the belts of Jupiter, and other astronomical topics. Lick soon abandoned his scheme to build a pyramid and decided instead to erect a telescope "superior to and more powerful than any telescope yet made," right on his favored city spot, the corner of Fourth and Market.

An in-town telescope was never built (fortunately), largely due to Davidson's intervention. As both an amateur astronomer and a geodeist, a profession that took him to towering mountain sites, he had long been convinced that astronomy would best be served by taking its instruments to the highest elevations possible, where a telescope's resolution would improve immensely in the clear, more rarefied atmosphere. Isaac Newton first pointed this out in the eighteenth century. "For the

Air through which we look upon the Stars, is in a perpetual Tremor," he wrote in his *Opticks*. ". . . The only remedy is a most serene and quiet Air, such as may perhaps be found on the tops of the highest Mountains above the grosser Clouds." And preferably in a region with a dry season, free of rain.

Over time Lick came to accept Davidson's compelling idea and in the fall of 1873 authorized the funds to construct a state-of-the-art observatory in the arid Sierra Nevada Mountain Range at an elevation of 10,000 feet. Caught up in the excitement of this novel venture, Lick pledged $1 million, a princely sum. No observatory had ever been established in such a remote and elevated locale. In that decisive shift, astronomy would soon change in a remarkable way, whisking the field away from its previous urban settings.

Over the next three years, Lick fiddled with the provisions of his trust, fired and hired assorted board trustees, reduced the price tag to a tightfisted $700,000, and changed his mind on the telescope's location. Once set for a spot near Lake Tahoe by the Nevada border, the site was eventually shifted to Mount Hamilton, a shorter peak (4,200 feet high) just to the east of San Jose, where Lick could look up and proudly view it from his property. Davidson, sorely disappointed by the lowered elevation and Lick's parsimonious ways, left the project and refused to speak to his former benefactor ever again.

Davidson's snub mattered little in the end, for Lick soon passed away. He died on October 1, 1876, at the age of eighty. Only then did construction of the mountaintop observatory, an arduous and unprecedented endeavor, truly get under way: Congress at last approved transfer of the public land, the local county built a road to the top, and the mountain peak was certified by an expert as exceptional for its atmospheric stability. Mount Hamilton's sharp, knife-edged profile causes minimal disturbance as air flows in from the west. Fulfilling Lick's decree, the largest refracting telescope in the world—one with lenses ten inches wider than the previous record holder at the U.S. Naval Observatory—was installed in a magnificent domed building, designed in the Italian Renaissance style and large enough to accommodate the scope's lengthy tube. Massive hydraulic cylinders allowed an astronomer to raise or lower the entire circular floor to keep him level with the telescopic eyepiece. The top thirty feet of the mountain had been blasted away to provide a level space for the rambling complex, which included housing, workshops, offices, and a library. The observatory operated as a small town, with families living on-site and supplies brought up daily by wagon from San Jose. A visitor dubbed it the "little Republic of Science."

Lick became the new republic's patron saint, for his egotism never completely disappeared with his noble gift to astronomy, even after his death. In January 1887, as soon as the telescope base was complete, Lick's remains were brought up the mountain and reburied, his body resting directly under the grand instrument he funded, in the very base of the pier supporting the giant refractor. Tour groups still visit the tomb today. Davidson claimed credit (as did others) for the interment idea, first voiced when Lick was alive. He was surprised the old man agreed. At the time Davidson had suggested a cremation and burial of the ashes, to which the former carpenter quickly replied, "No sir! I intend to rot like a gentleman."

The choice for director of the new Lick Observatory was Edward Holden, a graduate of West Point and an unaccomplished astronomer whose sole qualification seemed to be that his energy and initiative had once impressed Simon Newcomb, then America's most revered astronomer, while he was assisting Newcomb at the Naval Observatory. A proud and pompous man, Holden at least had a keen eye for talent. Aware of Keeler's outstanding work at Allegheny, Holden hired him in 1886 to get the new mountaintop observatory and its equipment up and running. Of all Holden's hires, James Keeler was by far the best trained. Bringing Keeler to the mountain was the best decision Holden ever made during his tumultuous directorship.

[2]

A Rather Remarkable Number of Nebulae

Keeler traveled to the Lick Observatory along a road that was a marvel of engineering in its day. Although Mount Hamilton is less than a mile high, the journey from its base to the top is more than twenty miles in length, with the roadway sinuously zigging and zagging as it gradually ascends. There are some 360 switchbacks in all, and some were even given special names, such as "the Tunnel," "Crocodile Jaw," and "Oh My Point," branded by the oft-heard refrain as people sat atop the stagecoach and looked down in horror at the point's steep drop-off. The serpentine route was installed to maintain a gentle gradient, so that stagecoach horses in the nineteenth century never needed to break their stride.

Upon reaching the top, Keeler was immediately enamored of the breathtaking scenery. "The view from the observatory peak is a very beautiful one, particularly in the spring, when the surrounding hills are covered with bright green verdure, and the eye looks down upon acres of wild flowers," he later wrote in a pamphlet for visitors. "To the west lies the lovely Santa Clara valley, shut in from the ocean by mountains somewhat lower than the Mt. Hamilton range. Sometimes the entire valley is filled with clouds, rolling onward under a clear sky and bright sun like a river of snow. . . . The surrounding mountain tops project out of the fog like black islands." Often the ocean fog arrives at sunset, rolling in from the Pacific at the Golden Gate, to the north, and Monterey Bay, to the south.

Not everyone on the mountain was enthusiastic about Keeler's arrival. The observatory's superintendent, Thomas Fraser, was initially wary of the newcomer. "If he has the right ring all will be right," said Fraser, "but if Stubern [sic] then things will go wrong and he will have to leave that is all there is to it." It didn't take long, though, for Fraser to be won over by the exceptional skill Keeler displayed as the telescope was being prepared for operation.

Its great lenses were finally installed on New Year's Eve 1887, but due to severe weather the staff could not test the telescope out until a few days later. Often in the wintertime, storms would sweep over the mountain with winds gusting more than 60 miles per hour, which would drift the snow about the dwellings more than ten feet high. Once the staff got back to the telescope, the trial run did not go well. To their horror the astronomers discovered that Alvan Clark, the telescope maker, had misstated the instrument's required length. Much like the Hubble Space Telescope's initial mishap a century later, they couldn't get it into focus. The telescope's tube should have extended fifty-six feet, but instead was six inches too long, forcing them to get out their tools and spend valuable days cutting the tube down to size. Clark's son, a partner in the telescope firm, was there for the trial, "a terrible old blow and grumbler," Keeler told Holden. While Clark insisted that his firm's glass was superb and the eyepieces "triumphs of art," he declared the dome "worthless."

With its tube shortened, the telescope was at last tried out on January 7, 1888, a cloudless night that was piercingly cold. With the dome frozen solid that evening, the handful of staff members and guests present could only passively observe the objects that happened to pass by the dome's slit, open toward the southeast. Yet, "no inconvenience was felt beyond the necessity of a little waiting," recalled Keeler. He was pleased to find the clock running smoothly and the mounting working well. The group first observed Rigel, a blue-white double star, followed by the Orion nebula, its great streamers making it one of the most spellbinding sights through a telescope. "Here the great light-gathering power of the object glass was strikingly apparent," Keeler noted. Then, just after midnight, Saturn came into view. Keeler reported that the planet was "beyond doubt the greatest telescopic spectacle ever beheld by man. The giant planet, with its wonderful rings, its belts, its satellites, shone with a splendor and distinctness of detail never before equaled." Everyone in the party took a look. Afterward Keeler spent some time studying Saturn more carefully, which led to Lick Observatory's first discovery. He spied a fine, dark line in Saturn's outer ring, "a mere spider's

James Keeler
(*Mary Lea Shane Archives of the Lick Observatory, University
Library, University of California–Santa Cruz*)

thread," as he described it. It was a breach (now best known for historic reasons as the Encke Gap, after an early-nineteenth-century German astronomer) that had never before been clearly seen. A superb drawing Keeler made of Saturn, based on his sketch of the planet that night, was displayed at the 1893 Chicago World's Fair.

Six feet tall with fair wavy hair, Keeler cut a fine figure. Despite his isolated upbringing in rural Florida, he became a keen judge of human nature and was often called upon to handle personnel and scientific crises at the observatory, which he carried out with the calm discretion of an international diplomat. "He was tolerant, amused and unwilling to take sides," said Keeler's biographer Donald Osterbrock. "He always sought to put the best construction he could on anyone's activities, to emphasize the positive, and never to criticize unless absolutely necessary. It was perhaps not the most courageous philosophy in the world, but it [took] him far."

And as an astronomer, Keeler was outstanding, studying a range of subjects from solar eclipses to planetary features. Photography was still in its infancy, so Keeler continued to make drawings that were praised by his colleagues as marvelous reproductions. "Beautiful and accurate," reported fellow Lick astronomer Edward E. Barnard in a notice to the

Royal Astronomical Society. ". . . [Keeler] has a real artistic ability such as very few observers possess." Keeler's real forte, however, was in using a spectroscope, which was a relatively recent addition to astronomy's instrumental arsenal. The scientific basis for it was established in the seventeenth century.

A young Isaac Newton, sitting in a darkened room in 1666, let a small stream of sunlight enter through a hole in his window shutter. He then passed it through a triangular prism of glass. Beholding a rainbow of colors on the wall behind him, an enchanting phenomenon observed with pieces of glass since antiquity, Newton clearly demonstrated that white light was a mixture of many hues: On one end was a band of red, followed by orange, yellow, green, and blue, until it reached a deep violet on the other end. He dubbed this multicolored display a spectrum, a word previously used to denote an apparition or phantom. By the early nineteenth century Joseph von Fraunhofer, a master Bavarian optician, cleverly combined a slit, a prism, and a small telescope—what came to be called a spectroscope—to examine the spectrum of the Sun more closely. Peering through the eyepiece, he was surprised to discern hundreds of dark lines in the spectrum, as if a series of black threads had been sewn across a rainbow. They resembled the ubiquitous bar codes now found on consumer products. But unfortunately, Fraunhofer died before he could pursue the origin of those mysterious dark slashes.

Answers arrived from the creative experiments being conducted in chemistry laboratories. Even before Fraunhofer's spectral tests, chemists had noticed that metals or salts, when heated to incandescence, emit certain colors. Salts containing sodium, for example, burn an intense yellow-orange when heated by a hot flame. When looking at the heated material through a spectroscope, the chemists saw that its spectrum was composed of discrete lines of color, resembling a picket fence with colorful posts. Whereas the solar spectrum was a continuous rainbow riddled with dark lines, these laboratory spectra were the exact opposite: thin bright lines of colorful emissions set against a dark background.

By 1859 the physicist Gustav Kirchhoff and the chemist Robert Bunsen (creator of the legendary lab burner) at last revealed the meaning behind these bright and dark lines. With the clear hot flame of Bunsen's improved instrument, free of the deceptive contamination that plagued earlier researchers, the two German colleagues were able to conclusively prove that each chemical element produces a characteristic pattern of colored lines when heated and viewed through a spectroscope. The elements weren't emitting an entire rainbow but rather just a few select colors. More consequential, the patterns were as unique and distinguishing as a fingerprint. Each element on the periodic table had its own personal set of emissions. Using their spectroscope one evening to peer at a distant fire in the port city of Mannheim, visible across the Rhine plain from their laboratory window, Kirchhoff and Bunsen were thrilled to detect the spectral signatures of barium and strontium in the roaring blaze. It didn't take long for them to fathom that they could analyze the Sun and stars in a similar fashion, as light knows no distance in the voids of space. Light can be sent through a spectroscope whether it originates from a distance of one foot, ten miles, or a billion light-years away. Before this revelation, astronomers only knew that a star shines, that it occupies a certain position on the celestial sphere, and in some cases moves. But now they were acquiring the means to determine a star's composition and temperature, information once thought impossible to glean.

When an element is hot and glowing, it radiates its distinctive pattern of spectral colors. But at other times it can absorb those same wavelengths, which explains the origin of the dark lines that Fraunhofer found in the solar spectrum. Each element in the Sun's cooler outer atmosphere absorbs its designated colors, robbing the sunshine of those selected wavelengths before they arrive on Earth. The bright lines are simply the reverse of this process—the elements emitting those very same wavelengths of light as they fiercely burn. Either way—dark or bright—the pattern of lines indicates the presence of the element. Not until the early twentieth century, with the advent of atomic physics, did scientists come to understand this behavior as arising from the electrons in an atom jumping from one energy level to another, the atoms emitting bursts of light when they lose energy and gaining energy when they absorb the photons.

Astronomers quickly realized that, along with revealing a star's composition, a stellar spectrum could also tell them how the star was moving. In the 1840s the Austrian physicist Christian Doppler had surmised that the frequency of a wave, such as the tone of a sound wave or the color of a light wave, would be altered whenever the source of the wave moved. We've all heard the pitch of a siren rise to a higher tone as a police car or ambulance races toward us. This is the very effect that

Doppler spoke of: The sound waves emitted by the screeching siren crowd together as they approach us, shortening their length and likewise raising the pitch. Conversely, as the police car pulls away, the sound waves stretch out, producing a lower pitch. In an analogous fashion, a light wave's length is shortened (gets "bluer") when the source of the light approaches and is lengthened (gets "redder") when the source moves away.

Astronomers, though, don't assess the overall color of a star or galaxy to measure its speed. That would be too difficult. They can more easily examine how the bright and dark lines in a celestial spectrum shift from their well-known laboratory positions. Depending on the object's motion, the lines can shift toward either the blue or red end of the spectrum. If a star or nebula, for example, is headed for us, its spectral lines move over toward the blue—that is, the lines get "blueshifted." If moving away, the lines swing over toward the red and hence become "redshifted." The exact velocity is pegged from the amount of shift in the spectral bands. Blueshifts and redshifts are nothing less than the speedometers of the universe.

Keeler had the eye of a hawk in measuring how the celestial light entering his spectroscope was separated into its component wavelengths, with each spectral line offering enticing clues. He was America's leading practitioner of this new technique, with some of his best work being done on measuring the speeds of nebulae within the Milky Way. In Latin, *nebulae* is the word for "clouds" or "mist," exactly what these extended objects look like through a telescope. Some are roundish and were dubbed "planetary nebulae" in the eighteenth century by British astronomer William Herschel, who thought they resembled planets through his telescope. Today, astronomers know that such circular nebulae are the result of aging stars casting off their outer envelopes. Other nebulae, such as the renowned Orion nebula, are more irregular and diffuse, made luminous by the new stars being born within these great cosmic oceans of gas.

By the late 1880s, as Keeler entered his thirties and continued these celestial explorations, he faced a career crisis. He was eager to marry Cora Matthews, the niece of Richard Floyd, the superviser of the observatory's construction and president of the Lick Trust. The couple had first met on the mountain but could not tie the knot right away because

Lick officials would not provide them adequate housing at the observatory once married. There was also Keeler's growing dissatisfaction with director Holden, a tyrannical and humorless man who often tried to share credit for some of Keeler's discoveries and at times ordered the young man to carry out observations he was not eager to do. It was said that Holden, given his West Point background, ran the observatory "as though it were a fort in hostile territory," barking out commands like a general under seige. On top of that, there was the tiresome isolation atop the mountain, with few opportunities to escape to the city and engage in a fuller social life. "I am a human being first and an astronomer afterwards," Keeler confessed to a friend.

Faced with these growing concerns, Keeler began networking among his astronomer contacts and in 1891 secured the directorship of the Allegheny Observatory, a return to his first place of employment. His old boss Langley had by then moved to Washington, D.C., where he served as secretary of the Smithsonian Institution and was beginning work on his lifelong dream to successfully launch a flying machine.

In terms of telescope power, Keeler's transfer to the Allegheny Observatory, situated on a hill across the river just north of America's steel capital, was a giant leap backward. The weather was poorer, the air was tainted with Pittsburgh's industrial smoke, the atmosphere was more turbulent for viewing, and the observatory's main telescope was a 13-inch refractor, far smaller than Lick's 36-incher. Yet, in some ways it was a blessing. The constraints forced him to focus his astrophysical studies on such objects as nebulae, a less trendy territory and hence riper for discovery. Because of their larger size, compared to stars, the fuzzy objects could still be adequately examined, even with a smaller scope. Moreover, astronomical photography had become more efficient and convenient, allowing him to build up exposures and see spectral details he could not see before with his eye alone. He doggedly tracked down every new advance in spectroscopic and photographic equipment in hope of offsetting Lick Observatory's advantages. The experience, though exhausting, only enhanced his astronomical abilities.

From his new post in Pennsylvania Keeler eventually made headlines worldwide. He had been using his spectroscope at Allegheny to measure how fast some of the major planets, such as Venus, Jupiter, and Saturn, were rotating. Based on a method already used to gauge the Sun's rotation, Keeler knew that a spectral line in light arriving from the edge of the planet rotating toward us would be shifted toward the blue end of the spectrum; this same line would shift equally the other way, toward the red, when emanating from the edge, or "limb," of the planet moving away. Along the way, Keeler cannily comprehended that he could also peg the velocity of Saturn's rings with the very same technique.

In 1856 the famous Scottish theorist James Clerk Maxwell had theoretically proven on paper that Saturn's rings were not solid, akin to a phonograph record, but rather composed of innumerable particles, little "moonlets" circling around in independent orbits. Saturn's immense gravitational pull, avowed Maxwell, would have torn apart any sort of solid disk. If true, then Newton's law of gravity would predict that the myriads of tiny chunks located in the outer part of the ring would be traveling slower than those closer in, nearer to Saturn's gravitational grip—just as Pluto, far from the Sun, orbits at a slower velocity than the solar system's inner planets.

A spectrum, taken on the night of April 9, 1895, gave Keeler the direct proof. The spectral lines indicated that the ring's particles were circulating around Saturn according to the rules of Sir Isaac. The ring was not a rigid plate after all. Within days, Keeler dispatched a report to the newly established *Astrophysical Journal*, and a torrent of newspaper and magazine articles about his triumph followed. His scientific reputation rose sharply, especially since he had devised such an elegant and simple test of Maxwell's conjecture, one that other astronomers knew they could have done years earlier, if only they had been so clever.

While Keeler was busy with Saturn, Lick director Edward Holden was scheming to expand his astronomical empire, by bringing the historic Crossley reflector to the observatory—a telescope first constructed by a Londoner, Andrew Common, in 1879. He had built it to test out some design ideas, even earning a gold medal from the Royal Astronomical Society in 1884 for the fine photographs taken with it, including the first image of a nebula, Orion. Its mirror was a glass disk, three feet wide, coated with a thin layer of silver, a relatively new development in reflector technology. Early telescopic mirrors had been made out of metal, which readily tarnished and easily got out of shape. Widespread use of reflecting telescopes did not occur until instrumentalists in the mid-nineteenth century learned how to cast large and sturdy glass mirrors, with the glass first ground and polished into an ideal shape for focusing the light and then its surface coated with a thin surface of metal for high reflectivity.

Satisfied with his design, Common was soon eager to make an even bigger scope and sold his award-winning instrument in 1885 to Edward Crossley, a wealthy textile manufacturer who moved it to his estate in Yorkshire. But after a few years, Crossley sadly deemed the English countryside unsuitable for decent astronomical observations and put

Original Crossley telescope at the Lick Observatory
(*Mary Lea Shane Archives of the Lick Observatory, University Library, University of California–Santa Cruz*)

the reflector (as well as the special dome he had built for it) up for sale in 1893.

Holden may have been a poor astronomer but he was a powerful persuader. He convinced the English tycoon to donate his entire assembly for free to the University of California, which now owned and operated the Lick Observatory. Once the parts for the scope and its dome arrived in 1895, Holden pushed mightily to get the system reassembled as soon as possible. As the dome was reconstructed on the edge of Ptolemy Ridge, a time capsule was inserted into its wall. The small zinc box, still hidden away, contains a letter from Crossley, the calling cards of the Lick astronomers then on staff, a Lick visitors pamphlet, and a set of U.S. postage stamps.

Lick astronomers, however, were not at all interested in this new addition to their astronomical arsenal. One disgruntled staffer declared the equipment "a pile of junk," after some halfhearted attempts were made to put the telescope back into working order. For many, the Crossley was the last straw in a battle that had been raging for a very long time: a face-off between the director and his workforce. Tired of Holden's militaristic commands, hogging of the spotlight, and endless interference, the staff eventually revolted. Holden (described by Lick employees behind his back as "the czar," "the dictator," "that humbug," "an unmitigated blackguard," and "the great I am") was forced to resign. The university regents had lost confidence in him. Holden took his final ride down "Lick Avenue," the mountain's dusty road, on September 18, 1897. Only one person, a young assistant, went out to say good-bye.

Keeler, by this time, was getting restless back in Pennsylvania. The mighty iron and steel mills in the Pittsburgh area were expanding, dirtying up his sky even further with the black soot of their coal fires. And, though he was noted as the country's most able spectroscopist, Keeler was more and more hampered by his tiny 13-inch refractor, a telescope originally built forty years earlier for amateur viewers. Its aging lens absorbed the higher wavelengths of light—blue and ultraviolet—which limited him to work primarily in the yellow-red region of the spectrum. To make matters worse, his former assistant at Lick, William Wallace Campbell, had arranged for Lick to get a new spectrograph (an instrument that not only disperses the light into its constituent colors but records the spectrum as well). It was being built in Pittsburgh, and Keeler had agreed to test it out before it was shipped to California. The experience made him realize that it would soon be impossible for him to compete with Lick, especially since a great economic turndown, a depression that started in 1893 and lasted for years, had dried up sources of funds to expand his facility and raise his salary. Holden's firing came at an opportune time for Keeler.

In the search for Holden's replacement, a number of names came into play, including the venerable Simon Newcomb, George Davidson, who had originally coaxed Lick to fund the observatory, and several senior Lick astronomers. Keeler was added to the mix as a dark horse but soon became a favorite among the more progressive university regents. They wanted someone young, someone with impressive credentials, who would help the University of California achieve first-class status. Keeler won the vote by 12 to 9, Davidson coming in second.

Hearing that they might lose their director, Allegheny Observatory supporters launched a last-minute effort to raise enough funds from the Pittsburgh elite to build a new edifice for Keeler, one equipped with an imposing 30-inch telescope. Poems were even written and printed in local newspapers to boost the cause:

> "Stay with us, Keeler," so they say,
> "And twice as much as Lick we'll pay."
> Wherefore perchance he'll not resign
> But stay and keep our stars in line.

If the full amount required had been raised, Keeler would likely have stayed, not wanting to be disloyal to a town he had come to love. But the campaign fell short (to the relief of his wife, who longed to return to the sunny climes of the West Coast). Yerkes Observatory, in Wisconsin, home to the newest record-holding telescope, a 40-inch refractor, also made him a job offer but could not guarantee a permanent staff position. Keeler, anxious to advance both his research and professional career, at last telegraphed his acceptance of the Lick directorship to University of California officials. It was a time when the United States was finally emerging from its deep economic depression. Hope and optimism were on the rise, as the nation was attaining status as a world financial power, at last surpassing Great Britain in overall worth. Highways were being paved with asphalt, and cities brightly glowed at nighttime, awash in electric light. Telephone and telegraph wires lined urban streets like thick, artificial spiderwebs. Keeler's vocation was carried forward on the swelling tide.

Keeler went back to Mount Hamilton, or the "hill," as it was affectionately known to its residents, on June 1, 1898, seven years after he had first departed for the East Coast. There he found his new duties resembling that of a small-town mayor. "It [was] like being shipwrecked on an island," recalled Kenneth Campbell, who had grown up on the mountain while his father, William, was on staff. "The Director of the Observatory was, I would say, . . . the czar. . . . He had to see that Mrs. MacDonald didn't break her leg on that back step, as well as worrying about spiral nebulae." By then the complex housed three senior astronomers, three assistant astronomers, a small group of workmen, and assorted spouses, servants, and children, some fifty people in all. If a hostess sent out an invitation for an evening gathering, it was plainly understood: no clouds in the sky, no party. Astronomy always came first. A new teacher for the one-room schoolhouse was hired nearly every year (as she often ended up marrying one of the astronomers). For relaxation, residents took some clubs over to the rudimentary golf course, eight holes laid out by one of the senior astronomers on a stretch of flat land just below the mountaintop. No need for man-made hazards; they were all natural—ditches, ridges, ravines, and rock formations; the "greens" were oiled dirt. Occasionally a ground squirrel would carry off a ball, mistaking it for a tasty nut.

A biologist visiting Mount Hamilton returned to the valley below feeling as if he had "dwelt awhile upon Mount Sinai, . . . watched the marshalling of the stars and the dividing of the constellations." Saturday nights were often held aside for visitors, with loaded stages and buggies coming up the mountain sometimes twenty to thirty in procession. Leaving San Jose, the wagons could take up to seven hours to traverse the twenty-five serpentine miles, passing first through orchards of figs, oranges, olives, and peaches. Always in sight during the slow ascent were the observatory's bright white domes. Not until 1910 did the automobile reduce the travel time to two hours.

Keeler resided with his wife and two children, Henry and little Cora, in part of a three-story residence known as the Brick House, just a stone's throw from the main building, where the telescopes were located. The move to Lick decidedly changed his routine. His research was now curtailed by innumerable administrative duties, especially correspondence with university officials, suppliers, prospective students, colleagues, and the general public. "There are no astronomical phenomena expected to accompany, or precede, the second coming of

Christ," he politely responded to one correspondent. In style and temperament, Keeler was the anti-Holden. "No member of the staff was asked to sacrifice his individuality in the slightest degree," said Lick astronomer W. W. Campbell. "No one's plans were torn up by the roots to see if they were growing. . . . Keeler's administration was so kind and so gentle, and yet so effective, that the reins of government were seldom seen and never felt."

Science, though, remained Keeler's prime objective in accepting the directorship. He once again had access to large telescopes situated in a premier environment for viewing, far removed from polluted industrial air. He completed his first paper, the spectral analysis of a peculiar star's outer envelope, within a month of his arrival. For this, he used the famous 36-inch refractor. As director, Keeler could have wielded his power and become the prime user of the 36-inch, but instead he made a daring and momentous decision. He decreed that Campbell, who had become Lick's main spectroscopist during Keeler's absence, would continue using the 36-inch to carry out an ambitious project Campbell had already begun, measuring the velocities of the stars. Keeler, to everyone's astonishment, chose to work on something completely different: getting the disreputable Crossley reflector up and running.

Keeler became interested in reflecting telescopes while he was still director of the Allegheny Observatory. He knew such telescopes would be particularly advantageous for carrying out his specialty—spectroscopy. The thick glass lenses in refracting telescopes tended to absorb certain wavelengths selectively (depending on the glass and lens construction), keeping that light from registering on either the eye or a photographic plate. This was a dismaying effect to a spectroscopist, who was devoted to collecting each and every light wave emanating from a celestial object. Mirrors, on the other hand, didn't have this problem. They shepherded all light waves equally, no matter what the color, right to the focus. Moreover, lenses were reaching their maximum size at the end of the nineteenth century; they couldn't be manufactured much bigger than forty inches without getting distorted by their own weight. Mirrors, on the other hand, could be made much larger. In Keeler's estimation, reflecting telescopes had acquired a stigma in the past because they had been placed in cheap, flimsy mounts.

Keeler had seen the power of reflectors firsthand while visiting England in 1896 and attending a meeting of the British Association for the Advancement of Science. There Isaac Roberts, a former businessman and accomplished amateur astronomer, displayed the eye-catching photographs taken with his 20-inch reflector. Roberts had pioneered many of the techniques for taking long-term exposures and was the first to reveal that the Andromeda nebula was a spiral. Photography was then having a tremendous impact upon astronomy, radically transforming its procedures. Holden, right before Lick opened, wrote that astronomers can now "hand down to our successors a picture of the sky, locked in a box." Observers were able to continue their research at their office desks, analyzing their images with mathematical precision, no longer dependent on crude drawings, hasty notes in a logbook, or the fading memory of their night at the telescope. Changes in a celestial object could at last be accurately monitored, from year to year and decade to decade.

After the palace revolt against Lick's former director, the Crossley had been abandoned. It was the mountain's white elephant. No Lick observer was interested in using the reflector, not a surprising turn of events given its dreadful reputation. Even before Holden left, a staff astronomer had written a long memorandum summarizing what sort of research could be done with the Crossley. The title of his paper broadcasted the answer with unforgiving bluntness: "No Work of Importance."

Keeler thought otherwise, even though he had never before used a reflecting telescope. He was interested because he was after rare game: the particular stars and nebulae that had eluded previous spectro-scopists due to their faintness, and the Crossley's special features were going to allow him to obtain a decent spectrum. The Crossley was not just any telescope mirror; it was the largest of its kind in America, but Keeler faced innumerable engineering problems, which he had to solve before the Crossley would be fully functional. For one, the spectrograph he inherited was so large that it had to be removed from the telescope each time the dome needed to be shut. And the telescope's mounting, originally set so it would correctly track the stars in England, had to be realigned to account for Mount Hamilton's more southerly location. Then there was the need for a new eyepiece, as well as a drive clock to keep the telescope in sync with the moving sky. Chemicals had to be gathered for silvering the yard-wide mirror—silver nitrate, caustic potash, ammonia, and a reducing solution composed out of rock candy, nitric acid, alcohol, and water—and telephone wires extended from nearby astronomers' cottages to the dome, so there would be electric light to illuminate the guidewires in the eyepieces.

Making improvements in fits and starts—three steps forward, two back—Keeler and his associates at last got the telescope operating tolerably in September 1898, just four months after he arrived back at Lick. On the fifteenth of that month he tried out his camera for the first time. His opening target, Altair, the brightest star in the constellation Aquila, was out of focus, but another exposure, east of the star, was better. "The fainter stars look pretty sound, but the brighter ones show irregularities," he wrote down in his logbook. Two weeks later he took a photograph of the Moon, then nearly full. "Plates are fairly good," he briefly noted. Inside the Crossley dome, the upper wall was painted black, in order to absorb stray reflections from the sky; the lower half, though, was colored bright red, so Keeler and his assistants could see where they were going in the dark. The whole interior was bathed in the faint glow from a lantern fitted with panes of crimson glass, as the photographic plates were not sensitive to red light. Such precautions were essential since the Crossley mirror was held by an open framework of iron rods instead of mounted within an enclosed tube.

In late fall Comet Brooks appeared in the sky. This led to Keeler's first research paper based on his observations with the Crossley. His images, taken over eleven consecutive nights with the help of his assistant Harold Palmer, displayed finer details than previous photographs of comets. They even captured a double nucleus. "On the negative of November 10, obtained with an exposure of 50 [minutes]," reported Keeler, "the head of the comet is made up of two clearly separated nebulous masses, surrounded by the nearly circular coma. . . . I am inclined to believe that the division of the nucleus was real." Keeler was not the first to discern such cometary structure, but it was exciting nonetheless.

He soon was observing the Pleiades, the impressive cluster of stars (the "Seven Sisters") situated near the constellations Taurus and Orion in the autumn nighttime sky. Taking a series of photographic exposures, sometimes lasting longer than an hour, he was able to show that the Pleiades is embedded in filamentary and diaphanous clouds of gas. "Nebulous wisps . . . are characteristic of the region," he reported. He later wowed astronomers with a spectacular photograph of the Orion nebula, convincing them of the capability of a reflecting telescope to bring out features formerly too faint to discern. The stunning image served as the frontispiece for an issue of the *Publications of the Astronomical Society of the Pacific*, and it amazed even him. "The photographic power of the Crossley reflector on a fine night is surprising," he wrote, "at least to one who has hitherto worked with refractors only."

Keeler went on to use the Crossley to record other arresting celestial sights, such as the sinuous and radiant strands of the Lagoon, Omega, and Trifid nebulae. "We know them so well today," Osterbrock pointed out, "that it is hard for us to realize how sensational his photographs were to the astronomers of his time. . . . They showed much more detail than even the best drawings of the earlier visual observers." Keeler was generating the Hubble Space Telescope pictures of his time.

Outside his duties as Lick director, Keeler was spending all his available time on Ptolemy Ridge, becoming the world's expert on nebulae. "The [Crossley's] workmanship is poor and the design is clumsy, but on a fine night the photographic power is quite extraordinary. It has seemed to me worth while to devote some time to ordinary photography of nebulae, as nothing that I have yet seen in this line comes up to what I can get with the Crossley," he told a friend.

The week before and the week after the new Moon, when the lunar orb was in inky shadow, were his best viewing times. Only then was the sky dark enough to photograph the faint nebulae he was beginning to detect, without interference from a bright lunar spotlight. When the night was clear and calm, he often had time to take several exposures. But then there would be stretches, even when the sky was cloudless, when the wind was so strong that the Crossley shook on its mount, ruining his observation.

Keeler at last got around to his first spiraling nebula on April 4, 1899. He started off with one succinctly named M81, situated in the constellation Ursa Major, just above the "pot" of the Big Dipper. He carefully tracked it from nine until eleven o'clock that evening, as two hours were needed to gather enough light to record an image on the plate. Once the plate was developed, he right away noticed a faint spiraling but considered it "valueless." A misalignment of the telescopic axis had unfortunately led to the stars appearing as small arcs.

His luck was better the following month. With the Crossley fixed, he took several photos of M51, known as the Whirlpool for its wondrous view of the spiraling, face-on. Keeler's four-hour exposure captured aspects of the nebula never before seen, largely due to the steady air above Mount Hamilton. Keeler sent a transparency of this exposure to his friend George Ellery Hale, director of the Yerkes Observatory, where it took Hale's breath away. "Everyone in the Observatory considers [this picture] to be far superior to anything of the kind they have ever seen or expected to see," Hale responded enthusiastically.

There was something even more consequential in the image, although Keeler didn't appreciate the import right away. Surrounding M51 in the picture were seven more nebulae—though smaller and fainter. In a brief note to the Royal Astronomical Society in London, he listed the exact locations of these nebulae and described them. Some were round, others spindle-shaped or elongated. And that was only the

Photo of Whirlpool galaxy (M51) taken by James Keeler in 1899 with the Crossley telescope. One faint nebula seen in upper left.
(*Copyright UC Regents/Lick Observatory*)

beginning. "Several other faint nebulae, the positions of which were not noted, were observed during the search," he wrote. "In fact, this region seems to be filled with small, apparently unconnected nebulae, large numbers of which would doubtless be revealed by long-exposure photographs." It was a fascinating find, but he just assumed it was an uncommon grouping of nebulae, likely confined to that sector of the sky.

When a selection from Keeler's growing archive of pictures was prominently displayed at the Third Conference of Astronomers and Astrophysicists, held at Yerkes in September 1899, it created great excitement. Astronomers formerly skeptical of a reflector's value, such as E. E. Barnard, began to change their opinion. Barnard, who had fled from Lick to Yerkes during the Holden debacle, just stood in front of Keeler's photographs for hours, taking in every scrumptious detail of the Orion nebula, the Pleiades, and the M51 spiral.

Media savvy, Keeler knew the value of a good pitch in helping both the observatory and his career. After a well-publicized solar eclipse, he had advised a fellow astronomer, who was about to convey his eclipse observations to a conference, to dwell "on the successes rather than on the failures. If you were to tell a reporter that three plates out of ten were failures, he would receive a totally different impression from what he would if you gave him the equivalent statement that seven out of ten plates were successes." Keeler sent copies of his best pictures to the Royal Astronomical Society, the New York Academy of Sciences, and the American Philosophical Society in Philadelphia, all institutions that could influence opinions within the scientific community. He also made sure that Crossley, the reflector's former owner, received a particularly nice print of the Orion nebula. "The finest I have ever seen," replied the English businessman. "It proves to me how important it is not only to have a powerful instrument but also a site where it can be used to the greatest possible advantage." Getting his results widely distributed seems to have paid off for Keeler. In 1900 he was elected to the National Academy of Sciences, a year after he had received its prestigious Henry Draper Medal for astrophysical research. He was now one of America's leading astronomers.

In late summer, right before the Yerkes conference, Keeler had started to examine the faint nebulae more closely. He took a one-hour exposure of NGC 6946, a fuzzy patch first noticed by astronomer William Herschel at the end of the eighteenth century and listed as the

6,946th object in the New General Catalogue, published by J. L. E. Dreyer in 1888. Upon developing his plate Keeler saw immediately that it was yet another spiral, similar to M51 and M81 but smaller in size. A few nights later he examined two more fuzzy nebulae. Again he found, in each case, spiraling arms wrapping around a brightened center. All these dim nebulae appeared to be flattened disks, much like the Andromeda nebula, but they were set in different orientations.

And something more surprising developed as this work progressed. Each time Keeler took a photograph, he found even fainter nebulae loitering in the background of his image. At the start of his venture, when he first saw the seven nebulae on his plate of M51, he thought it "a rather remarkable number of nebulae to be found on a plate covering only about one square degree." That's a segment of the sky the size of two full Moons. But he soon discovered that this celestial flock wasn't so remarkable after all. With each additional picture Keeler took, he detected more and more nebulae arrayed over the heavens. Throughout the fall of 1899, whenever the nighttime sky was clear and moonless, he made his way to the Crossley and kept adding to his count. He took a four-hour exposure of NGC 891, a spiral seen edge-on, and the plate revealed thirty-one new nebulae, scattered around the central spiral like background extras in a movie scene. On a photograph of NGC 7331 he saw twenty more and "there are nearly as many on several other plates," he reported. "Besides these new nebulae . . . the plates contain a considerable number of objects which are probably nebulae so small that the resolving power of the telescope is insufficient to define them in their real form and to bring out their true character."

Keeler was dumbfounded. Space was awash with tiny nebulae, and most of them displayed a conspicuous spiral form, though seen from assorted angles. "There are hundreds, if not thousands, of unrecorded nebulae within reach of our 36-inch reflector," reported Keeler. By assuming that there were three new nebulae in each square degree (a number he admitted was far too conservative), he estimated that "the number of new nebulae in the whole sky would be about 120,000." He was positive there were more. Before this, about nine thousand nebulae had been cataloged by astronomers but only seventy-nine were identified as spirals, less than 1 percent. The Yerkes Observatory, in Wisconsin, by then had opened to great fanfare with a bigger telescope, one with a lens forty inches in width, but it still could not compete with Keeler's reflector. Even Barnard conceded that his new home at Yerkes,

Keeler's 1899 image of NGC 891 with background nebulae marked
(*Copyright UC Regents/Lick Observatory*)

situated at a more lowly thousand feet above sea level, was "a mirey climate for a great telescope and discoveries are few and far between."

In an article in *Astronomische Nachrichten*, a highly respected German astronomical journal, Keeler drew attention to his baffling finds:
"The spiral nebula has been regarded hitherto as a rara avis—a strange
and unusual phenomenon among celestial objects, to be viewed by the
observer with special interest, and marked in catalogues with exclamation points. . . . But so many other nebulae also proved to be spirals that
the classification . . . soon lost its significance. . . . The same form occurs
over and over again, on a smaller scale, among the fainter nebulae."
Spirals were now the norm, not the exception, in the celestial sky.
Keeler figured they must be an important constituent of the universe,
ranging in size "from the great nebula in *Andromeda* down to an object
which is hardly distinguishable from a faint star disk."

But what in blazes was a spiral nebula? No one knew for sure, solely
because there was as yet no way to determine the distance, a recurrent
problem for astronomers. If the spirals were nearby, part of the Milky
Way, then they would be relatively small given their size in the sky, each
possibly a new star forming. But if the spiraling patches were very far

away, then they would have to be huge to appear as they did in telescopic photographs, as big as the Milky Way itself.

To Keeler, the whirling shape seemed to indicate that the object, whatever its nature, was rotating. And like many of his contemporaries, he speculated that the spirals were somehow linked to star formation. "If . . . the spiral is the form normally assumed by a contracting nebulous mass," he pondered, "the idea at once suggests itself that the solar system has been evolved from a spiral nebula." Given this view, each spiral then marked the spot where a new star and its planetary companions were hatching. The idea that our solar system condensed out of a rotating nebula of gas had already been introduced by both Immanuel Kant and Pierre-Simon de Laplace decades earlier. In a lecture at Stanford University, Keeler made this very point: "The heavens are full of beautiful illustrations of the views of Laplace . . . [in] photographs of great spiral nebulae in various stages of condensation, taken recently with the Crossley reflector at the Lick Observatory."

Much as Einstein's relativity inspired numerous works of art and literature since its inception, so too did the nebular theory in the nineteenth century, as seen in this stanza from "The Princess," by Great Britain's poet laureate Alfred Lord Tennyson in 1847:

> This world was once a fluid haze of light,
> Till toward the centre set the starry tides,
> And eddied into suns, that wheeling cast
> The planets . . .

It's interesting to contemplate how far Keeler might have gone in this line of research. With his phenomenal skill at the telescope, he had a good shot at obtaining spectral data that forced him to consider other explanations for the nature of the spiral nebulae. "Keeler . . . was a far better trained, more experienced spectroscopist than any [other astronomer of his time]. No doubt he would have reached the conclusion that the spirals were galaxies of stars," contends Osterbrock, himself a Lick Observatory director seven decades after Keeler. Keeler might have also noticed, far earlier than others, that the spirals were racing away from the Milky Way at high velocities. He had the smarts, and he had the equipment. He had already obtained the velocities of myriad planetary nebulae and had a plan to move on to the spirals. His friend Hale had that impression; he was sure that Keeler intended to "follow up his remarkable beginnings with the Crossley reflector, cataloging the new nebulae, and doing something with their spectra."

But we will never know, for Keeler died unexpectedly on August 12, 1900, one month shy of his forty-third birthday. Throughout the spring and summer of 1900 Keeler had been suffering from what he called "a hard cold." An entrenched cigar smoker since his college days, he had already been experiencing heart problems. His doctor also diagnosed pleurisy of the lung, "nothing very serious," Keeler told friends, but he was likely afflicted with either emphysema or lung cancer. He couldn't manage walking the steep rise from the Crossley reflector back to his home without stopping several times short of breath. With his doctor forbidding him to continue observing, he left the mountain at the end of July for a short rest with his family. He was expecting to return to use a new spectrograph, just completed for the Crossley, and begin examining spiral nebulae. But within weeks Keeler died in San Francisco, after experiencing two strokes. The setback for astronomy, said his friend and colleague Campbell, was "incalculable." Harvard College Observatory director Edward Pickering wrote that the "loss cannot be overestimated. . . . There was no one who seemed to me to have a more brilliant future . . . or on whom we could better depend for important advances in work of the highest good." The journal *Science* ran a tribute to Keeler on the first page of its September 7, 1900, issue.

On Mount Hamilton, the memory of Keeler became sacrosanct and remains so to this day. He was the ideal director, an astronomer without equal cut down in his prime. But Keeler's acclaimed reputation beyond the Lick Observatory grounds gradually faded. In encyclopedias he is primarily remembered (if he is mentioned at all) for his work on Saturn's rings, with only a brief reference to his pioneering use of a reflecting telescope at high altitude, which allowed him to record the myriad spiral nebulae. Yet his tenacious pursuit of the nebulae with the Crossley reflector is truly his most lasting legacy. "The day of the refractor was over," said Osterbrock, "and although a few more intermediate-sized ones were built, no American professional astronomer ever thought seriously of building a very large telescope as anything but a reflector, after Keeler's work with the Crossley."

With his innovative spirit and success in restoring a once-despised instrument, Keeler pushed reflectors to the forefront of astronomical research. Campbell, who had been carrying out his program to map the motions of the stars, knew that Lick needed a second telescope in the southern hemisphere to complete the observations. Chosen as Keeler's successor to the directorship, he decided to build another 36-inch reflector, similar to the one that Keeler so successfully got working. In 1903 this telescope was erected on a site outside Santiago, Chile, where it was in operation for twenty-five years. The refractor at Lick had cost hundreds of thousands of dollars; Campbell built his Chilean scope for a thrifty $24,000.

In the fall of 1901, just a year after Keeler's death, the Yerkes Observatory assembled a trial reflector of its own in one of its small domes. With a mechanical system far superior to the Crossley, which allowed the mirror to be highly stable, this Yerkes reflector yielded photographs of nebulae that were even better than Keeler had obtained, despite its smaller 24-inch aperture. "The results obtained with the two-foot reflector show that very fine atmospheric conditions are necessary for the best results," reported the telescope's builder, George Ritchey. "It is interesting to think of the photographic results which could be obtained with a properly mounted great reflector in such a climate and in such atmospheric conditions as prevail in easily accessible parts of our country, notably in California."

Keeler not only turned reflectors into astronomy's instrument of choice, he inspired astronomers to take a new look at the universe. Was the cosmos defined as simply the Milky Way, or was there more to the universe than met the unmagnified eye? Keeler took a problem previously tackled by amateurs, for the most part—the spiral nebulae—and turned it into a prime concern for professional astronomers. He gave traditional astronomy a good shake at the end of the nineteenth century and in the process reinvigorated a debate that had been going on for centuries. What was the true nature of those irresolvable nebulae—so mysterious, so enthralling—that pervaded the celestial sky? Could the universe possibly be far larger?

[3]

Grander Than the Truth

Contemplating a universe of magnificent vastness has not been a recent affair. In the first century B.C., the Roman poet and philosopher Lucretius approached the question with cunning logic: "Let us assume for the moment that the universe is limited," he posed. "If a man advances so that he is at the very edge of the extreme boundary and hurls a swift spear, do you prefer that this spear, hurled with great force, go whither it was sent and fly far, or do you think that something can stop it and stand in its way?" For Lucretius and a few Greek thinkers before him, it was hard to imagine that an impenetrable cosmic barrier existed. It seemed ludicrous.

But Lucretius's reasoning never flourished. It was overshadowed by the authoritative cosmology espoused by Aristotle in the fourth century B.C. The noted Greek philosopher preferred a motionless Earth poised in the center of a celestial sphere of set dimensions, a concept of such influence that it endured for centuries. Over that time scholars only occasionally reflected on the possibility of a universe significantly bigger. In the sixteenth century, for example, Thomas Digges in England imagined the stars scattered throughout a boundless space, while in Italy Giordano Bruno presciently declared that "the center of the universe is everywhere, and the circumference is nowhere." Even Isaac Newton had a good scientific reason to prefer a cosmos without end. If the universe had a border, gravity would gradually draw all its matter inward, and ultimately the universe would collapse. To keep the cosmos stable—immutable and immovable—required that the stars be spread infinitely outward in all directions. "If the Matter was evenly disposed throughout an infinite Space," wrote Newton to a friend, "it could never convene into one Mass."

Yet most found such enormity difficult to grasp and horrifying to ponder. A character in Thomas Hardy's nineteenth-century novel *Two on a Tower*, an astronomer named Swithin St. Cleeve, gave splendid voice to this apprehension: "There is a size at which dignity begins; further on there is a size at which grandeur begins; further on there is a size at which solemnity begins; further on, a size at which awfulness begins; further on, a size at which ghastliness begins. That size faintly approaches the size of the stellar universe. So am I not right in saying that those who exert their imaginative powers to bury themselves in the depths of that universe merely strain their faculties to gain a new horror?"

Even as late as the eighteenth century, most celestial observers still backed away from questions of the universe's true size and nature, for professional astronomers at this time were primarily mathematicians who used Newton's laws to predict the motions of the Moon, planets, and comets. Stars themselves, as distinct celestial objects, were not yet as interesting or provocative to them as determining with utmost precision their coordinates (in essence, their heavenly latitude and longitude) for celestial atlases. As a result, cosmological conjectures on the universe's size, shape, and destiny were largely thrashed out by those on the fringe, such as Thomas Wright, a dilettante and schemer who clawed his way up the social ladder from a rather modest background as a carpenter's son. After serving as a watchmaker's apprentice, seaman, and then teacher of mathematics and navigation, he went on to make a comfortable living in England giving private lessons on architecture and science to noble families. He tutored Lord Cornwallis's daughters (sisters of the Revolutionary War general), hunted with the Earl of Halifax, and dined regularly for a time with the Duke and Duchess of Kent.

With the backing of his wealthy benefactors, Wright published a lavish book in 1750 titled *An Original Theory; or, New Hypothesis of the Universe*, which attempted to explain the structure of the Milky Way. Then thirty-nine years old, the Englishman applied his self-taught expertise in surveying and geometry to the question he had been pondering, off and on, for many years: Why does the Milky Way appear as a misty streak that stretches across the celestial sphere? Galileo with his

Thomas Wright of Durham
(*From Thomas Wright's* An Original Theory; or,
New Hypothesis of the Universe, *1750*)

telescope had revealed that this cloudlike band was composed of innumerable stars, but why should the stars arrange themselves in such a streamlike fashion?

Limited in formal education, Wright filled his book with arcane theological digressions, as was the style of his time, but in the midst of his ramblings he introduced the startling idea, now deemed obvious, that our position in space affects how we perceive our celestial environment. He proposed that the Milky Way could be "no other than a certain Effect arising from the Observer's Situation, I think you must of course grant such a Solution at least rational, if not the Truth; and this is what I propose by my new Theory." Hedging his bets, he offered a couple of explanations for the Milky Way's appearance. One model pictured the stars moving in a vast ring, much like the rings of Saturn, around a central point. But, strongly guided by his religious views, he preferred to

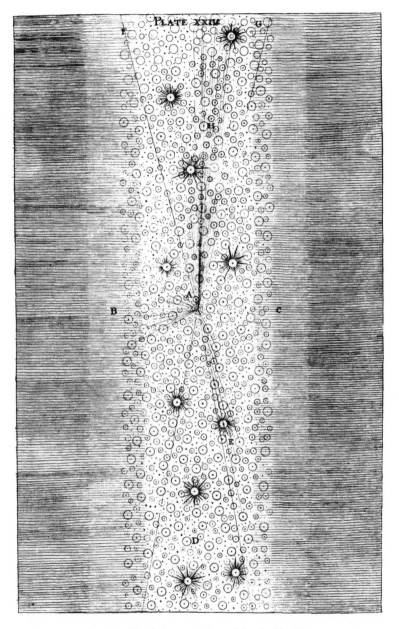

Thomas Wright's engraving of the Milky Way,
depicting it as a disk of stars (*From Thomas Wright's
An Original Theory; or, New Hypothesis of the Universe, 1750*)

think of the Milky Way as a thin spherical shell of stars—essentially a bubble—with the solar system on the surface and the Eye of Providence, the "agent of creation," residing in the center.

Wright included a number of lush illustrations, thirty-two in all, which conveyed his seminal ideas better than the text itself. One engraving—the one still found in textbooks today—displays the Milky Way as a flat layer of stars. This was a first step in imagining his huge spherical shell. "I don't mean to affirm that [the disk] really is so in Fact," he wrote, "but only state the Question thus, to help your Imagination to conceive more aptly what I would explain." Looking along the plane of Wright's big, gently curving shell, in which the Sun is embedded, Earth's inhabitants would readily perceive a disklike structure. The Milky Way appears as a band, mused Wright, because we observe this thin layer of stars edge-on; when looking away from the plane, stargazers see fewer stars.

Wright went on to consider whether certain cloudy spots, then being observed in the heavens in greater numbers, might be additional creations, bordering upon us but "too remote for even our telescopes to reach," countless spheres with many "Divine Centres." He seemed to be echoing the Swedish philosopher Emanuel Swedenborg, who in 1734 also wondered if "there may be innumerable other spheres, and innumerable other heavens similar to those we behold, so many, indeed, and so mighty, that our own may be respectively only a point."

If left there, Wright's imaginative ideas and dazzling illustrations would have likely generated hardly a footnote in astronomical history. He even reverted to a more medieval cosmic model, outrageous in its fires-of-hell imagery, some years later. But as British historian Michael Hoskin first pointed out, Wright managed to achieve a degree of acclaim when others widely disseminated what they *thought* he meant. A few months after the publication of An Original Theory, its key ideas were summarized in a Hamburg journal. The review selectively stressed Wright's concept of the Milky Way as a flat ring, rather than a sphere. This ring was compared to our solar system, with the stars moving around much like the planets circling the Sun. Inspired by this brief journal account, a young Prussian tutor in 1755 wrote his own book on the subject. Like Wright, he described the nebular patches in the nighttime sky as "just universes and, so to speak, Milky Ways. . . . These higher universes are not without relation to one another, and by this mutual relationship they constitute again a still more immense system." These words were virtually ignored until the author—Immanuel Kant—achieved fame as one of the world's great philosophers. Even then his ideas on the universe's design almost didn't survive. Kant's manuscript was destroyed when his printer went bankrupt. Fortunately, a shorter version was tucked away in the appendix of another book that he published in 1763.

Kant, trained in science, imagined that Wright's ring of stars was actually a continuous disk. This was more than wishful thinking; he was inspired by the latest astronomical evidence. Pierre-Louis de Mauper-tuis in France had been observing dim objects in the sky, what he called "nebulous stars," that appeared elliptical in shape, the very way a disk would appear when tipped at an angle. "I easily persuaded myself," wrote Kant, "that these stars can be nothing else than a mass of many fixed stars. . . . On account of their feeble light, they are removed to an inconceivable distance from us." With such reasoning, Kant arrived at the correct image of a galaxy's basic structure. Kant was astonished that previous observers of the heavens had not figured out the structure of our galaxy earlier. The Milky Way resembled a flat plate. Moreover, it was just one of many star-worlds scattered throughout the heavens. The German scientist Alexander von Humboldt later dubbed them Kant's "island universes," a phrase that would resonate throughout the astro-nomical community like a mantra—some championing Kant's vision, others deriding it. Johann Lambert, a former tailor's apprentice in Alsace who had learned some science on his own, independently arrived at a similar conclusion in 1761 with his *Cosmological Letters on the Arrangement of the World-Edifice*. With the publication of these works, the "mystery of the nebulae" came to vex both philosophers and astronomers for more than a century.

From the days of Ptolemy, astronomers talked about certain stars in the sky that appeared "cloudy" to the eye. The most famous is in the north-ern constellation Andromeda, the mythical princess situated in the sky near her parents, Cassiopeia and Cepheus, and her husband, Perseus. At her waist is an oval patch of light, best seen on the darkest of nights. As early as the tenth century, astronomer Al-Sufi of Persia noted it as a "little cloud" in his catalog of the heavens. With the invention of the telescope more nebulae were sighted, and by the early 1700s Edmond Halley (of comet fame) counted six in all. To some observers, these pale entities were breaks in the celestial sphere, through which the light of the Empyrean—the highest heaven—came shining down. Others suggested that they were the hazy atmospheres surrounding distant stars. Halley, however, thought of them as unique celestial objects, unlike anything else in the heavens. They "appear to the naked Eye like small Fixed Stars," he wrote, "but in reality are nothing else but the Light coming from an extraordinary great Space in the Ether; through which a lucid *Medium* is diffused, that shines with its own proper Lustre."

Gradually found in greater numbers, these celestial objects took on even more importance in 1781 when the celebrated comet hunter Charles Messier published in France his list of more than one hundred nebulae, a directory that is still used today. The Andromeda nebula, for example, is commonly known as M31 because it's the thirty-first nebula in Messier's catalog. Messier, though interested in the nebulae themselves, primarily wanted to let his fellow observers know that these celestial regulars, the most prominent of their kind, should not be mistaken for comets. He was putting up cosmic road signs for his colleagues, pointing out those nebulae visible above the horizon from the latitude of Paris.

No one was more excited by Messier's list than William Herschel, England's soon-to-be crown prince of astronomy. As soon as Herschel received a copy of Messier's catalog, he immediately aimed a telescope at the celestial clouds. "I . . . saw, with the greatest pleasure, that most of the nebulae, which I had an opportunity of examining in proper situations, yielded to the force of my light and power, and were resolved into stars," he wrote a few years later. He was the first to make this discovery, using a telescope twenty feet in length with a mirror then twelve inches wide. It was the most powerful in its day, allowing him to see that many of the nebulae (what we now call open clusters and globular clusters) were actually comprised of hundreds and thousands of stars. This led him to the belief that *all* nebulae were far-off systems of stars. Any nebula still appearing cloudlike through his eyepiece, he figured, was simply too distant to behold individual stars clearly.

Herschel promptly initiated a grand hunt for nebulae, literally sweeping the heavens with his giant reflector. Previous endeavors to spot nebulae paled beside this effort. By 1786 he had sighted a thousand new nebulae and star clusters; three years later he added hundreds more. "These curious objects, not only on account of their number, but also in consideration of their great consequence, [are] no less than whole sidereal systems," he wrote. He even boasted at one point that he had discovered fifteen hundred new universes. Each, he excitedly reported, "may well outvie our milky-way in grandeur."

Herschel had come late to this pursuit. Raised in the Duchy of Hanover (now part of Germany) within a family of musicians, he fled as a teenager to England, Hanover's ally, in the midst of war. There he supported himself by copying musical manuscripts, composing, giving private lessons, and performing in local concerts. Eventually he obtained financial security by becoming a choral director in the city of Bath. Yet he was restless for more intellectual stimulation.

Inspiration arrived on May 10, 1773. On that day Herschel, then thirty-four years old, bought a copy of a popular astronomy textbook. "When I read of the many charming discoveries that had been made by means of the telescope," said Herschel, "I was so delighted with the subject that I wished to see the heavens and Planets with my own eyes thro' one of those instruments." By the autumn he was beginning to handcraft metal mirrors for a reflecting telescope. He became obsessed with his new hobby, soon shifting his interests from the music of the Earth to the music of the heavens. So passionate was his commitment to astronomy that his younger sister, Caroline, who had earlier joined him in England, fed him morsels of food by hand, so that he would not have to pause while grinding and polishing. Pointing his home-built instruments toward the sky, he came to memorize the heavens and in 1781 climactically spotted Uranus, the first planet discovered since the dawn of history. He was promptly elected a fellow of the Royal Society and procured an annual stipend from England's King George III, a pension that at last allowed Herschel to devote himself to his astronomical interests, especially building ever-larger telescopes (the largest he ever constructed was forty feet long).

Herschel was far ahead of his time, as he used his telescope to examine the universe much the way an astronomer would today. While other astronomers in his day focused solely on the motions of the stars and planets, he was determined to discern nothing less than the "construction of the heavens," the title of one of his most notable papers. He wanted to reach out into distant space, far beyond the realm most studied by his contemporaries. Wright and Kant had done the same, but they were merely theoretical speculators, not practicing astronomers. Herschel insisted that his ideas be "confirmed and established by a series of observations." Photography was still decades away, so to do this

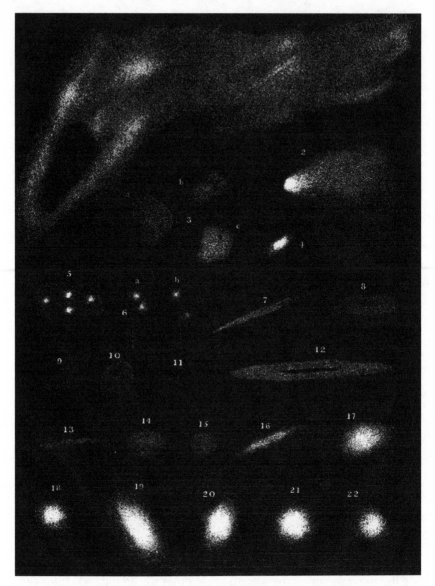

Drawings of nebulae by astronomer William Herschel, 1811
(*From* Philosophical Transactions of the Royal Society of London
101 [1811]: 269–336, Plate IV)

he had to spend hours at his eyepiece, awkwardly perched on a platform at the top of his telescope. So skilled did he become at fashioning telescopes that his instruments were the only ones at the time capable of seeing out to cosmological distances. His tireless assistant Caroline was often with him, jotting down the positions and descriptions of the many nebulae he came upon during his scans of the heavens.

"I have seen double and treble nebulae, variously arranged; large ones with small, seeming attendants; narrow but much extended, lucid nebulae or bright dashes; some of the shape of a fan, resembling an electric brush, issuing from a lucid point, others of cometic shape, with a seeming nucleus in the center; . . . when I came to one nebula, I generally found several more in the neighbourhood," he reported. At one point, Herschel even imagined other beings residing within those nebulae, looking back at us: "The inhabitants of the planets that attend the stars which compose them must likewise perceive the same phænomena. For which reason they may also be called milky-ways by way of distinction." He seemed to be confirming the Wrightian and Kantian visions: that the universe is vastly larger and more complex than previously imagined. The Milky Way was a cohesive system of stars and beyond that was a limitless universe, populated by other stellar systems, comparable to our own.

Astronomers might have become quite comfortable with and accepting of the idea that other galaxies existed, more than a century before Hubble proved it conclusively, if not for the fact that Herschel abruptly changed his mind about those hundreds of "new universes." A new observation forced him to reconsider his previous assertions. It happened on a cold November evening in 1790 when Herschel came upon an eighth-magnitude star that was surrounded by a faintly luminous atmosphere of considerable extent. "A most singular phænomenon!" he jotted down in his notebook. He called this haze a "planetary nebula" because of its resemblance to a planetary disk (as noted earlier, now known to be an aging star shedding its outer envelope of gas). "Cast your eye on this cloudy star," he wrote, "and the result will be no less decisive . . . that *the nebulosity about the star is not of a starry nature. . . .* Perhaps it has been too hastily surmised that all milky nebulosity, of which there is so much in the heavens, is owing to starlight only." In Herschel's mind, nebulae had to be comprised of either stars or a "shining fluid"—not both. So he decided that any nebulae that remained unresolved through his telescope were no longer distant stellar systems, but instead collections of luminous matter, likely the stuff out of which stars ultimately condensed.

Herschel's telescopes were so much better than the equipment of any other astronomer at the time that his colleagues trusted his judgment on this matter. They simply didn't have the telescopic power to confirm his findings. As a result, Herschel's pronouncement became the accepted wisdom. The universe swiftly shrank back to the borders of the Milky Way. We were alone in the universe once again . . . at least for a while.

Throughout the nineteenth century, the two explanations for the unresolved nebulae went through a relentless tug-of-war, one side winning the hearts of astronomers for a time, then the other. Some insisted they were nearby clouds of gas, while others championed them as far-off islands of stars. Each faction was seeking a solitary explanation, simple and elegant—and that meant choosing between the two possible options.

Cosmology at this time continued to be of more interest to independent astronomers than the professionals who toiled at university- or government-sponsored observatories, and it was one of these self-directed observers who gave renewed hope to those who favored the idea that the dim nebulae were similar to the Milky Way, separate galaxies whose individual stars over the vast distances melted into a uniform pool of light. The excitement arose when William Parsons, the third Earl of Rosse, constructed a giant telescope on the grounds of his ancestral home, Birr Castle, in central Ireland, seventy miles west of Dublin. So big was the telescope tube that at the observatory's opening ceremony, a dean of the Church of Ireland walked right down the huge cylinder wearing a top hat and an open umbrella.

Young Rosse (then Lord Oxmantown, prior to succeeding his father in the earldom) served in the British Parliament, but his passion was telescope-building, with his decided aim, according to those who knew him, "to make a telescope of the largest dimensions possible with the resources of his time." In 1834, at the age of thirty-four, Rosse left politics to devote himself to a newfound career as a gentleman scientist. He had long wanted to surpass Herschel's instruments in size and devised the methods himself for casting and polishing the metal mirror in his own workshops, personally training the laborers on his estate to assist him. Though an aristocrat, he put on no airs; a British reporter once caught him working at a vise, his shirtsleeves rolled up, displaying brawny arms. The mirrors he constructed were made out of a tin and copper alloy, a blend that resulted in a reflectivity almost as high as silver. Rosse's first big success was a three-foot-wide mirror mounted in a tube twenty-six feet long. "It is scarcely possible to preserve the necessary sobriety of language in speaking of the moon's appearance with this instrument," reported a friend.

The triumph gave Rosse the confidence to construct a mirror twice the size, taking no notice that the Irish weather was more infamous for rain than clear skies. First put into operation in 1845, this reflector, when erect, was said to resemble one of the ancient round towers of Ireland and was dubbed the "Leviathan of Parsontown." "Sweeping down from the moat towards the lake, stand two noble masonery walls," reported a houseguest. "They are turreted and clad with ivy, and considerably loftier than any ordinary house. As the visitor approaches, he will see between those walls what may at first sight appear to him to be the funnel of a steamer lying down horizontally." It was the telescope's immense wooden tube, which was more than fifty feet long and held a polished metal mirror six feet in diameter. This mirror provided fourteen times more surface area for collecting light than Herschel's most productive telescope. A pulley system, attached to the top of the tube, allowed the telescope to be pointed by two men on the ground. A series of staircases and galleries provided the observer access to the mouth of the great tube. It was an astounding telescope size for its time and wouldn't be matched for another seven decades.

The Leviathan's prime targets were the "strange stellar cloudlets that fleck the dark vault of the heavens." Rosse was determined to see if he could resolve the nebulae—those that remained stubbornly cloud-like—into stars. But what he turned up was something even more intriguing.

In the spring of 1845, Rosse and his assistant Johnstone Stoney began to study the fifty-first nebula, M51, in Messier's famous catalog. When William Herschel viewed it years earlier, he saw only a bright round nebula; his son later observed it as a ring with two branches. But Rosse, to his amazement, detected a distinct coiling, arms of gas wrapped around M51's center like a whirling pinwheel. No one had ever anticipated something like this. Some nebulae were shaped like spirals, "a structure and arrangement more wonderful and inexplicable than anything which had hitherto been known to exist," reported Great Britain's Royal Astronomical Society.

In these days before astrophotography, Rosse sketched a picture of the configuration with painstaking care. "With each successive increase of optical power, the structure has become more complicated and more unlike anything which we could picture to ourselves," Rosse reported.

A drawing of Lord Rosse's Leviathan (*From* Philosophical Transactions of
the Royal Society of London *151* [*1861*]: *681–745, Plate XXIV*)

"That such a system should exist, without internal movement, seems to
be in the highest degree improbable." This is when M51 came to be
called the Whirlpool because of the striking swirl of its appearance.
Rosse went on to discern more than a dozen such spiral nebulae in the
celestial sky.

Despite Rosse's gorgeous drawings, a few believed the spiraling lanes
of nebulous matter "existed only in the imagination of the astronomer."
Rosse's mirror was so large—its light-gathering power so great—that
no other telescope could verify his find. But for others, the discovery
revived and energized Herschel's earlier speculation that other systems of stars resided outside the borders of the Milky Way. Scottish
astronomer and science popularizer John P. Nichol was certainly
thrilled, for he had long been pushing the idea that "numerous firmaments, glorious as ours, float through immensity, doubtless forming one
stupendous system." He was a Kantian. Nichol thought of a galaxy
(what he called a "grand group") as the chief feature in the universe. "It
is indeed wholly unlikely that our group, as a single instance of a
species, should rest alone and forlorn amidst desert untenanted Space,"

Lord Rosse's drawings of M51 (top) and M99 (bottom),
which in the mid-1840s were the first nebulae found to have a spiraling
structure (*From* Philosophical Transactions of the Royal Society of
London *140* [*1850*]: *499–514, Plate* XXXV)

he wrote. The universe, to Nichol, was *"thronged with similar clusters,*
separated far from each other as islands in the great Sea." Some "are situated so deep in space," he went on, "that no ray from them could reach
our Earth, until after travelling through the intervening abysses, during
centuries whose number stuns the imagination." He even imagined
some so far distant that their light left "at an epoch farther back into
the Past than this momentary lifetime of Man, by at least THIRTY
MILLIONS OF YEARS!" This was a brave estimate for someone to
make in 1846, a time when many in the public still held to a biblical
age for creation of only six thousand years and scientists over the previous fifteen years were just beginning to find evidence (then still controversial) that it was much longer.

It was said that Rosse's telescope was "poised so skilfully that a child
could guide its movements." By one astronomer's reckoning, it could

gather twenty thousand times the light of the unaided eye. But the Leviathan did possess one blatant shortcoming: "It does not present objects in a perfectly distinct manner," said Richard Proctor, a contemporary of Rosse's who once had the opportunity to peek at the sky with the giant telescope. "It used to be remarked of the great four-feet reflector of Sir William Herschel, that it 'bunched a star into a cocked hat.' " Proctor believed the same was true for Rosse's great instrument. The sheer weight of the telescope's mirror—a truly massive four tons—distorted its images at times. Views of planets through the Rosse scope, judged Proctor, were "perfectly wretched." Although the metal reflector had its good days as well as bad, criticism like this dampened enthusiasm for further advancement on mirrored telescopes. Herschel and Rosse had made great strides with their big reflectors, but most astronomers still preferred gathering their celestial light with lenses. Not until James Keeler got the Crossley reflector up and running at Lick Observatory in the 1890s did astronomers at last change their minds on their instrumental preference.

Rosse, an engineering wizard, was always more attracted to constructing a telescope than to using it. His astronomical work continued for some twenty years, but most of the measurements were carried out by associates. His greatest contribution to astronomy was his discovery of the spirals, revealed when the Leviathan first went into operation. In doing this, he introduced an entirely new celestial creature, a novel species of nebula that would tantalize and frustrate astronomers for decades to come.

Popular interest in astronomy grew immensely in the nineteenth century, likely fueled by the rising use of photography, which at last allowed the general public to view and admire gorgeous pictures of the celestial heavens at their leisure. The first known daguerreotype of a celestial object, the Moon, was taken by the American physician John Draper in the 1840s. Later, the brightest stars were imaged. But the process became more routine with the introduction of more sensitive plates in the 1870s, which allowed fainter and more diaphanous objects, such as nebulae, to be photographed.

At the same time, the invention of the spectroscope offered a novel means for astronomers to pursue the mystery of the nebulae. Widely known as the "new astronomy" or astrophysics, spectroscopy was particularly favored by the enthusiasts who lacked formal mathematical training in classical astronomy. Professional astronomers were slow to appreciate the power of the new instrument. Indeed, they were distraught to see their telescopes, once set up like grandiose metal sculptures within towering domes, now surrounded by a chaotic array of chemical and electrical contraptions required to carry out spectroscopic work. But nonprofessionals astutely perceived that spectroscopy, despite its inelegance, opened up virgin astronomical territory. Rather than dully peg the positions of stars to stupefying accuracies, they were going to discern the very nature of celestial objects—*what* they are instead of *where* they are.

No one was more dedicated or persistent in this new enterprise than William Huggins. At the age of thirty he sold his textile business in England and erected a private observatory at Tulse Hill, then a rural area about four miles south of central London. Soon tiring of routine astronomical observations, he was reinvigorated when he heard about the latest spectroscopic discoveries. He compared it to "coming upon a spring of water in a dry and thirsty land." By 1862 he was able to show that the elements found both on the Earth and in the Sun also dwelled in the distant stars. "The chemistry of the solar system prevailed," said Huggins, "wherever a star twinkled."

Then, on the evening of August 29, 1864, he shifted his attention from stars to nebulae. He aimed his telescope at a bright planetary nebula in the Draco constellation. He recalled in a memoir years later that he felt "excited suspense, mingled with a degree of awe" as he put his eye to the spectroscope. The spectrum he beheld was a surprise: "A single bright line only!" he noted. "At first I suspected some displacement of the prism, and that I was looking at a reflection of the illuminated slit. . . . This thought was scarcely more than momentary; then the true interpretation flashed upon me. . . . The riddle of the nebulae was solved. The answer, which had come to us in the light itself, read: Not an aggregation of stars, but a luminous gas." A star was simply too complex to be emitting a single spectral line; the emitter had to be a gaseous cloud, he thought, readying itself for stellar construction. In light of this and other findings, it became more popular to think of *all* nebulae as embryonic stars and planetary systems in the making. This idea was strengthened in 1888 when the English celestial photographer Isaac Roberts captured a full picture of the Andromeda nebula, an astounding feat at the time because of its faintness. When it was displayed at a

Royal Astronomical Society meeting, murmurs could be heard in the audience: "The nebular hypothesis made visible!" The photo displayed a bright core surrounded by a wide, hazy cloud. When Huggins saw the image, he exclaimed that it had to be "a planetary system at a somewhat advanced stage of evolution; already several planets have been thrown off."

With the great weight of his opinion, Huggins helped force the pendulum the other way. The island-universe theory was no longer a viable contender; it became passé. In his late-nineteenth-century *A Text-book of General Astronomy for Colleges and Scientific Schools*, a classic in its day, astronomer Charles Young stressed that astronomers no longer considered a spiral nebula as "a 'universe of stars,' like our own 'galactic cluster' to which the sun belongs. . . . In some respects this old belief strikes one as grander than the truth even. It made our vision penetrate more deeply into space than we now dare think it can." To Young, the Milky Way was some 10,000 to 20,000 light-years wide. "What is beyond the stellar system, whether the star-filled space extends indefinitely or not, no certain answer can be given," he said.

The island-universe theory had already been shaken in 1885 when a nova—a new pinpoint of orange-yellow light—was sighted near the center of the Andromeda nebula. At its brightest, around the sixth magnitude, this nova was nearly as luminous as the *entire* nebula. "This strange and beautiful object has broken silence at last, though its utterance may be difficult to interpret," said the Greenwich Observatory astronomer E. Walter Maunder.

If Andromeda were a distant external universe, it was reasoned, the nova had to be shining with the energy of some fifty million suns, "a scale of magnitude such as the imagination recoils from contemplating," said Agnes Clerke, a nineteenth-century historian of astronomy. That was actually a stupendous underestimate of the nova's power, but even that tally was too preposterous to consider in any serious fashion in 1885. The idea that a star could totally obliterate itself as an explosive supernova was not even a fantasy at the time. There was no physics to explain it. Stars were regarded as stable and enduring. It seemed more likely that the nova was an infant sun condensing and turning on within a vast collection of luminous matter on the edge of the Milky Way or perhaps a dark star running into nebulous matter and provoking an incandescent outburst.

Astronomer Edwin Frost, then nineteen and entering his senior year at Dartmouth College when the nova appeared, recalled the event with great vividness: "[The nova] was in the heart of the Great Nebula . . . and was a star of about the seventh magnitude. It thus became the only individual star distinguishable in this nebula, which at that time we supposed to be a purely gaseous body. . . . The distance of the nebula was then not regarded as greater than that of the stars in our portion of the Milky Way. . . . Among astronomers, as well as the public generally, it was thought that we might be observing the sudden transformation of the nebula into a star," and perhaps a planetary system as well. Another great nova, dubbed Z Centauri, appeared in the spiral nebula NGC 5253 a decade later, reinforcing the belief that spiral nebulae were relatively close by. Given what astronomers then knew about stars, there was no other explanation.

So, by the turn of the twentieth century, most astronomers had settled on this common story for the spiral nebulae—that they were new stars and planets emerging. This idea gained momentum when Thomas Chamberlin, a respected geologist, joined up with Forest Ray Moulton, an expert on celestial mechanics, on modeling how the solar system came to be formed. The Chamberlin-Moulton theory suggested that a nomadic star passed near our Sun long ago, drawing out streams of gas. This material eventually became a rotating nebula with spiraling arms, from which the planets slowly condensed. Chamberlin, while working on this idea at the University of Chicago, had heard about the amazing images of spiral nebulae that James Keeler was obtaining with his reflector atop Mount Hamilton, which seemed to suggest that he and Moulton were on to something: The spirals might be the gas, just recently torn off and ready for condensation into the planets that would eventually orbit the star, the bright center of the spiral nebula. Chamberlin wrote Keeler, saying, "[I would deem] it a very great favor to be able to make use of your great harvest of new forms." Keeler obliged.

"The question whether nebulae are external galaxies hardly any longer needs discussion. It has been answered by the progress of discovery," declared Clerke with confident finality in her influential book *The System of the Stars*. "No competent thinker, with the whole of the available evidence before him, can now, it is safe to say, maintain any single nebula to be a star system of coordinate rank with the Milky Way." To Clerke, such contemplations were "grandiose" and "misleading." Our galaxy and the universe were one and the same—synonyms in the dictionary of the heavens.

But soon after Clerke wrote her comments, new observations were beginning to suggest something very different. At the Potsdam Observatory, in Germany, Julius Scheiner spent seven and a half hours in January 1899 gathering a spectrum of the Andromeda nebula. What he saw was unexpected. The spectrum did not resemble a cloud of gas, such as the Orion nebula, at all. Instead, it resembled the light emitted by a vast collection of stars. "That the spiral nebulae are star clusters is now raised to a certainty," reported Scheiner. He began to imagine that the Milky Way itself was a spiral nebula, very similar to Andromeda. But at that point Scheiner was effectively a lone voice in the cosmic wilderness. At the Lick Observatory, Keeler took special note of the German's finding but died before he could follow up.

A further investigation was not undertaken until 1908, when Edward Fath, a graduate student at Lick, used the Crossley telescope to confirm Scheiner's findings on the spectrum of Andromeda (M31), as well as several other spirals, for his dissertation. It was wearying work, as Fath had to sustain a photographic exposure over several nights. One plate was exposed for a total of eight hours and forty-seven minutes. Another took more than eighteen hours. But the tediousness of the procedure paid off. To Fath, the results were unmistakable: from its spectral signature, Andromeda appeared to consist of myriad stars, many of them similar to our own Sun. As a double check, he took the spectra of some globular clusters, which were known to be assemblies of stars. Each spectrum looked exactly like Andromeda's.

"The hypothesis that the central portion of a nebula like the famous one in *Andromeda* is a single star may be rejected at once," reported Fath, "unless we wish to modify greatly the commonly accepted ideas as to what constitutes a star." He suspected the spirals were very remote, as the stars could still not be resolved into individual pinpoints of light, but he had no definitive proof—no slam dunk—to back up that guess. In 1908 there was as yet no way to measure the distance out to Andromeda directly.

As a result, Fath didn't promote his conclusion, perhaps because he was still a lowly graduate student and not in a position of authority to overturn the nebula-as-new-solar-system belief. Or perhaps because Lick Observatory director W. W. Campbell, a careful and conservative man of science, instructed him to downplay his speculations. Whatever the reason, Fath took a particularly cautious tone in the close of his official report. He said that his interpretation "stands or falls" on the question of determining a true distance to a spiral.

The response to Fath's report was like the sound of one hand clapping. Aside from a few outliers, hardly anyone else cared. Fath was soon offered a post at the Mount Wilson Observatory, where he did some follow-up work for a number of years but arrived at no breakthroughs. He eventually settled into a teaching job at Carleton College in Minnesota.

And that's where the matter stood until a man, whom Keeler had once rejected for a Lick graduate fellowship, took over the Crossley reflector in 1910 and continued the groundbreaking work of both Keeler and Fath. And in doing so, Heber Curtis challenged the conventional wisdom with single-minded determination. With great industry and zeal, he took on the problem of the spiral nebulae and made it his own.

[4]

Such Is the Progress of
Astronomy in the Wild and Wooly West

s Lick Observatory entered its third decade, life on Mount Hamilton continued to be a rustic adventure. Residents hiked the mountain trails, staged amateur theatricals, and read aloud around the roaring fireplace on frosty evenings. As long as the weather was favorable, the telescopes were scheduled for use nearly every night of the year. The lone exception was Christmas Eve, when operations were shut down for the holiday and the graduate students would sneak quietly into the cavernous dome to hang their stockings on the gear of the giant telescope.

There was a new addition to the observatory grounds, a tennis court, where, on Saturday afternoons, as one onlooker described it, "a spectacular performance is kept up, consisting of wild up-bursts of tennis balls, a la Roman candle, followed by hot chases down the canyons." A molasses jug served as the loving cup for the annual Fourth of July tournament.

Those wanting to go into town often hitched a ride with one of the lucky few, such as Heber Curtis, who owned a car. The astronomer would load people into his Mitchell automobile, nicknamed Elizabeth, making sure to stash a bag of flaxseed in the trunk, so he could pour the seeds into the radiator whenever it started leaking. Pictures of Curtis in his later years, taken after an illness, typically depict a small and stern-looking man. But while he was at Lick, the students knew him as a "wonderfully kind, jolly person, always smiling, always happy." His genial composure was only broken when he had to sneeze, a feat once described as "remarkable."

By the 1910s the island-universe theory, dormant for many years, was slowly reemerging among a select group of scientists in both the United States and Europe. These astronomers were specifying that the spirals' sizes and the brightness of their novae only made sense if they were milky ways at great distance. The highly respected English astrophysicist Arthur Eddington was captivated by the vast breadth of this idea; it engaged his theoretical fantasies. "If the spiral nebulae are within the stellar system [the Milky Way], we have no notion what their nature may be. That hypothesis leads to a full stop," he noted. "If, however, it is assumed that these nebulae are external to the stellar system, that they are in fact systems co-equal with our own, we have at least an hypothesis which can be followed up. . . . [It] opens up to our imagination a truly magnificent vista of system beyond system . . . in which the great stellar system of hundreds of millions of stars (our galaxy) . . . would be an insignificant unit." For Eddington, the heavens just seemed to make more sense viewed from this grander perspective.

The epicenter of this resurgence was located right at the Lick Observatory, where its director, W. W. Campbell, was at last persuaded by the mounting evidence and openly declared that thinking about the spirals as enormous distant bodies was "in best harmony with known facts." And those facts were largely being gathered by Curtis, one of his most able staff members. Campbell was still focused on his monumental campaign, a virtual assembly line of stellar measurements systematically proceeding from target to target, to catalog the velocities of stars within the Milky Way. The survey was being done in hope that the data would reveal new clues on stellar evolution. It was left to Curtis to get back to the Crossley telescope and revive the observatory's investigation of the spiral nebulae, a program that had not been a top priority since Keeler's death. The compact reflector, however, was still one of the best tools around for imaging and analyzing the hazy celestial clouds.

Curtis, a gifted mechanic, right away made significant improvements to the telescope. First off, he erected a new observing platform that could be raised and lowered by an electric motor, installed a powered dome shutter, and devised a better mechanism for driving the telescope. The mirror had already been remounted in 1904 into a thick metal tube, whose rivets along the side made it resemble a beam on a naval battleship. This telescope remains in operation, now searching for extrasolar planets. It's possibly the oldest reflecting telescope still in use for professional research.

Heber Curtis standing by the renovated Crossley telescope
(*Mary Lea Shane Archives of the Lick Observatory, University
Library, University of California–Santa Cruz*)

When Curtis rekindled Keeler's pursuit of the spiral nebulae, the island-universe theory was regarded as just a good guess, an intuitive suspicion. Curtis was after more concrete proof. He started to dig deeper into the problem, in the same way that Keeler would likely have proceeded. But Keeler had a mere two years to work with the Crossley before his death; Curtis, fortunately, had more time, which allowed him to extend astronomy's knowledge of the spirals throughout the 1910s. This celestial quest became, in the words of a fellow astronomer, Curtis's "magnum opus."

No one was more surprised perhaps at the zigs and zags in Curtis's career path than Heber Doust Curtis himself, who went to college at the University of Michigan in Ann Arbor, at the same time that Campbell happened to be teaching there. But their paths never crossed, for Curtis was a dedicated student of the ancient languages—Latin, Greek, Hebrew, Sanskrit, and Assyrian—earning first a bachelor's degree, then a master's. At this stage, Curtis voiced no interest whatsoever in science and never set foot inside an observatory. After teaching high school briefly in Detroit, he moved to California in 1894 to become a professor of Latin and Greek at Napa College, a small institution north of San Francisco. Curtis seemed destined for a life of quiet scholarship in the classics, until he came upon a small telescope at the college and on an impulse began to tinker with it.

His tiny college later merged in 1896 with the University of the Pacific, situated in the San Jose area, and he moved there, opportunely within the shadow of Lick Observatory. He continued his astronomical activities and got so caught up in his newfound hobby—and so adept at observing—that he was chosen to teach mathematics and astronomy at the small college. He was even able to spend some time on Mount Hamilton during the summers of 1897 and 1898 as a special student. The experience convinced him that he wanted to make astronomy his life's work. He hoped to continue at the Lick Observatory as a graduate student, but his inadequate preparation in science put up roadblocks. Keeler, Lick's director at the time, was looking for someone more professionally skilled in spectroscopy. Curtis was finally offered a fellowship at the University of Virginia, where for his PhD he reluctantly focused on a more mathematical topic, celestial mechanics, although along the way he made sure to get as much instrumental experience as possible. It was a risky move. He was resigning from a college professorship to start anew as a student in a field in which he had no prior training—and with a growing family to support as well.

Serendipity offered an assist. Just as Curtis was headed east in 1900 to begin his doctoral studies, Lick astronomers William Campbell and Charles Perrine were traveling to Georgia to scrutinize a solar eclipse, whose shadow was scheduled to cut across the southeastern United States. Curtis signed on as a helper, saying he was "ready and glad to be put at anything from a shovel up." Given this opportunity, he proved to the Lick men that he could handle a telescope and spectrograph as if he had been using them all his life. Campbell took notice. As soon as Curtis finished his degree at Virginia in 1902, Campbell, by then Lick's director, hired him on as an assistant. To Curtis, having lived on a small mountain in Virginia was simply good training for a life on Mount Hamilton, where kids hunted rattlesnakes for fun in the summertime.

Curtis arrived at Lick covered in thick yellow dust from the long stagecoach ride, raring to begin his research straightaway. For the first few years, he focused on traditional Lick specialties, such as measuring stellar velocities, computing the orbits of binary stars, and going on solar-eclipse expeditions. Life was fairly routine, until one memorable April morning in 1906 when the mountain experienced a minor temblor. Damage was minimal at the observatory—a few coal-oil lamps overturned, loosened bricks on some of the buildings—but looking toward San Francisco, Lick residents saw an enormous tower of black smoke. They didn't realize how serious the disaster was until noon, when the daily stage from San Jose, which normally ran like clockwork, never showed up. By evening the astronomers turned Lick's 12-inch telescope completely horizontal and aimed it toward the Golden Gate, the strait connecting San Francisco Bay to the Pacific Ocean. Through the scope they saw three miles of fire-front, burning fiercely. "And, naturally, the lens inverted everything, so we saw buildings fall up and flames sweep down—which was a weird, weird sight. . . . It reminded me of . . . Dante's *Inferno*," said Douglas Aitken, who had lived on the mountain at the time as a young boy.

Curtis missed all the excitement because two months earlier he had arrived in Chile to head up Lick's southern station on the summit of San Cristobal, on the outskirts of Santiago. With him were his mother, wife, and three small children. After a few years they became so comfortably settled in Chile that they contemplated an extended stay, having become fluent in Spanish and grown fond of the South American lifestyle. "Queer how completely we seem to have taken root here," noted Curtis. But in 1909 Curtis received an unexpected invitation to return to Lick, not as an assistant or associate, but as a senior astronomer. Short on staff, the observatory needed an experienced hand to work with the Crossley reflector. In accepting the post, Curtis became Keeler's anointed successor, the next in line to tackle the mystery of the spiral nebulae.

Curtis first spent time getting to know the Crossley's strengths and weaknesses: What were the faintest stars it could photograph? How many hours of exposure were required? He had the good fortune to start his new venture just as a famous celestial visitor, Halley's Comet, visibly reappeared in the skies in 1910, as it did every seventy-six or so years, providing a superb target to test out the Crossley's photographic abilities. The comet this time passed relatively close to Earth, creating quite a stir throughout the world, so by the time it completely disappeared from telescopic sight in 1911, the Crossley and other Lick telescopes had taken nearly four hundred pictures of its spectacular passage.

With the Crossley checked out, Curtis at last turned his attention to the mysterious nebulae. Keeler and others at Lick had previously amassed a photographic library of around one hundred nebulae and clusters using the Crossley. By the summer of 1913, Curtis boosted that number to more than two hundred. "Many of these nebulae show forms of unusual interest," he jotted down in his observatory report. "The great preponderance of the spiral form becomes more and more striking with the progress of the survey." He was beginning the process of identifying and cataloging the nebulae, particularly the spirals, in hope of detecting patterns that would lead to revealing what they were. His descriptions conveyed the rich diversity in their appearance: A spiral could be either "patchy," "branched," "irregular," "elongated oval," or "symmetrical." For the moment, he was merely recording what he saw, not venturing to discuss what they might be.

It was tiring work. "Crossley still has its old reputation of using up more energy than any other instrument on the hill," Curtis told a colleague. Despite the improvements he had made on the telescope, it was still difficult to reach the eyepiece at certain positions. "If you got a little bit sleepy at night, it was dangerous, because it went down a great many feet [from the observing platform] to a floor in the basement," said one of the telescope's later users. One wisecracker suggested the only way to observe with the Crossley in comfort was to fill the dome with water and observe from a boat.

When he first started his study, Curtis assumed that the spirals were comparable to the size of a modest cluster of stars, spanning no more than several hundred light-years in width. It was a reasonable assumption. Over at the Mount Wilson Observatory, with its new 60-inch reflector, George Ritchey had begun to photograph the spiral nebulae and was concluding they were a mix "of smooth nebulous material and also of soft star-like condensations or nebulous stars." He surmised he was seeing a collection of developing stars—a good-sized cluster but certainly not an entire "island universe."

But Curtis began to doubt this viewpoint as he gathered more evidence with the Crossley. Some of the first hints surfaced when he

An edge-on galaxy photographed by Heber Curtis in 1914,
showing the dark lanes of dust and gas within the disk
(*Copyright UC Regents/Lick Observatory*)

rephotographed a number of nebulae that Keeler had previously
imaged. By comparing his most recent spiral pictures with those gathered years earlier, he hoped to see how the swirling clouds had rotated.
The amount of motion measured was going to help him judge their
distance. But Curtis didn't detect any sign of movement, not a smid-
geon "rotatory or otherwise," he reported. "As the spirals are undoubt-

edly in revolution—any other explanation of the spiral form seems impossible—the failure to find any evidence of rotation would indicate that they must be of enormous actual size, and at enormous distances from us." It would simply be impossible to measure a shift by sight alone if the spiral were considerably larger and at the same time pushed far off into space.

Even earlier Curtis started reporting that some of the spirals he photographed—the ones so tilted they were seen edge-on—resembled "the Greek letter Φ . . . for lack of a better term": an oval ring crossed by a straight dark line. He expressly mentioned them in his research notes: NGC 891 "shows dark lane down center," he jotted down. And NGC 7814 was described as small but with a dark lane "beautifully clear."

This was at a time when Yerkes astronomer E. E. Barnard was also acquainting astronomers with myriad "dark nebulae" within the Milky Way. Barnard was gathering exquisite photographic evidence that the coal-black regions within the Milky Way that appeared to be devoid of stars ("holes in the heavens," Herschel called them) were actually clouds of cosmic gas and dust—colossal streams of inky darkness without the hint of a glow. Curtis immediately connected this finding to his work: The dark lanes he was sighting in the spirals had to be "due to the same general cause that produces certain occulting effects in our own galaxy. . . . " The dark bands were almost certainly matter—but matter that wasn't glowing.

This also explained why no spiral was ever seen in certain areas of the celestial sky, aptly named the "zone of avoidance." Spiral nebulae were very exclusive objects; they tended to huddle around the north and south galactic poles, as if shunning the long white swath of the Milky Way. Astronomers had long scratched their heads over this peculiar distribution. If spirals were truly the birthplaces of new stars, why weren't they found in the richest star fields? Why were the spirals found in only those sectors of the sky where stars were scarce? Not one spiral had ever been spotted in the thick of the Milky Way. Curtis cleverly deduced that this cosmic quarantine was only an illusion: If his dark-banded spirals were truly distant galaxies, then the Milky Way, too, must have its own dark band. All the dark gaseous clouds within the Milky Way were collectively acting like an opaque wall, making it impossible to see the spirals that resided beyond this obstruction, keeping the spirals hidden. "[The] great band of occulting matter in the plane of our galaxy . . .

serves to cut off from our view the distant spirals lying near the projection of our galactic plane in space," explained Curtis. And that couldn't happen unless the spirals were very far off.

To Curtis this argument made perfect sense, but he was presenting the idea at a time when most astronomers still thought of the vast expanses between the stars as a pristine emptiness and the Milky Way as transparent as a glass window. His reasoning wasn't as readily accepted as he had hoped.

Curtis spent much of the 1910s on this fight—gathering data, giving lectures, coming up with fresh new arguments. He gathered clues as if he were a cosmic sleuth. "Were the Great Nebula in *Andromeda* situated five hundred times as far away as at present," reasoned Curtis, "it would appear as a structureless oval . . . with [a] very bright center, and not to be distinguished from the thousands of very small, round or oval nebulae found wherever the spirals are found. There is an unbroken progression from such minute objects up to the Great Nebula in *Andromeda* itself; I see no reason to believe that these very small nebulae are of a different type from their larger neighbors." But his mounting certainty that the spirals he photographed, both large and small, were all distant galaxies strewn through space was based solely on circumstantial evidence. He had convinced his colleagues at Lick, which came to be identified as a stronghold of island-universe supporters, but the majority of astronomers still preferred to think of all the stars and nebulae as inhabiting one great system, the Milky Way. Curtis was absolutely right, but convincing the wider community of astronomers was an entirely different matter.

And then something interesting . . . and very unusual . . . happened.

On July 19, 1917, some three hundred miles southeast of Mount Hamilton, George Ritchey was taking a routine photograph of a spiral nebula with the 60-inch reflector at the Mount Wilson Observatory. It was the fourth in a series of long-exposure photos he had been taking of NGC 6946 over the previous seven years. This time, though, he noticed a new pinpoint of light in the spiral's outer region. It had to be a nova, for this "new star" wasn't in any of his previous pictures. More important, this nova was distinctly different from the dazzling one that had flared up in Andromeda thirty-two years earlier. This one was very, very *faint*.

The unforgettable nova that briefly blazed within Andromeda in 1885 had reached a brightness that could be discerned by the naked eye (just barely); the nova in NGC 6946, on the other hand, was about sixteen hundred times dimmer. Ritchey knew he had caught the nova fairly early in its burst; a plate taken just a month earlier with another telescope showed no extra speck of light whatsoever. Telegrams announcing the new find were quickly sent out to other observatories.

Curtis likely received the report with a sinking heart, for he had sighted similar novae months earlier. On the very day that Ritchey's telegram reached Lick, Curtis was actually at his desk, drafting a paper on three faint novae he had discovered in other spiral nebulae. He had been sitting on the news since March, when he first observed the flare-ups. He was being very careful, holding off any announcement until he was sure that the outbursts were not simply variable stars reaching their maximum brightness. His caution kept him from the prize of first announcement.

The first nova that Curtis spotted was in NGC 4527, an elongated spiral located in Virgo. By checking plates of this region made earlier at the Harvard, Yerkes, and Lick observatories, Curtis confirmed that no star had been visible in the spiral over the previous seventeen years. The tiny dot on his photo reached around fourteenth magnitude (some sixty thousand times dimmer than the stars in the Big Dipper). And in the course of his plate search, he came upon two additional faint novae: this time in M100 (also known as NGC 4321), a spectacular spiral in Coma Berenices viewed face-on. One of these novae had flared in 1901, the other in 1914. "That both these novae should have appeared in the *same* spiral is especially worthy of note," reported Curtis. By the time Curtis announced his finds in July 1917, though, all three of these novae had completely disappeared. But he made sure to point out in his bulletin that the new stars "must be regarded as having a very definite bearing on the 'island universe' theory."

With such startling news from both Mount Wilson and Lick, nova hunting spread like wildfire among the top U.S. observatories. Going to old astronomical plates and searching for novae became the craze, and new candidates were found right away. The list was getting longer week by week. "Such is the progress of Astronomy in the wild and wooly West," joked one Mount Wilson astronomer. Curtis was tremendously excited by all the discoveries. Every time he found a new nova in a spiral, he'd go through the observatory and show off the plate, like some proud papa in a hospital maternity ward.

Curtis soon had a big enough sample of novae to make a judgment

1901, April 16   1914, March 2

Arrows point to the novae discovered by Heber Curtis
in photos of NGC-4321 taken in 1901 and 1914.
(*Copyright UC Regents/Lick Observatory*)

call: He suspected that the 1885 outburst in Andromeda, as well as the
1895 one in Centaurus, were rare and exceptional celestial events. Curtis guessed that their spectacular radiance had misled astronomers into
thinking the novae's host nebulae had to be close by. He suggested that
nova bursts actually came in two varieties: The rarer ones were big and
spectacular (now known to be stars blowing apart), while the ones seen
more often were less energetic (determined later to be a flaring off the
surface of a white dwarf star). And since the majority of the novae being
sighted in the spirals more resembled the ordinary novae seen periodically within the Milky Way, he concluded that the spiral nebulae had to
be *millions* of light-years distant, in order for those novae to appear so
dim. He said as much to the Associated Press. He boldly told its reporter
that the nova bursts he had discovered occurred some 20 million years
in the past, meaning the nebulae had to be 20 million light-years distant
for the light to be reaching us now. (With 1 light-year equaling about six
trillion miles, that's more than a hundred million trillion miles.) For
Curtis the faint novae were bona fide proof that the nebulae resided far
beyond the borders of the Milky Way. But Curtis was championing this
idea too early, before the physics could explain it. Many of his fellow
astronomers were still fairly skeptical, unwilling to conjure up new
celestial creatures willy-nilly. For them "Occam's Razor" prevailed, the
long-standing rule of thumb established by the English philosopher
William of Occam in the fourteenth century. *"Pluralitas non est
ponenda sine necessitate,"* declared Occam, which can be translated as

"plurality must not be posited without necessity." Best to choose the simplest interpretation over an unnecessarily complex one—unless forced to do otherwise. One type of nova was far more preferable than two.

Despite the lack of support for his creative hypothesis, Curtis was still gaining appreciable momentum on his endeavor, at least until World War I intervened. Just months after the United States officially joined the fight in 1917, Curtis went first to San Diego and then to Berkeley to teach officer recruits navigation. Afterward, he proceeded to Washington, D.C., to work for the Bureau of Standards on the design and development of military optical devices. Before taking his leave, though, Curtis made sure to compile a master list of the spiral nebulae he had photographed with the Crossley, by now more than five hundred. And as before, each of his photographs revealed ever more nebulae, pale and murky, surrounding the more consequential spirals that he officially cataloged. On one plate alone he counted 304 additional spirals. Keeler had estimated that 120,000 spiral nebulae were within observational range of the Crossley. Another Lick astronomer later upped that number to 500,000. Now Curtis was raising the figure even higher. "The great numbers of small spirals found on nearly all my plates of regions distant from the Milky Way, long since led me to the belief that [an earlier] estimate of half a million was likely to be under, rather than in excess of, the truth," he reported. "[I] believe that the total number accessible with the Crossley Reflector with rapid plates and exposures of from two to three hours may well exceed 1,000,000." This was an astounding hike in the spiral estimate.

While still in the U.S. capital wrapping up his work after the 1918 armistice, Curtis was invited to deliver a semipopular lecture on the spiral nebulae before the Washington Academy of Sciences and the Philosophical Society of Washington. His expertise on the topic was getting noticed. "Get up a collection of about 40 classy slides and send to me at once," he wrote Campbell at Lick in great excitement. Curtis was thrilled at the opportunity, his first actually, to lay out before an influential scientific conclave all his hard-won evidence in support of the island-universe theory. He planned to use a lantern slide—the early-twentieth-century version of PowerPoint—to display the various types of spirals he had come across, to point out the dark lanes running through them, and to reveal the many fainter nebulae lurking in the background of his photographs of the spirals.

On the appointed day—March 15, 1919—a large audience gathered to hear Curtis in the new lecture room at Washington's prestigious Cosmos Club (then located at Lafayette Square), the traditional meeting place for the city's intelligentsia. Curtis opened with a tip of the hat to William Herschel. "The history of scientific discovery affords many instances where men with some strange gift of intuition have looked ahead from meager data, and have glimpsed or guessed truths which have been fully verified only after the lapse of decades or centuries," he said. "We have now, as far as the spiral nebulae are concerned, come back to the standpoint of Herschel's fortunate, though not fully warranted deduction . . . that these beautiful objects are separate galaxies, or 'island universes,' to employ the expressive and appropriate phrase coined by Humboldt." With these words, Curtis became the most outspoken and identifiable advocate of the island-universe theory.

Curtis at the time estimated that our stellar home was some 30,000 light-years wide and contained about a billion stars, with the Sun nicely situated right near the center. He was wrong about that: The Milky Way's dimensions were even then being revised upward, and the Sun was losing its front-row seat on galactic affairs. But Curtis was right about the spiral nebulae being far-off galaxies.

Over the course of that March evening, Curtis laid out his arguments point by point. First, there was the peculiar distribution of the spiral nebulae. If they are stars in the making, he asked, why are there no spirals in the very place where stars are most numerous, the Milky Way? "Occulting matter," he answered, was masking our view, making it only appear as if the spirals were avoiding the plane of the Milky Way. And then there was the very light of a spiral to consider: The spectrum of a spiral revealed that its light was emanating from a massive assembly of stars, not just a cloud of gas.

His logic was impeccable. Going through the historical records, Curtis determined that nearly thirty "new stars" had made an appearance within the Milky Way over the last three hundred years, each suddenly rising to great luminosity and then sinking back into obscurity once again. But half that number had already been sighted in spiral nebulae in just a few years, making it all the more likely that the "spirals are themselves galaxies composed of hundreds of millions of stars." Moreover, with the novae being so faint, they had to be situated millions of light-years away. "This is an enormous distance," admitted Curtis, "but, if these objects are galaxies like our own stellar system, this is about the order of distance at which we should expect them to be placed."

Curtis was fully aware of the magnitude and complexity of the new cosmic scheme he was proposing. "We know that the relative space in this, our galaxy, occupied by . . . our solar system . . . is about the same order as that occupied by a single drop of water in Chesapeake Bay," he told his audience. "To go still beyond such a concept, the island universe theory forces us to consider a still mightier whole, a space containing hundreds of thousands of stellar universes like our own, each containing millions upon millions of suns. . . . Awe-inspiring as are the concepts of astronomy, this newer concept surpasses them all; it staggers the imagination." Curtis was plainly carried away by the allure of this astounding idea. His audience was enthralled as well. At the end they applauded with great enthusiasm and kept him long afterward for further discussion.

At the close of the war, officials at the Bureau of Standards had hoped that Curtis would stay with the agency, but he refused. "As to my staying here permanently, I have no idea whatever of doing that," he assured Campbell. "[I'm] anxious to get back to my hill and the Crossley, and *stay* there. . . . I am more than ever of the opinion that men like you [and] Hale . . . get all the hard knocks, and not half the fun out of life that those of us lower down get." An observer at heart, he was eager to return to his nebulae. By May 1919, he was back on Mount Hamilton gathering more evidence in support of distant galaxies.

Curtis already had a few converts to his cause. Astronomer Andrew Crommelin of the Royal Observatory in Greenwich favored the island-universe theory as well, but voiced caution: "The hypothesis of external galaxies is certainly a sublime and magnificent one," he said. "[But] our conclusions in Science must be based on evidence, and not on sentiment." His fellow astronomers were setting the bar high. Curtis had to provide more than logical arguments to win his case. He needed concrete evidence. Additional clues had been arriving, but they did not originate with Curtis. They instead turned up at the Lowell Observatory, Lick Observatory's long-standing competitor located in northern Arizona.

[5]

My Regards to the Squashes

Roman god. Bringer of War. Fourth planet from the Sun. Astronomers eager to solve the spiral nebulae dilemma had Mars, strangely enough, to thank for a further step toward an answer—at least in a roundabout way.

The red planet, with its vivid ruby luster, has fascinated stargazers for millennia, but interest grew even more intense after the invention of the telescope. With the extra magnification astronomers could at last discern markings on the surface of Mars. Bright patches around its poles, similar in appearance to our own planet's arctic and antarctic regions, were seen to wax and wane with the Martian seasons. So Earthlike was this behavior that by 1784 William Herschel was reporting that Mars "is not without a considerable atmosphere . . . so that its inhabitants probably enjoy a situation in many respects similar to ours."

Scrutiny of Mars was particularly favorable in the fall of 1877, when Earth and Mars were at their closest, approaching in their orbits to within thirty-five million miles of each other. The superb viewing conditions allowed the Italian astronomer Giovanni Schiaparelli to catch sight of numerous dark streaks crossing Mars's reddish ochre regions, then known as "continents." In his native language, he called these thin shadowy bands *canali,* or "channels," which many figured arose from natural geographical processes.

But Schiaparelli's term was translated inaccurately, a gaffe that encouraged many fanciful conjectures. The most controversial, by far, was the assumption that the "canals" were irrigation works built by advanced beings, who were directing scarce resources over the surface

Percival Lowell (*Lowell Observatory Archives*)

of their planet for cultivation. "Considerable variations observed in the network of waterways," wrote French astronomer Camille Flammarion in 1892, "testify that this planet is the seat of an energetic vitality. . . . There might at the same moment be thunderstorms, volcanoes, tempests, social upheavals and all kinds of struggle for life." No one championed this idea more avidly than Percival Lowell, a wealthy businessman whose crusade generated a Mars mania among the public, so much so that the *Wall Street Journal* in 1907 reported that evidence for the existence of Martian folk surpassed that year's financial panic as the news story of the year.

Lowell, the oldest of five children, came from a well-established New England family. He was one of the Boston Brahmins, upper-crust Bostonians who had made their fortunes creating the American cotton industry. A few years after graduating from Harvard in 1876, Lowell began to travel extensively, especially to the Far East, which led to his writing several well-received books on the region and its religions. By the 1890s, though, restless and searching for individual expression, he renewed a childhood interest in astronomy. "After lying dormant for many years," recalled his brother, "it blazed forth again as the dominant one in his life." Independently wealthy, Lowell decided to establish a private observatory atop a pine-forested mesa nestled against the small village of Flagstaff, Arizona (then still a territory of the United States). His initial aim was to observe the particularly close approaches of Mars occurring in 1894 and 1896. Later, the entire solar system became his celestial playground. He was taking to heart his family's motto—*occasionem cognosce*, "seize your opportunity." It was a daring venture for an amateur astronomer with no professional experience, especially since he found himself competing with the new and larger astronomical outposts then being built by universities and research institutions. But in this rivalry, Lowell became the outsider, dedicating his observatory to the pursuit of questions that interested him and him alone. Given his obsession with the red planet, the high perch on which the observatory rested, 7,250 feet above sea level, was soon dubbed Mars Hill.

Lowell devoted the rest of his life to this infatuation. A rugged individualist and showman, he once listed his address as "cosmos" in a friend's guestbook. Though often charming when necessary, the patrician Bostonian could easily become enraged if either his opinions or scientific credentials were challenged. He eventually fired one charter member of his observing staff for continually insisting that the canals on Mars might be illusory after all.

Lowell installed a 24-inch refractor on Mars Hill. Though a modest-sized telescope (by then several others in the world had lens widths of thirty inches or more), it was still perched more than three thousand feet higher than the giant scope at the venerable Lick Observatory, and Lowell sought to outdo his competitor at every turn. Sometimes he tried too hard. Lowell and his staff occasionally reported on sightings—certain elusive stars or markings on planets—that simply weren't there. Lick staffers rolled their eyes in exasperation at the dubious announcements coming out of Flagstaff and hinted that there were defects in Lowell's scope (or with his eyesight). Before long a battle of the observatories ensued—California's top instrument versus Arizona's best. One newspaper headlined the unceasing skirmish as "The Strife of the Telescopes."

In 1900 Lowell upped the stakes when he ordered a custom-built

spectrograph that was an improved version of the one already in use at the Lick Observatory. He directed its manufacturer to make it "as efficient as could be constructed." To operate it, Lowell hired a recent graduate of the Indiana University astronomy program, Vesto Melvin Slipher, who was grateful to be posted at one of the few observatories in the United States with a large telescope, along with high altitude, clear air, and good "seeing," minimal blurring from atmospheric activity.

Lowell originally thought of Slipher's job as temporary ("I . . . take him only because I promised to do so," Lowell told one of Slipher's professors at Indiana), but the young astronomer ended up remaining until his retirement in 1954, serving as the observatory's director for thirty-eight of those years. Lowell chose well. Slipher took a spectrograph intended for planetary work and with great skill and extraordinary patience eventually extended the observatory's celestial scans far beyond the solar system. Instead of discerning new features on Mars, the observatory's raison d'être, he found himself revealing a surprising facet of the cosmos, previously unknown. He detected the very first hint—the earliest glimmer of data—that the universe is expanding, although it took more than a decade for astronomers to fully recognize just what he had done.

In the nineteenth century, with rural farms in the United States often miles apart, lit by only candle or kerosene, and no interfering glow from a nearby metropolis, the nighttime sky was breathtaking in its appearance. The Milky Way streaked across the celestial sphere like a ghost on the run. This sublime stellar landscape must have been a powerful lure, for many of America's greatest astronomers a century ago were born on Midwest farms, including Slipher. "V.M.," as he was best known to friends and colleagues, was one of eleven children, and at his school in Indiana he displayed a keen knack for mathematics. Going off to Indiana University at Bloomington at the age of twenty-one, he earned a degree in both mechanics and astronomy. He must have had qualms upon arriving at Flagstaff in the summer of 1901. Before coming to the Lowell Observatory, the biggest telescope he had ever operated was a tiny 4½-inch reflector. He had certainly never handled a spectrograph as large and complex as the one he was expected to operate. It was a daunting task for a beginner. The young man struggled for a year to handle the spectrograph with ease. He even confused the red and blue ends of

the spectrum initially, a scientific faux pas of the first magnitude. In distress, Slipher asked Lowell if he could go to Lick to get some instruction, but his boss firmly said no. Given the animosity between the two observatories, Lowell didn't want Lick knowing that one of his staff needed help. "When you shall have learnt all about the spectroscope and can give them as much as you take it will be another matter," asserted Lowell.

Slipher and Lowell were an intriguing mesh of personalities, like a harmony created from two different notes. Flamboyant, aggressive, and driven in his passions, Lowell hated to share the spotlight, especially when it came to announcing a discovery made at *his* observatory. Slipher was fortunately Lowell's opposite in character, a man who, it was said, "kept himself well insulated from public view and rarely attended even scientific meetings." He was a peacekeeper at heart and knew it wasn't wise to steal Lowell's thunder. More than that, he didn't

Young Vesto Slipher (*Lowell Observatory Archives*)

want to. An unassuming and dignified man who always wore a suit and tie to work when not observing, Slipher was markedly deliberate and cautious in his pronouncements. A picture of him at the observatory, fresh from the Midwest, reveals a handsome, dark-haired lad with a gaze and smile like that of Mona Lisa. He preferred to correspond with his peers rather than travel and often had others present his findings. Director and underling, consequently, got along famously.

Frequently away from the observatory, either traveling or taking care of business in Boston, Lowell remained in contact with Slipher via a steady stream of letters and telegrams. While Slipher stood in as the observatory's effective director, Lowell offered his pronouncements from afar on matters astronomical ("Don't observe sun much. It hurts lenses"), administrative ("Permit nobody whatever in observatory office"), and personal ("Will you kindly see if shredded wheat biscuit are to be got at Haychaff"). They consulted each other on hires, equipment, budgets, and even vegetables. Lowell doted on his observatory garden and insisted on news of its condition whenever he was away. "How fare the squashes?" asked Lowell one year as fall harvest approached. His letter the following week closed with, "My regards to the squashes." And finally, "You may when the squashes ripen send me one by express."

Slipher did not respond. "Why haven't I received squashes? Express at once if possible," Lowell anxiously telegraphed right after Christmas. Slipher reluctantly had to answer that the poor gourds, alas, had shriveled up and died.

All was forgiven, though, by next spring. "Thank you for taking so much pains with the garden! Just keep on planting and you will get something," wrote Lowell. Slipher did; by July he was sending Lowell his latest bounty. "Your vegetables came all right and delighted me hugely," replied Lowell. More were sent in October.

As with his gardening, Slipher made progress on the spectrograph as well, eventually becoming a virtuoso at its operation. He first used it to verify the rotation periods of Jupiter, Saturn, and Mars. Next was Venus. The planets in the solar system were always Lowell's first priority. Slipher was then directed to use the instrument to analyze planetary atmospheres, an assignment that got him into the thick of Lowell's battles with the astronomical community when he tried to measure whether water vapor was present in the Martian air. Slipher believed he had detected a slight signal, which Lowell immediately publicized as boosting his vision of a watery Mars. But Lick astronomer W. W. Campbell, after conducting the same observation, saw no sign at all of water vapor in the red planet's atmosphere.

Despite the disagreement, Slipher was gaining confidence and improving the sensitivity of the spectrograph through trying out different kinds of prisms and photographic plates. By 1909 he was able to

Boston,Feb. 8,1909.

Dear Mr. Slipher:-

    I would like to have you take with your red sensitive plates the spectrum of a white nebula - preferably one that has marked centres of condensation.

    Always sincerely yours,

Percival Lowell

* *Continuous spectrum*
* *but I want its outer parts.*

Mr. V.M.Slipher.

Percival Lowell's 1909 letter directing Vesto Slipher to get a white nebula's spectrum (*Lowell Observatory Archives*)

confirm that some gas existed in the seemingly empty space between the stars, a triumph that later won praise from astronomers around the world. These pursuits eventually led Slipher to his greatest discovery of all, an unanticipated revelation that involved the spiral nebulae.

It began innocently enough. On February 8, 1909, Lowell in Boston sent a typed letter to Slipher with concise instructions: "Dear Mr. Slipher, I would like to have you take with your red sensitive plates the spectrum of a <u>white</u> nebula—preferably one that has marked centres of condensation." By "white," Lowell meant a spiral nebula, which in 1909 was still generally understood to be a new planetary system under construction. In a handwritten footnote at the bottom of his note, Lowell stressed that he wanted "its outer parts." He longed to see if the chemical elements found at a spiral nebula's edge, as revealed by the fingerprints of its spectral lines, matched the composition of the giant planets situated far from our solar system's center. A connection would mean the spirals could indeed be baby solar systems under way.

Slipher balked at first. "I do not see much hope of our getting the spectrum of a white nebula," he told Lowell. He knew that it would take at least thirty hours to get just a plain old photograph of the nebula with the observatory's 24-inch telescope. Nebulae were extremely faint through its lens. To get a spectrum, with far less light hitting the photographic plate after its passage through the spectrograph, seemed impossible.

But Slipher had something to prove. Campbell at the Lick Observatory had recently written yet another article critical of the Lowell Observatory. It was the latest volley in the observatories' ongoing war over whose refractor could get the better results. Lowell had earlier asserted that the superior air on Mars Hill allowed his 24-inch refractor to see 173 stars in a given field of the sky, where Lick's 36-incher could see only 161. Slipher, deeply loyal to his astronomical home, wanted to settle the matter once and for all. He was eager to set up a challenge between the two observatories, comparing photos of stars taken at the same time on similar plates, but Lowell nixed the idea. To reclaim some honor, Slipher decided to focus on the difficult task of getting the spiral nebula spectrum. "I have come to the conclusion," he had written John A. Miller, his former astronomy teacher at Indiana, just a few months earlier, "that where we can defend ourselves . . . we shall have to do it or otherwise everything we publish will be discredited."

Though Slipher considered the spectral task hopeless, he persisted and by December 1910 was able to wrench some feeble data from the Great Nebula in Andromeda. "This plate of mine," he informed Lowell by letter, "seems to me to show faintly peculiarities not commented upon." He was going to say "to show faintly, perhaps" but had scrawled out the "perhaps." He was now convinced he had captured something on the spectrum previously unseen by other spectroscopists, such as Scheiner in the 1890s.

By trial and error, coupled with an astute technical mind, Slipher started making improvements to the spectrograph. Instead of using a set of three prisms, which better separated the spectral lines, he decided to use just one. Though this made the spectrum more congested and difficult to read, it vastly increased the amount of light available since there was less glass to absorb the incoming photons. More important, he understood that increasing the speed of the camera was vital and so bought a very fast, commercially available camera lens. The entire spectrograph, which included counterweights to keep the telescope balanced, weighed 450 pounds and resembled an oversize nutcracker attached to the bottom of the telescope. The celestial light, instead of going into the eyepiece, was directed to the prism, which separated the beam into its constituent wavelengths. A small photographic plate was suitably positioned to record the spectral lineup, from red to violet.

Planet studies, reports on the return of Halley's Comet, and administrative duties diverted Slipher's attention for a while. He could not get back to the question of the spiral nebula until the fall of 1912. But by then his refashioned spectrograph was operating two hundred times faster than the instrument's original specifications, allowing him to slash his long and tedious exposure times. With his modifications in place, he could at last try for the spectrum that he had so long sought. It was a tantalizing goal, not only scientifically, but personally as well. Two years earlier Lick director Campbell had spoken at Yale University and specifically observed that "there is no more pressing need at present than for a greatly increased number of nebular radial velocities." To beat Campbell at his own specialty—radial velocities, how fast celestial objects move either toward or away from us as if traveling along a radius—would be sweet triumph indeed for Lowell Observatory loyalists. No one had yet gauged the speed of a spiral nebula. That required a spectrum with more detail than had ever been previously obtained or even deemed possible.

Slipher carried out his first measurement on September 17. It took a total of six hours and fifty minutes for the extremely faint light to fully register. "It is not really very good and I am of the opinion that we can do much better," he soon relayed to Lowell, "but in view of the results got elsewhere of it generally with much longer exposures, it seems to me encouraging and I mean to try it again." The spectrum was very tiny, a mere centimeter long and a millimeter wide. The photographic plate itself was barely eight centimeters long, but there was just enough room for Slipher to write "Sept 17 And Neb" on the top of the glass to indicate his target had been the Andromeda nebula.

Gale's Comet required his attention for most of October, so he was not able to get back to Andromeda until November 15. The weather was fair with some clouds, but the wind was strong. He started the measurement at seven that night. Being winter, it was already fully dark, and he worked into the early-morning hours. The plate was exposed for eight hours and was left in the spectrograph, shutters closed, so the following night he could align the telescope once more upon his target and continue the observation for another six hours. By taking the longer photographic exposure and narrowing his slit, he saw some improvement in the spectrum when compared with the one taken in September.

He returned to the problem on December 3 and 4, when the Moon no longer rose at night to interfere with his observation of the dim nebula. This time, Slipher scribbled in his workbook that the transparency of the air was "very good," underlining it for emphasis. Over the two nights he was able to gather his sparse photons for a total of thirteen and a half hours. The only problem that arose was a troublesome clock drive that took fifteen minutes to fix.

When carrying out these observations, the interior of the wooden dome at times resembled the movie version of a mad scientist's laboratory, with high-voltage induction coils sparking and sputtering by the side of the telescope. A row of old-fashioned Leyden jars provided the ignition. It was a wonder that Slipher didn't electrocute himself. This Rube Goldbergian contraption vaporized samples of iron and vanadium, whose light then served as a calibration for Slipher's measurement. The spectrum of these elements, at rest within the dome, could be compared to the spectrum of the nebula rushing around in space; the difference between the spectra determined the nebula's speed.

Since each spectrum that Slipher produced from Andromeda was so tiny, he needed a microscope to measure how much the spectral lines

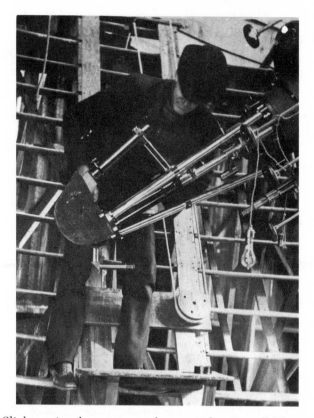

Vesto Slipher using the spectrograph mounted on Lowell Observatory's
24-inch refracting telescope (*Lowell Observatory Archives*)

had shifted, compared to their positions on the calibrated standard. The
more the shift, the higher the velocity of the nebula. The microscope
had been with Lowell in Boston temporarily, and Slipher didn't get it
back until mid-December. But once the scope arrived he couldn't resist
taking a quick peek at the Andromeda plates he had so far. There were
"encouraging results or (I should say) indications," Slipher reported to
Lowell, "as there appears to be an appreciable displacement of the nebular lines toward the violet." A shift of the lines toward the blue-violet
end of the spectrum meant Andromeda would be moving *toward* Earth.
"I congratulate you on this fine bit of work," Lowell wrote back.

But Slipher felt he needed to acquire an even better spectrum to peg
the exact speed. It was an endeavor, he told Lowell, that "would doubtless impress all these observers as a quite hopeless undertaking, and
maybe it is, but I want [to] make an attempt."

He started the final measurement on December 29 at 7:35 p.m. and stayed with it until some clouds rolled in near midnight. On a scale from 1 to 10—1 being the worst, 10 the best—Lowell Observatory astronomers often joked that at 10 you can see the Moon, at 5 you can still see the telescope, and at 1 you can only feel the telescope. Fortunately, the sky was clear the following night, and he was able to collect additional light for nearly seven hours. Perhaps pressing his luck, he went into a third night, New Year's Eve. This time the weather was poor, and he had to finish up just before 1913 rang in. Yet, the additional attempt allowed him to squeeze one more hour of data onto his photographic plate.

Slipher had no time as yet to accurately measure this last plate, but he did a speedy check and right away knew that something was up. "I feel safe to say here that the velocity bids fair to come out unusually large," he wrote Lowell right away. For Slipher to make such an impetuous claim at such an early stage was downright radical for a man normally so cautious. He must have been thrilled at what he had found.

Throughout January he focused on measuring all four of his plates more carefully, in order to gauge the velocity of Andromeda precisely. He did this by placing the plate of the nebula's spectrum in a "spectro-comparator," which measured it against the standard spectrum—the rest frame. By turning a screw, he shifted one plate relative to the other. When the spectral lines at last matched, he recorded how much he had to shift the nebula plate to get it in line with the standard. The amount of shift established the velocity of the nebula. His calculations to convert the measured shift into a velocity filled page after loose page, with his figures neatly recorded in pencil. He started on January 7 and ended on the twenty-fourth.

The final result astonished Slipher. The Andromeda nebula was rushing toward Earth at the ridiculous speed of 300 kilometers per second (or around a million kilometers per hour), about ten times faster than Slipher had been expecting, given the average speed of a star in the Milky Way. Nebulae weren't supposed to act like this. Astrophysicists at the time generally believed that nebulae were rather slow cosmic creatures, plodding along at speeds far lower than stars. Instead, spiral nebulae seemed to be in a special class all to themselves. Andromeda was setting a cosmic speed record. In present-day terms, it's nearly forty times faster than a space shuttle in orbit.

Slipher, prudent as always, remeasured the plates he had just taken to make sure there was no error. He also sent a print of the spectrum to Edward Fath to obtain an independent check that the shift was real. In 1908, when Fath had taken his own spectrum of the Andromeda nebula at the Lick Observatory, he too had discovered a shift in its spectral lines. But at the time he simply wrote off the unexpected change as a likely malfunction of his spectrograph. It was the accepted wisdom that celestial objects simply did not move that swiftly. He heedlessly decided to brush aside the anomaly because, as he reported, "the shift has no direct bearing on the question for which an answer was sought." Again, the hapless Fath missed his chance at making astronomical history. One can imagine his chagrin at receiving Slipher's print. He had seen the same spectral message as Slipher four years earlier, only to ignore it and not follow up.

By February Slipher came to trust both his instrument and his expertise (which in hindsight was truly incredible; today, with far better equipment, astronomers measure Andromeda approaching us at 301 kilometers per second, a difference from Slipher's rate of less than a third of a percent). Slipher informed Lowell that the plates "agree as closely as could be expected and I can not doubt the reality of the displacement." Andromeda had to be moving at an astounding clip. Instead of announcing the result in a major astronomical journal, though, Slipher chose to publish his brief account—just nine paragraphs—in the *Lowell Observatory Bulletin*. True to form, Slipher held off on any grander statement until he had secured some confirmation.

Yet even one spiral nebula velocity was an exceptional accomplishment. Many were thrilled for Slipher. "It looks to me as though you have found a gold mine," wrote Miller, "and that, by working carefully, you can make a contribution that is as significant as the one that Kepler made, but in an entirely different way."

Max Wolf at the Königstuhl Observatory in Heidelberg admired the spectrum's "beauty." Edwin Frost, then editor of the *Astrophysical Journal*, wrote his sincere congratulations at the revelation of such an "incredible" velocity. "It is hard to attribute it to anything but Doppler shift," he said. "Your success on this object indicates the value of elevation above the sea. . . . It is a pity that someone cannot try other objects of this sort at elevations of 12,000 to 15,000 ft." Astronomers would, but only decades later.

Then there were others, such as Campbell at Lick (predictably), who were highly skeptical. "Your high velocity for [the] Andromeda

Nebula is surprising in the extreme. I suppose . . . the error of [your] radial velocity measurement may be pretty large. I hope you have more than one result for velocity."

To be fair to Campbell, an extraordinary finding like this needed extraordinary proof, and Slipher knew that as well. He had already put out the call for others to try to confirm it. Within a year, Wolf was able to follow up. His spectrum was cruder but still in fair agreement. Soon after, even persnickety Lick Observatory came to confirm Andromeda's fleetness. Lick astronomer William H. Wright obtained a velocity that nearly matched Slipher's. "I had planned to get at this work years ago when Fath got his big displacement . . . but you seem to have beaten me to it," Wright told Slipher.

Lowell was enormously pleased. "It looks as if you had made a great discovery," he wrote, right after Slipher's initial finding. And then the director added, "Try some more spiral nebulae for confirmation." Slipher took up the challenge with great enterprise, for he was better at following directions than initiating his own scientific pursuits.

Working on Andromeda, though, was a holiday compared to gathering the spectral light from other spirals. Though its center is barely discernible to the naked eye, Andromeda is still the biggest and brightest spiral in the nighttime sky. The others only get progressively smaller and dimmer, which made it even harder for Slipher to obtain their velocity. "Spectrograms of spiral nebulae are becoming more laborious now because the additional objects observed are increasingly more faint and require extremely long exposures that are often difficult to arrange and carry through owing to Moon, clouds and pressing demands on the instrument for other work," he noted in his work papers. The job for him was "heavy and the accumulation of results slow."

Slipher's first target after Andromeda was M81, a spiral that is brighter than most, and then he looked at a peculiar nebula situated in the Virgo constellation known as NGC 4594. In his notes, he described it a "telescopic object of great beauty." It's now popularly known as the Sombrero galaxy for its distinctive resemblance to a Mexican hat viewed from the side. Slipher eventually saw that NGC 4594 was moving at a speed "no less than three times that of the great Andromeda Nebula." This time, however, the nebula was not traveling *toward* Earth but instead was whisking *away* at some 1,000 kilometers per second. Slipher was greatly relieved. Finding a nebula that was racing outward rather than approaching removed any lingering doubts that the velocities might not be real. "When I got the velocity of the Andr. N. I went slow for fear it might be some unheard-of physical phenomenon," he wrote his mentor Miller. Now, by the spring of 1913, he was reassured that the spectral shifts on his plates reliably meant *movement*.

At this stage, with just a few measurements in hand, Slipher began to think of the nebulae as drifting by the Milky Way—coming toward us on one side of the galaxy, and wandering away on the other side. He was reluctant to speculate publicly on what the spiral nebulae might be, but he did share some of his pet theories in private correspondence with his astronomer friends. At first he thought they might be dust clouds illuminated by reflected starlight, much the way he had already proven, to great acclaim, how the famous Pleiades star cluster shines. Or maybe, he went on to muse, the spirals were very old stars "undergoing a strange disintegration, brought about possibly by their swift flight through stellar space." But, even then, he was beginning to have reservations about such interpretations. If the spirals were indeed single stars surrounded by fine matter, Slipher posed in one 1913 letter, why are spirals not "more numerous in, rather than outside, the Galaxy?" That was the very same question Curtis was starting to ask over at the Lick Observatory.

Throughout the succeeding months Slipher kept expanding his list, one spiral at a time. His accomplishment was all the more amazing, considering the relative crudeness of his instrument. Lowell Observatory's 24-inch telescope had only manual controls, ones that weren't yet sophisticated enough for fine guiding. Yet he had to hold the tiny image of each spiral nebula on the slit of the spectrograph with utmost care and steadiness for hours on end as the heavens progressively rotated above him. When asked years later how he was able to do this, Slipher replied dryly, "I leaned against it." Given the faintness of his targets, his exposures often ran twenty to forty hours, which meant they extended over several nights, even weeks if there was unfavorable weather. And nothing could be done when the Moon was brightly shining. "With such prolonged exposures the accumulation of plates is not very rapid," he informed Lowell, "but the results are worth while and encouraging," so much so that Slipher was beginning to feel uncharacteristically possessive of his findings. "It is our problem now and I hope we can keep it," he told his boss.

Slipher need not have worried. No one else could catch up to him. By the summer of 1914 he had the velocities of fourteen spiral nebulae in hand. And with this bounty of data, an undeniable trend was at last emerging: While a few nebulae, such as Andromeda, were approaching us, the majority were rapidly moving away.

For island-universe devotees this was great news. "My harty [sic] congratulations to your beautiful discovery of the great radial velocity of some spiral nebulae," wrote Danish astronomer Ejnar Hertzsprung. "It seems to me, that with this discovery the great question, if the spirals belong to the system of the milky way or not, is answered with great certainty to the end, that they do not." The speeds were simply too great for them to stay put within our home galaxy. But Slipher at this stage was still on the fence. "It is a question in my mind to what extent the spirals are distant galaxies," he responded.

For most of his career Slipher published few detailed papers of his work, outside of his observatory's in-house bulletin. He either sat on his data until he was absolutely sure of the results or generously sent his findings to others to use in their analyses. Part of this might have been a reaction to the rumpus the observatory faced whenever Lowell defended his more sensational findings. Slipher inwardly feared that the unwelcome publicity was affecting astronomers' opinions on all other research coming out of Flagstaff. So, he preferred to keep his head down, out of the line of fire, adopting the philosophy, Let the work speak for itself. The singular exception for Slipher was his work on the spiral nebulae velocities. He had worked on so many stellar and planetary spectra that he was absolutely confident of what he was seeing—so confident that he for once overcame his homebound nature and traveled to Northwestern University in Evanston, Illinois, to present his results in person.

In August 1914 sixty-six astronomers from around the United States gathered at Northwestern for their annual meeting, four days of scientific talks, official business, concerts, and social excursions to Lake Michigan. It was the conference when the astronomers unanimously voted to change their title from the Astronomical and Astrophysical Society of America to simply the American Astronomical Society. At the same time, a young man named Edwin Hubble, a graduate student at the Yerkes Observatory, in Wisconsin, was elected for membership.

The presentations were made in the lecture room of the university's Swift Hall of Engineering. Slipher's paper, one of forty-eight read at the meeting, was titled "Spectrographic Observations of Nebulae." At the start of his talk, Slipher told the audience that he began his investigations simply to obtain a spiral nebula's spectrum, but went on to say that the exceptional velocity of the Andromeda nebula made him shift his attention to the velocities themselves. The average speed of the spirals, he reported, was now "about 25 times the average stellar velocity." Of the fifteen spiral nebulae he had observed so far, three were approaching Earth, the rest were moving away. The velocities ranged from "small," as it was recorded on his list, to an astounding 1,100 kilometers per second. That was the greatest celestial speed ever measured up to that time.

When Slipher finished delivering this remarkable news, his fellow astronomers rose to their feet and gave him a resounding ovation. No one had ever before witnessed such a spectacle at an astronomical meeting. And with good reason: Slipher had alone climbed to the top of the Mount Everest of spectroscopy. Even Campbell, his relentless competitor, came to both accept the finding and respect the tremendous effort behind it. "Let me congratulate you upon the success of your hard work," he wrote Slipher after the meeting. "Your results compose one of the greatest surprises which astronomers have encountered in recent time. The fact that there is a wide range of observed velocities—some of

Astronomers at the 1914 American Astronomical Society meeting
in Evanston, Illinois. Vesto Slipher is circled on the left,
Edwin Hubble on the right. (*From* Popular Astronomy,
"*Report of the Seventeenth Meeting*," 1914)

approach and some of recession—lends strong support to the view that the phenomena are real."

Soon after, Slipher was notified that the National Academy of Sciences in the United States was about to begin publication of a periodical titled *Proceedings*, aimed at displaying the nation's best scientific work. Slipher was asked to contribute an account of his groundbreaking research. "I am . . . glad to have your kind offer to present my papers to the Academy," he replied. "It only remains for me to do something worth sending." Slipher, as usual, was being modest to a fault.

Over the next three years, after he had gathered more spectra, Slipher at last came around to Hertzsprung's view. He, too, began to envision the Milky Way as moving among other galaxies just like itself. He first made this view public before the American Philosophical Society, when he was invited to give a key address at its 1917 annual meeting, one of the nation's most important scientific gatherings. Keen to report on his most up-to-date findings, Slipher even enlisted the help of a mathematician—Elizabeth Williams, in Boston, who had long worked as Lowell's top computer—two weeks before the lecture to help him double-check the direction and magnitude of his full complement of spiral nebulae, now numbering twenty-five. She telegraphed her results in the nick of time.

"It has for a long time been suggested that the spiral nebulae are stellar systems seen at great distances," said Slipher at the April conference in Philadelphia. "This is the so-called 'island universe' theory, which regards our stellar system and the Milky Way as a great spiral nebula which we see from within. This theory, it seems to me, gains favor in the present observations." With all but four of his twenty-five spiral nebulae racing outward, Slipher speculated at one point that the spirals might be "scattering" in some way, a precocious intimation of the cosmic expansion that took many more years to fully recognize.

Though other astronomers were confirming a few of Slipher's results, the Lowell Observatory astronomer was the absolute ruler in this new celestial realm. He dominated the field for years. By 1925, forty-five spiral nebulae velocities were pegged with assurance, and it was Slipher who had measured nearly all of them. As early as 1915, researchers in Germany, Canada, the United States, and the Netherlands began to look for a pattern in Slipher's growing mound of data. It was an extremely difficult task, though, as the speeds measured for the spiral nebulae were entangled with other velocities, such as Earth's orbital travels and the Sun's journey through the galaxy. It was like trying to determine the exact speed of a train off in the distance, while you yourself are in a car racing down a highway.

The investigators began by subtracting out the extra factors—first the Earth's motion, then the Sun's—to see how fast the spiral nebulae were truly moving. Once these secondary velocities were removed, the astronomers saw that the nebular speeds continued to be enormous, far higher than the average velocity of a star within our galaxy. More important, they confirmed that the mistlike disks were indeed generally headed away from us. A few nebulae, such as Andromeda, were exceptions (they didn't yet know that Andromeda and the Milky Way were gravitationally bound together and so wouldn't be flying away from each other), but all in all the spiral nebulae were primarily moving outward into space in all directions. The German astronomer Carl Wirtz went even further in 1922 by looking at a nebula's size and luminosity to roughly judge which of the nebulae were closer to us and which were farther out. By making this assumption, he noticed a particular progression to the stampede outward: The more distant the nebula, the faster it was receding. That was intriguing.

But perhaps this relationship between speed and distance was a false impression. Maybe the effect would disappear as the velocities of more and more nebulae, especially those found in the southern celestial sky, were measured. It could all average out: half of the nebulae moving toward us, the others away. Astronomers began to worry that what looked like an overall recession might turn out to be a temporary illusion. To take care of this, they began to insert a special component into their equations, a term they labeled K, which kept track of the trend. Maybe this term would eventually fade away, but maybe not.

Despite these loose ends, by the time of the 1917 American Philosophical Society meeting, the island-universe theory was rousing from its slumber. Heber Curtis had begun to publish his findings on the spiral nebulae in the major journals, and his cogent arguments in support of distant galaxies were already convincing the top astronomers who counted, such luminaries as Eddington at Cambridge University, in England, Campbell at Lick, and Hertzsprung, then at the Potsdam Observatory, in Germany. The swift velocities that Slipher was finding only strengthened the idea that the spirals were indeed situated far beyond the Milky Way's borders. But success could not be fully grasped until astronomers figured out a method for determining how far away Andromeda and its sister spirals truly were. Nothing could ever be settled in this ongoing debate until someone determined the distances to these exasperating nebulae, in a way that every astronomer had confidence.

What Slipher and Curtis did not yet know was that a novel way to carry out such a celestial measurement had been budding even as they were beginning their researches on the spiral nebulae. It involved a gifted woman with a keen eye, who came upon some intriguing stars while examining photos of an alluring feature in the southern nighttime sky.

[6]

It Is Worthy of Notice

First-time travelers to the southern hemisphere might mistake the clouds for high cirrus formations, somehow made luminous in the dark of night. Ancient Persians called the biggest one Al Bakr, or the White Ox. Europeans were introduced to the "two clouds of mist" from accounts of the first circumnavigation of the globe by Ferdinand Magellan and his crew in the early sixteenth century. And so the hazy pair came to be named in honor of the Portuguese-born explorer. The Large and Small Magellanic Clouds are each a chaotic collection of stars, richly diffused with glowing gas.

Such novel and fascinating sights were a compelling reason for European and American astronomers to set up observatories in the southern hemisphere. The Harvard College Observatory did just that in the 1890s, when it established a southern station in the highlands of Peru, just above the town of Arequipa. Before this, for more than a decade, Harvard had been carrying out a formidable task: to catalog every star in the northern sky and accurately gauge its color and brightness. Presented with a sizable endowment for a program in spectroscopy, observatory director Edward C. Pickering resolved to photograph and classify the spectra of all the bright stars as well. The Peruvian observatory allowed Harvard to extend the reach and sweep of this endeavor to the southern sky. By doing this, Pickering was helping astronomy move beyond just tracking the motions of stars across the sky to figuring out their basic properties. Though tedious and wearying, such astronomical surveys can often reveal a few surprises along the

The Small and Large Magellanic Clouds (top left, bottom left) as seen from
Cerro-Tololo Inter-American Observatory, in Chile. The Milky Way
is on the right. (*Roger Smith/NOAO/AURA/NSF/WIYN*)

way. The Harvard survey was no exception, but it took many photographs to get there.

With the huge number of photographic plates of the northern and southern skies stacking up at the observatory on Garden Street in Cambridge, Massachusetts, Pickering shrewdly recognized the value of smart young women yearning to make contributions in an era that generally denied them full access to scientific institutions. Here was a ready workforce, he noted in one annual observatory report, entirely "capable of doing as much and as good routine work as astronomers who would receive much larger salaries. Three or four times as many assistants can thus be employed, and the work done correspondingly increased for a given expenditure."

Williamina Fleming (standing) directs her "computers" while
Harvard Observatory director Edward Pickering looks on
(*Harvard College Observatory*)

These women "computers," as they were called, many with college degrees in science, were situated in two cozy workrooms, pleasantly decorated with flowered wallpaper and star charts. Working at mahogany writing tables, crammed together, each woman through the day might peer through a magnifying glass at her selected plate or industriously record her findings in a notebook. Resembling assembly-line workers in a factory, with their plain, unadorned dresses, these dedicated women—swiftly, accurately, and cheaply—numbered each star on a given plate, determined the star's exact position, and assigned it either a spectral class or photographic magnitude. Annie Jump Cannon, who established the stellar classification system adopted internationally in the course of this work, praised Pickering's modern outlook. "He treated [the computers] as equals in the astronomical world," she claimed (somewhat Pollyannishly), "and his attitude toward them was as full of courtesy as if he were meeting them at a social gathering." He was their gallant Victorian gentleman.

Pickering's first hire was his housekeeper, Williamina Fleming, who had displayed a keen intelligence in carrying out her duties. Frustrated one day by a male assistant's ineptitude, Pickering had declared that his maid could do a better job, and he found out she could. From the 1880s until Pickering's death in 1919, some forty women came to be employed in "Pickering's harem," as it was jokingly known. One of his most brilliant choices was Henrietta Leavitt, who first began work at the Harvard Observatory as a volunteer.

Leavitt grew up in Massachusetts, within a big and supportive family (she was the oldest of five surviving children) that cherished education. Her father held a doctorate in divinity from the Andover Theological Seminary. The Leavitts moved to Cleveland when Henrietta was a teenager, where she eventually began her undergraduate studies at Oberlin College. In 1888, at the age of twenty, though, she returned to Massachusetts and entered the Society for the Collegiate Instruction of Women in Cambridge (what later became Radcliffe College). She primarily took courses in the arts and humanities but in her fourth year enrolled in an astronomy course. It must have been an inspiration, because after receiving her certificate in 1892, which stated that she had undertaken an education equivalent to a Harvard bachelor of arts degree, Leavitt remained in Cambridge to take some graduate courses and work as an unpaid helper at the college observatory.

According to those who knew her, she was a serious-minded woman, devoted to her family circle and to her friends. A photograph of her reveals a woman of quiet beauty with soulful eyes. "For light amusements, she appeared to care little," said Harvard astronomer and colleague Solon Bailey. Yet, he went on, she was still "possessed of a nature so full of sunshine that, to her, all of life became beautiful and full of meaning." Her good-natured disposition remained even after she experienced, sometime after her graduation, a serious illness that left her severely deaf.

As a volunteer she became an expert in stellar photometry, gauging the magnitude of a star by assessing the size of the spot it imprints upon a photographic plate. The brighter the star, the larger the spot made dark by the star's light upon the negative. In carrying out this work, she was also instructed to keep an eye out for variable stars, stars that regularly increase and decrease in brightness over a fixed span of time. These variables were found by comparing photographs of the same region of the sky taken at different times. Leavitt would place the nega-

Henrietta Leavitt at her Harvard College Observatory desk
(*AIP Emilio Segrè Visual Archives*)

tive of a photograph taken on one date directly over a positive photograph of the same region of the sky taken on another date. If the black and white images of a star did not exactly match, the star was likely changing its intensity and so was suspected to be a variable.

After writing up a draft of her initial research, Leavitt left Harvard in 1896 for a while, first traveling through Europe for two years and then moving to Wisconsin, where her father had a new ministry. But in 1902 she wrote Pickering for information on new job opportunities, at either Harvard or elsewhere. She obviously wanted to get back into astronomy and must have been overjoyed when Pickering made her an offer within three days for full-time employment. "For this I should be willing to pay thirty cents an hour in view of the quality of your work, although our usual price, in such cases, is twenty five cents an hour," he wrote her. She replied that it was a "very liberal offer," what today (taking inflation into account) would be a little over the U.S. minimum wage. A man would have gotten nearly twice that.

Not until the spring of 1904, though, did variable stars come back into her life in full force. Peering through a magnifying eyepiece at two photographic plates of the Small Magellanic Cloud, taken at different times, she noticed that several stars in the cloud had changed in brightness. On one plate a particular star was relatively luminous; on another plate that same star had turned far dimmer. It was as if the star were undergoing a slow-motion twinkle. Over the following year, she looked at additional images of the cloud and found dozens more. With each new delivery of plates from Harvard's station in Peru (and checks on old ones going back to 1893), she readily and meticulously updated her count, so much so that a Princeton astronomer described her as a "variable-star 'fiend.' " Soon she included the Large Magellanic Cloud in her tally, and by 1907 she found a record-setting total of 1,777 new variable stars residing within the prominent, mistlike clouds (before that, only a couple of dozen variable stars had been detected in the Magellanic Clouds). She dutifully reported her findings in the 1908 *Annals of the Astronomical Observatory of Harvard College*, with thirteen pages taken up with listing every new variable she had discovered, its exact position in the sky, as well as its minimum and maximum brightness.

More intriguing was what she wrote at the end of this paper. Over the course of her painstaking examination of the Small Magellanic Cloud, she came to notice a special group of variable stars, sixteen in number. They were later identified as Cepheid variables, stars that are thousands of times more luminous than our Sun. Their name was derived from one of the first and brightest discovered, δ Cephei, located in the constellation Cepheus the King, a major landmark in the northern sky. These stars regularly vary their brightness in a matter of days or months. The shortest cycle Leavitt measured for these Magellanic variables was 1.2 days, the longest 127 days. Yet no matter if the Cepheid had a long or short period, each was as regular as a metronome in its variation. "As a rule, they are faint during the greater part of the time," reported Leavitt, with the period of maximum brightness being fairly brief. The variable δ Cephei, for example, goes from dim to bright in just a day, then gradually fades back to its faintest magnitude over the next four days, until it suddenly brightens once again.

But it was the next sentence in Leavitt's report that turned into its most venerated statement. "It is worthy of notice," she continued, "that . . . the brighter variables have the longer periods." Since all her Cepheids were situated in the same celestial cloud, Leavitt could assume they were all roughly the same distance from Earth. And that meant she could trust that the Cepheids' periods were directly associated with their actual emission of light. Leavitt was in fact getting a first glimpse at astronomy's celestial Rosetta stone, a means for astronomers to solve the mystery of the spiral nebulae. The key was that link between a Cepheid's period—the steady rhythm of its oscillation—and its luminosity. She was on the brink of finding a new cosmic yardstick, one that would allow astronomers to determine the distances to far-off celestial objects that were formerly immeasurable by more traditional means.

Leavitt had chanced upon the celestial equivalent of lighthouses on Earth. A sailor, if he is familiar with the amount of light a particular lighthouse emits, could roughly estimate how far he is from land, given how bright the beacon appears to him from offshore. Similarly, a Cepheid's period labels it as having a particular brightness. The distance to the Cepheid is then obtained by figuring out how far away it must be to be viewed as the faint point of light we see from Earth. In this way, the Cepheid becomes a valuable "standard candle" (as astronomers call it) for gauging distances deep into space, when all other methods fail.

Bright and dim, bright and dim goes a Cepheid's cycle, but not endlessly. It was long believed that a Cepheid was an eclipsing binary star—one star regularly circling another like the Earth going around the Sun. But by 1914 it was recognized that this type of variable was actually a single, pulsating star, its atmosphere regularly ballooning out and then shrinking back in, over and over again. When the origin of a star's power was at last understood, astronomers came to see that a Cepheid is a star far more massive than our Sun, anywhere from five to twenty solar masses, that has reached a particular stage in its evolution. Having used up its main supply of hydrogen, the Cepheid becomes unstable for a while (about a million years) as it adjusts to burning new sources of nuclear fuel. When the star is compact, pressures build up, causing the star's outer atmosphere to expand and thus become more luminous. But once stellar pressures are reduced, gravity takes over and causes the star to contract back and become dimmer—that is, until stellar pressures build up once again. In this way, the Cepheid comes to pulsate in a regular fashion. More important, the brighter and more massive Cepheids oscillate more slowly than the fainter and smaller ones.

In 1908 Leavitt was wary that her initial sample of sixteen Cepheids was too small to secure a firm and predictable "period-luminosity" law. She needed more, but chronic illnesses and the death of her father delayed her a few years. Moreover, though very bright, allowing them to be seen over long distances, Cepheids are also very rare. Not until 1912 was Leavitt able to add nine more Small Magellanic Cepheids to her list. With twenty-five in hand, she could at last establish a distinct mathematical relationship between a Cepheid's blinking and its perceived brightness.

Science often involves discovering patterns, spotting regularity and order where none before had been noticed. And the pattern that Leavitt made plain, with such patient care and shrewd insight, in time opened up the universe. The connection was immediately apparent when Leavitt plotted her data on a graph. "A remarkable relation between the brightness of these variables and the length of their periods will be noticed," she wrote, with a decided animation rare in scientific discourse. On a logarithmic scale, the visible brightness of her Cepheids rises steadily as the stars' periods get longer and longer. Her variable stars huddle along a sure, straight line from the bottom left to the upper right of the graph paper. This historic finding was published as *Harvard College Observatory Circular*, No. 173, a three-page paper titled "Peri-

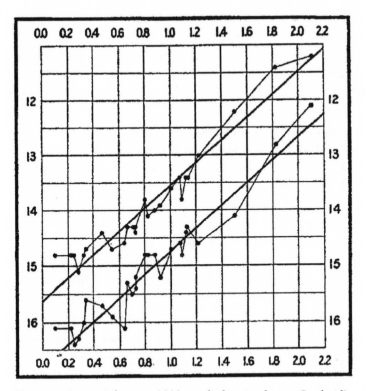

Henrietta Leavitt's historic 1912 graph showing how a Cepheid's brightness increases as the variable star's period gets longer (*From Harvard College Observatory Circular, No. 173* [1912], *Figure 2*)

ods of 25 Variable Stars in the Small Magellanic Cloud" and now considered a "masterpiece" of scientific literature.

Cepheids stood ready to be the perfect standard candles, but first she
needed to know the *true* brightness of at least one, the luminosity she
would observe if she were essentially right next to the star. If she could
determine the brightness of just one, her graph would let her know all
the others. Once her graph was calibrated in this way, an astronomer
could pick out a far-off Cepheid anywhere in the sky, measure its
period, and infer its actual luminosity. The distance to the Cepheid
then followed: By measuring the Cepheid's *apparent* brightness in the
sky (a much fainter magnitude), you could figure out how far away it
must be to appear that dim. Cepheids held the promise of being astronomy's handiest cosmic measuring tape. Astronomers could at last gauge
the distance to celestial objects farther out than they ever conceived
possible. Leavitt knew this, but she wasn't one to state things so daringly.
Besides, Pickering chose his women computers "to work, not to think,"
according to one Harvard astronomer. So, in a far quieter tone, Leavitt
simply wrote at the close of her paper, "It is to be hoped, also, that the
parallaxes [essentially, distances] of some variables of this type may be
measured."

What was needed was an indisputable distance to a bona fide
Cepheid. But Leavitt's going to a telescope to pursue an answer was out
of the question, not only because women were denied access to the best
telescopes at the time (generally considered man's work) but because of
her frail condition. Given her deafness and frequent illnesses, she had
been advised by her doctor to avoid the chilly night air, an environment
habitually faced by observers. She came to believe the cold aggravated
her hearing condition. If she had the know-how, she could have carried
out a calculation from her desk, using stellar data from previously published work, but Pickering held the strong conviction that his observatory's prime function was to collect and classify data, rather than apply it
to solve problems. The accumulation of facts was Pickering's prime
directive, so he quickly assigned Leavitt another task, a project on stellar
magnitudes that he considered far more important. She respectfully
carried out her boss's directive without objection for a number of years.
Back in her room at the observatory, she continued to work on the stars
photographed by others. Cecilia Payne-Gaposchkin, who came to the
Harvard Observatory in the 1920s, called this "a harsh decision, which
condemned a brilliant scientist to uncongenial work, and probably set

back the study of variable stars for several decades." Yet Leavitt's effort was not wasted. In the end her delegated work served as the basis for an internationally accepted system of stellar magnitudes.

But her desire to pursue the variables never left her; it only awaited the proper time to act on it. Soon after Pickering's death, Leavitt at last divulged her most cherished interest to the observatory's new director, Harlow Shapley. Once he arrived at Harvard in 1920, she lost no time in asking his advice on advancing her research on the stars in the Magellanic Clouds. By then Shapley had already calibrated the Cepheids, but he told Leavitt he would like to see a deeper investigation of the short-period variables, stars that pulse over a matter of hours instead of many days. "[It's] of enormous importance in the present discussions of the distances of globular clusters and the size of the galactic system," he said. Moreover, does the same period-luminosity law also work for stars in the Large Magellanic Cloud? he asked. He wished her success on tackling these questions.

But just as she was on the verge of completing her prolonged stellar magnitude project—possibly when she would have at last gone back to her work on the Cepheids—Henrietta Leavitt passed away at the age of fifty-three. She had faced a long and grueling struggle with stomach cancer. By the time of her death, on December 12, 1921, she had discovered some twenty-four hundred variable stars, about half the number then known to exist. Her contributions at Harvard had been unique, making it difficult for them to replace her. "Miss Leavitt had no understudy competent to take up her work," Shapley told a colleague the day after her death. Unaware of her passing, a member of the Royal Swedish Academy of Sciences four years later contacted the Harvard Observatory to inquire about her discovery, intending to use the information to nominate her for a Nobel Prize in Physics. But by the rules of the award, the names of deceased individuals could not be submitted.

Exploration

[7]

Empire Builder

In 1914 the world was plunged into turmoil as the Allied and Central powers rapidly faced off in the War to End All Wars, the four-year conflict that demolished old empires and reshaped the modern world. And yet, in this time of devastating upheaval, astronomy experienced some of its greatest discoveries. Vesto Slipher was measuring the fleeing spirals, Heber Curtis was ferreting out new ones, and Harlow Shapley was gearing up to move our Sun from its hallowed position at the center of the known universe. While the landscape of global politics was being redesigned, so too was our cosmos.

The Milky Way had long been pictured as relatively small, at most around 20,000 to 30,000 light-years wide (estimates at this time varied), but in 1918 Shapley radically increased our galaxy's girth to some 300,000 light-years. Moreover, he declared that our solar system was situated a good 65,000 light-years from the galaxy's heart. Barely recovered from its Copernican shift from the center of the solar system, Earth was demoted once again. The Milky Way's overall width was later amended, adjusted downward to some 100,000 light-years when better calibrations were undertaken, but even then it was far vaster than anyone had previously imagined.

Shapley would never have had this opportunity were it not for the astounding foresight and boundless fortitude of George Ellery Hale. A noted solar astronomer, Hale discovered that there were magnetic fields in sunspots, a sensational finding in its day, for it was the first magnetic field detected beyond Earth. He also cofounded the *Astrophysical Journal* (along with James Keeler) and helped transform the Throop College of Technology into the California Institute of Technology. But Hale made his most valuable contributions to astronomy as an administrator. It was largely through his focused efforts over several decades that America wrenched the baton from Europe in astronomical leadership. Hale nearly single-handedly orchestrated the construction of four great telescopes in the United States, each larger and more advanced than the one before. In carrying out this colossal endeavor, he allowed Shapley to revamp the Milky Way and the astronomers who followed to reveal the true vastness of the universe and the amazing diversity of its celestial inhabitants. Astronomer Allan Sandage of the Carnegie Observatories is convinced that astronomers "owe *all* to Hale and his dreams and positive actions to put those dreams into glass and steel. Where would world astronomy be today if Hale had not been an 'empire builder'?"

Hale took unique advantage of the magnificent productivity of his era. It was once jokingly noted that American astronomy became pre-eminent at this time because of two discoveries: Pickering discovered women and Hale discovered money. American industrialists were amassing great fortunes, capital that was just waiting to be tapped for philanthropic undertakings in an era before the federal income tax was permanently established. Of all the sciences in the Gilded Age, astronomy was the most popular destiny for private support in the United States. One reason is that astronomy held out the promise of a shiny white dome on a mountain, for all to look up and admire. Hale, too, commented that the public regards "astronomical research with a feeling of awe which is not accorded to other branches of science [because of] its power of searching out mysterious phenomena in the infinite regions of space."

Hale himself was the very personification of this union of money with science at the turn of the twentieth century. Hale's father, William, had secured sizable riches as the manufacturer of hydraulic elevators, produced for the many skyscrapers that began to dot the Chicago urban landscape after the Great Fire of 1871. His company also supplied them for Paris's Eiffel Tower. Some of the capital from these enterprising ventures offered Hale as a teenager sufficient funds to construct his own spectroscopic observatory in the attic of the family mansion in the Hyde Park section of Chicago, where he avidly studied the Sun's spectrum, alongside his books, laboratory equipment, and fossil collection. He was a precocious boy with a formidable power of concentration—always curious and always devising new ways to study the natural world. He chose the Sun as his target of interest because, as the closest star, he hoped it might better reveal the secrets of stellar evolution. Shortly after his twentieth birthday in 1888, he confirmed that the element carbon resided in the Sun, a matter then in great debate. Before Hale even graduated from college, he developed a new instrument—the spectro-heliograph—that enabled astronomers to photograph the surface of the Sun and its fiery prominences as never before. It imaged the Sun in one chosen wavelength of light, a spectral band being emitted by a specific chemical element of interest. Science was still reeling in the late 1800s from the magnificent discoveries in geology and biology that so beautifully demonstrated the gradual changes that occurred over Earth's history: new species evolving and landscapes continually sculpted by natural forces. Hale was seeking evidence of a similar dynamic within the universe itself.

Upon graduating from the Massachusetts Institute of Technology, class of 1890, Hale married his childhood sweetheart, Evelina Conklin, and took an extended honeymoon trip to Niagara Falls, Colorado, San Francisco, and Yosemite. But he was most excited, while out in California, to get a personal tour of Lick Observatory. There, he had the opportunity to work one night with James Keeler, as the Lick astronomer was observing planetary nebulae. Hale was mightily impressed and never forgot his first glimpse of the 36-inch refracting telescope, then the world's largest, its long tube "reaching up toward the heavens in the great dome," he later recalled. Hale primarily studied the Sun, Keeler the stars and nebulae, but both were fervent advocates of spectroscopy. They became fast friends.

Within two years of his return to Chicago, Hale became an associate professor at the newly reorganized University of Chicago. With the university's promise of future funding for a larger telescope, he allowed the university to use his personal observatory, grandly christened the "Kenwood Physical Observatory." The complex was built right next door to the Hale family mansion and housed both a 12-inch refractor, paid for by his father, and his revolutionary spectroheliograph. "I would not consider [joining the faculty] *for a moment* were it not for the prospect of some day getting the use of a big telescope to carry out some of my pet schemes," he told an acquaintance.

That prospect arrived sooner rather than later, due solely to Hale's resourcefulness. After attending the latest meeting of the American

George Ellery Hale at his spectrograph
(*Courtesy of the Archives, California Institute of Technology*)

Association for the Advancement of Science in Rochester, New York, in the summer of 1892, he went out to cool off on his hotel's veranda and overheard a conversation about two 40-inch telescope lenses that had unexpectedly become available. The glass disks had been made for a planned observatory in southern California that was aimed at surpassing Lick in telescopic power. A real estate boom had brought sudden wealth to the Los Angeles area, and for the sake of regional pride developers were eager to erect their own grand astronomical monument— until the promoters went broke when the land bubble burst. For Hale, so eager to acquire a large telescope for his solar investigations, ready access to such lenses was a stroke of luck. A lens forty inches wide had nearly 25 percent more surface area than the Lick's 36-inch lens and so would gather 25 percent more light, a huge and treasured gain for any

astronomer. Given the brush-off by a bevy of Chicago's wealthiest busi-
nessmen to sponsor the purchase of the lenses, Hale at last convinced
Chicago's streetcar magnate, Charles Tyson Yerkes, to fund construction of the giant instrument. Hale's work as an astronomer provided the
scientific arguments for this bold step; his family's wealth and position
gave him the self-confidence, even though he was only twenty-four
years old, to win over Yerkes in financing such a grand scheme.

The university had been pursuing Yerkes for months to pledge a gift
and offering to have his name attached to the world's largest telescope
was powerful enticement (as it had been with James Lick years earlier).
Hale was not shy in playing up that angle. "The donor could have no
more enduring monument," he wrote Yerkes. "It is certain that Mr.
Lick's name would not have been nearly so widely known today were it
not for the famous observatory established as a result of his munifi-
cence." Yerkes snapped at the bait: "Build the observatory," he told Hale
at a meeting. "Let it be the largest and best in the world and send the
bill to me."

Yerkes was a curious target for Hale's entreaties, as he had a shadier
pedigree than other robber barons of his time. A suave and colorful
character, Yerkes had made a fortune dealing in Philadelphia's municipal securities, until some questionable dealings led to his serving a
prison term for misappropriation of public funds. It earned him the reputation as "the embodiment and representative of corruption in municipal affairs." Having lost his wealth in the City of Brotherly Love, he
quickly made the money back once he moved to Chicago and bribed
the appropriate politicians to gain control of the city's streetcar system.
Right before this move west, Yerkes had also divorced his wife, who had
borne him six children, and married a much younger woman, Mary
Adelaide Moore, renowned for her beauty. So memorable was Yerkes'
life that author Theodore Dreiser immortalized it in his fictional works
The Financier and *The Titan*. Dreiser described this era in his autobiography as the moment in American history when "giants were plotting,
fighting, dreaming on every hand."

Yerkes clearly savored having his name attached to a big telescope. It
gave him class (not to mention an improved credit rating with local
bankers, which may have been his aim all along). The newspaper
Chicago Daily Inter Ocean reported that "Mr. Yerkes, when he took the
matter in hand, simply stipulated that the observatory and its telescope

should beat everything of its kind in the world." The *Chicago Tribune* chimed in as well, smugly writing that "the Lick Telescope will shortly be licked."

With great pomp and circumstance, the Yerkes Observatory officially opened in 1897. Hale, at the age of twenty-nine, was appointed its director. Seventy miles from Chicago, the observatory was situated in the resort town of Williams Bay, Wisconsin, a tiny hamlet next to Lake Geneva, where several University of Chicago trustees just happened to have summer homes. The shift to the "new astronomy" was highly evident at the new establishment, a fact noted at its dedication. Keeler, the keynote speaker, told the distinguished audience that "there may be some who view with disfavor the array of chemical, physical and electric appliances crowded around the modern telescope, and who look back to the observatory of the past as to a classic temple whose severe beauty has not yet been marred by modern trappings." The Yerkes Observatory was forging a new pathway for studying the heavens. Hale, a committed astrophysicist, made sure that there were photographic darkrooms, spectroscopic labs, and instrument shops specifically devoted to the astrophysicist's concerns. Hale was changing how an observatory worked.

A restless and anxious man, Hale was the astronomical equivalent of an industrial entrepreneur, always on the lookout for new technologies and new methods for obtaining cutting-edge results. Before the magnificent Yerkes dome even rose above Lake Geneva, Hale had already convinced his wealthy father to buy the materials for yet another telescope, this time a large reflector. A mirror blank, *sixty* inches in width, was cast for him by the Saint-Gobain glassworks, a French maker of wine bottles. By the time astronomers from around the world arrived at Yerkes for its opening, telescope maker George Ritchey was busy in the observatory's optical shop grinding the mirror and designing its support system. The descendant of Irish immigrant craftsmen and a former high school teacher in woodworking, Ritchey was employed by Hale as a skilled optician, having had some training in astronomy before he dropped out of college. He was legendary for his artistry in designing and fashioning new telescopes, but his obsession to achieve technical perfection led to his reputation as a cantankerous cuss.

Tests on the Yerkes 40-inch lens had made it obvious that lenses were reaching their limit. If the lens were any larger, the glass would sag under its own weight, distorting the image. To go bigger, both Ritchey and Hale knew that they had to use a mirror rather than a lens, bringing back the reflecting telescopes pioneered by Herschel and Rosse. Hale and Keeler, who was just beginning to operate the Crossley reflector at the Lick Observatory, had many discussions on this topic. A 60-inch reflector was going to vastly increase—more than double—the amount of light gathered by the 40-inch refractor at Yerkes. It held the promise of accelerating the observatory's output: Photographic exposures could be shortened, and the spectra of faint stars, previously too feeble to image, at last acquired. New celestial vistas were certain to open up, with millions of new stars being revealed. Ritchey and Hale formed a close bond over their joint vision that reflectors were the instruments of astronomy's future.

Hale had long had his eye on the West Coast to erect his 60-inch scope. "The possibility of having you for a neighbor in California is quite too delightful," wrote his friend Keeler, upon hearing of Hale's interest. "It seems to me that somewhere in the coast range, perhaps farther south than we are, would be the best site." Hale agreed. He knew the air was more arid, the weather more suitable, in the southern region of the state. And within a few years, he was able to check it out for himself.

At the end of 1903, Hale temporarily moved his family to Pasadena, then still a small town with many of its roads unpaved. His daughter, ill with asthma, required a warmer and drier climate than the one found by chilly Lake Geneva in Wisconsin. California's plentiful sunshine helped Hale, too, lifting him out of the depression he occasionally experienced. Once settled in, Hale became convinced that "Wilson's Peak," which he could see from his bedroom window on Palmetto Street, was "*the* place" to continue his astronomical work. He had actually been thinking about the site for quite a while, ever since Harvard briefly considered setting up a permanent telescope there in the late 1880s.

Some thirty miles from the Pacific Ocean, Mount Wilson rises abruptly from the valley floor. It is one of many peaks of the San Gabriel Mountain Range, which runs west to east and forms a barrier between the Los Angeles metropolitan area and the Mojave Desert to the north. Upon his first venture to the top of the mile-high peak burgeoning with scrubby live oak and commanding spruce, Hale felt as if he had arrived at the edge of the world, with its stunning views of the town directly below and the dark blue sea in the distance. He had certainly reached the end of his pursuit for the perfect location to carry out his observations.

With the University of Chicago unwilling to fully finance Hale's dream for a California outpost, the young astronomer sought other funding sources. Fortuitously, Andrew Carnegie had just established the Carnegie Institution of Washington, generously endowed with $10,000,000 "to encourage investigation, research and discovery in the broadest and most liberal manner." Carnegie, who had made his fortune in steel, made an even bigger name giving his money away. For Hale, an enterprise solely founded to support scientific investigation was nirvana and "seemed almost too good to be true," for Carnegie's gift surpassed the funds then endowed for research at all American universities combined. Hale immediately lobbied for a grant, laying out his plans with great vigor, but not surprisingly Carnegie was inundated with requests from around the country and slow to respond.

That didn't hamper Hale one bit. Even before he had any promise of money for a full-fledged observatory, Hale used funds allotted for a "University of Chicago Expedition of Solar Research" to Mount Wilson to start work on the mountain in the summer of 1904. When that grant ran out, he dipped into his own pocket to pay his men, gambling that his investment would pay off. What came to be known as the "Monastery" was built at this time on the edge of the south ridge, to serve as guest housing for the male astronomers. By the time the quarters were finished at the end of the year, complete with a huge granite fireplace built from native rock, Hale received word that the Carnegie Institution had at last agreed to sponsor his plans, which called for both a solar telescope and the 60-inch reflector atop the mountain. It was difficult to refuse a man with such energy, magnetism, and relentless dedication to a goal. His mistress, the Los Angeles socialite Alicia Mosgrove, described Hale as having "an inner excitement—a higher degree of interest—a higher degree of suffering." Hale resigned as Yerkes director to devote his full attention to the new Mount Wilson Solar Observatory of the Carnegie Institution of Washington. A deal had been arranged to lease the observatory's land for ninety-nine years, free of rent.

The shop for assembling the 60-inch telescope was set up in town on Santa Barbara Street. The stately headquarters of the Carnegie Observatories remains there today. The street is now crowded with residential dwellings, but then Pasadena was sparsely settled, with only a few farmhouses and barns nearby. At the time of Hale's first reconnaissance of Mount Wilson, two trails went up the mountain: an old Indian path and a road built by the Mount Wilson Toll Road Company, although this "road" was just a few feet in width. Mules—with names such as Jasper,

Pinto, Duck, and Maude—were on hand to either carry baggage or transport the weary. Strands of their hair, finer than human hair, were sometimes used as cross wires for the guiding telescopes, a smaller, parallel scope that helps astronomers keep their celestial objects on target during an observation.

Getting the 60-inch telescope set atop Mount Wilson was a herculean effort. All in all, hundreds of tons of material were hauled up by either mules or mule-assisted electric carts to construct the building and steel dome. To do this the nine-mile trail had to be first broadened and improved by pick and ax, foot by foot. The tensest moment was the transport of the mirror itself, which had been polished over a grueling four years to a smooth, parabolic shape so fine that no imperfection extended farther than two millionths of an inch. One misplaced wheel and the entire cargo could have suddenly plummeted to the canyon floor. To everyone's relief, the mirror arrived in the summer of 1908 without a scratch and within three months was finally cradled in its

Transporting the 60-inch telescope up Mount Wilson
(*Courtesy of the Archives, California Institute of Technology*)

mount (which miraculously survived the great 1906 earthquake in San Francisco, where it was built). Once the telescope was in operation, astronomers could see stars up to one hundred million times fainter than the brightest stars in the sky.

Hale decided he would not follow the Lick Observatory model, which involved erecting a self-contained village to house staff members and their families. For Mount Wilson only essential personnel stayed full-time on the mountain to maintain the observatory and its equipment; the astronomers now traveled up from the observatory's headquarters in town whenever scheduled to observe. "Hale was never so happy," noticed one onlooker, "as when, like a boy on a vacation, he could pack a knapsack and start on the eight-mile climb over the old trail to the summit." Hale also gave staff members the freedom to work on their individual research interests. In the past, observatories were often set up as data factories, conducting long-term surveys and gathering extensive libraries of plates for others to consult in solving problems. But for Hale astronomy was now experimental physics; telescopes were to be used like the instruments in a laboratory, to answer carefully chosen questions and to develop theories from the gathering facts. A firm believer in the potential of American science, Hale wanted his country's scientists to evolve beyond mere fact gathering and produce more fundamental discoveries—as he put it, to see "the woods" instead of the trees. It was a radical departure from the way astronomy had been done over the centuries, which largely applied Newtonian gravitation to the workings of the universe. Hale sought to embrace the new physics.

Some old habits from the Victorian era persisted, though. For dinner at the Monastery, linen cloths and napkins covered the table, with all astronomers required to wear coat and tie, which was a trial during hot weather, but they'd be barred from the dining room if they violated the dress code. Not until after World War II did this prim social atmosphere begin to crumble. Today, as soon as twilight beckons, T-shirt-clad astronomers on Mount Wilson dash off to their telescopes to glean every possible second of observing. But back then the dapper astronomers continued to eat dinner, even as it got dark, and got up to amble leisurely over to the scope only when the meal was finished.

Hale took nearly all the best Yerkes men with him to his new astronomical Shangri-la. He had a magnetic aura that drew in people and kept them in awe of him. His second in command at Mount Wilson, Walter Adams, once admitted that he stuck with Hale and astrophysics "partly because of the strong influence of Dr. Hale's remarkable personality. . . . A very slight change in circumstances might equally well have led me to follow the teaching of Greek as a profession."

Others on Hale's charter staff did not have a formal education in astronomy but instead trained on the job, bringing with them valuable skills from such fields as photography and mechanical engineering. These included Ritchey and Ferdinand Ellerman, who was first hired to assist Hale at his private observatory in Chicago.

Hale eventually extended his search for employees beyond his Yerkes loyalists. When he heard about a PhD student from Princeton who was impressing everyone, he arranged to meet the young man in New York City. Harlow Shapley showed up fully prepped to discuss all the latest astronomical discoveries. Instead, the two men ended up talking about the operas Shapley had time to catch the day before. The conversation went on for a while, but then Hale abruptly remarked, "Well, I must be going." Not one word on astronomy had passed between them—and no mention of a job. The Princeton grad assumed he had not passed muster, but to his surprise he soon received a letter from Hale. The message was all he had hoped for: "Please come to Mount Wilson."

[8]

The Solar System Is Off Center and Consequently Man Is Too

Upon arriving at Mount Wilson, Harlow Shapley had no immediate investigative plans, only a developing interest in variable stars. He had told his Princeton mentor, Henry Norris Russell, that he would probably work on odds and ends. But Shapley's wife, also skilled in astronomy, soon came upon an interesting set of stars while examining photos of a globular cluster. "I have looked at some cluster plates a little," Shapley wrote Russell, "and found five new variables in the middle of a cluster . . . or to tell the truth my hausfrau found them but I plan to take the credit for it." It helped him latch onto a focus; he was going to study the Milky Way's globular clusters, each a dense ball of stars that gleams like a cosmic sparkler frozen in time.

Shapley worked at Mount Wilson from 1914 until 1921, and the best research of his career was accomplished during this time. Shapley was a risk taker. As Hale later noted, "He is much more venturesome than other members of [the Mount Wilson] staff and more willing to base far-reaching conclusions on rather slender data." Counting up his scientific publications over those seven years, Shapley was the sole author of or a contributor to some 150 notes and papers. Years earlier, none of his childhood friends would have bet on that career outcome, given that Shapley's first job was being a hard-nosed reporter, covering crime and corruption in the Midwest.

Born in 1885, Shapley and his fraternal twin brother, Horace, along with an older sister, Lillian, and a younger brother, John, grew up on a Missouri farm a few miles from the town of Nashville, on the edge of

Ozark country not far from where Harry Truman, the thirty-third president, was also born (in 1884). Shapley's father, Willis, was a hay dealer. Young Harlow attended a one-room schoolhouse for a few years but was mostly taught at home. When milking cows, he recited poems by Tennyson to "keep the rhythm going."

"The St. Louis *Globe-Democrat* was our chief contact with the outside world," recalled Shapley. That may be why, at the age of fifteen, he became a reporter for the *Daily Sun* in Chanute, Kansas, a rough-and-tumble oil town about sixty miles northwest of his family's homestead. He later moved back to Missouri to work on the police beat for the *Joplin Times*. All the while he spent his free time reading in the local libraries, for Shapley's ambition right from the start was to save enough money to go to college. He eventually applied to the local high school, in order to work toward the diploma he vitally needed to matriculate, but, refused admission due to his meager educational record, he paid out of his own pocket to catch up on his academics at a collegiate prep school. Finishing up in 1907, at the age of twenty-one, he at last qualified for admission to the University of Missouri, just as his schoolteacher mother had always desired.

Given his years of experience reporting on midwestern mishaps, Shapley had always intended to major in journalism, but upon arriving on campus he discovered that the promised opening of the university's School of Journalism had been delayed. "So there I was," said Shapley later in life, "all dressed up for a university education and nowhere to go. 'I'll show them' must have been my feeling. I opened the catalogue of courses and got a further humiliation. The very first course offered was a-r-c-h-a-e-o-l-o-g-y, and I couldn't pronounce it! . . . I turned over a page and saw a-s-t-r-o-n-o-m-y; I could pronounce that—and here I am!" Shapley, a lover of tall tales since he was a child, was just joking around. He actually was in need of a job, and an offer from Frederick Seares, head of the university's astronomy department, to work for him at 35 cents an hour was likely the deciding factor. In whatever way Shapley came to major in astronomy, the choice suited him to a tee. Seares was mightily impressed by the former reporter, especially the fact, as he put it, that Shapley "*thinks* about what he is doing." Within two years Seares had Shapley teaching the introductory astronomy course. Although starting out with little training in physics and mathematics, Shapley ended up in 1910 graduating with honors.

Shapley spent another year at Missouri to obtain his master's degree

Young Harlow Shapley (*Photo by Bachrach,*
courtesy of AIP Emilio Segrè Visual Archives)

and chose to go to Princeton for his PhD when he won one of its distinguished fellowships. One of his recommenders had warned Princeton officials to accept this rising star before their competitors had a chance to steal him away. There in the idyllic midlands of New Jersey, under the guidance of Henry Norris Russell, the eminent astronomer and theoretician, Shapley specialized in eclipsing binaries—two stars positioned in such a way that, as they circle, one periodically passes in front of the other when viewed from Earth, causing the binary's light to dim for a while and then rise back. Shapley became a whiz in handling a slide rule and consulting mathematical tables to compute the stars' orbits, as well as their densities and size, Russell's special area of interest. Such work was immensely valuable in confirming the wide range of stellar types, including the existence of giant stars.

It was an odd pairing of adviser and advisee: Russell, with his stiff and aristocratic demeanor, the son of a Long Island clergyman, coupled with the "wild Missourian" with the round face and farmboy haircut, who once attended two New York City theater performances in one day and judged the experience as "worse than log tables." But they came to appreciate each other's professional expertise and industriousness. According to Russell's biographer David DeVorkin, the two were often seen strolling the campus together, with Russell using "his cane to sweep the undergraduates out of their path."

The connections Shapley had made at Missouri proved crucial for his next career step. Seares, his undergraduate professor, had moved to Mount Wilson in 1909 and helped open doors for Shapley to become a staff astronomer at the celebrated observatory. Soon after Hale offered the position in 1912, at a salary of $90 a month plus free board on the mountain. Shapley delayed his start date in order to do some travel in Europe and stay with Russell a bit longer to complete their "crusade" on eclipsing binaries but at last journeyed to Mount Wilson in the spring of 1914. Along the way, he stopped off in Kansas City to marry his University of Missouri sweetheart, Martha Betz, a gifted scholar and linguist he had met in a mathematics class. She took an interest in astronomy once they started dating and even helped him reduce the piles of data he had collected for his doctoral dissertation. On their honeymoon train ride out to California, they together happily computed eclipsing binary orbits. In less than a decade, Shapley had gone from fledgling newsman to professional astronomer, about to look through the eyepiece of what was then the largest telescope in the world.

Conditions at the mile-high observatory were still fairly primitive when Shapley arrived. "Just killed a 3 ft. rattlesnake with 8 rattles lying by our back door," reported one pioneering staff member. "We had to be rugged in those days," Shapley later recalled. "We would go up the mountain, a nine-mile hike, sometimes pushing a burro, sometimes not. The new road had not [yet] been put in." When not on Mount Wilson, Shapley spent his time at the observatory's offices and workshops in Pasadena, a town then in the process of transforming from an agricultural community of lush citrus groves and vineyards to a winter resort town filled with flowers and wealthy visitors from the East.

A sociable fellow, Shapley forged friendships with several colleagues right away, including solar astronomer Seth Nicholson and Dutch astronomer Adriaan van Maanen, the latter of whom first arrived at Mount Wilson in 1911 as a volunteer assistant and remained on as a staff member for thirty-five years. Among these friends and colleagues, Shapley was an incorrigible raconteur. "A discussion with him was like a rousing game of ping-pong, ideas flashing back and forth, careening off at unexpected angles and often coming to earth in a breathless finish," said Cecilia Payne-Gaposchkin, who knew Shapley later at Harvard. An enormously vain man, Shapley also liked to be flattered and got along best with those who fawned over him. Moreover, he never forgave a slight. "A generous supporter, a stimulating companion, he could also be an implacable enemy," added Payne-Gaposchkin.

The one person Shapley couldn't sway with his gee-whiz midwestern charm was Walter Adams, the effective leader at Mount Wilson. Hale, prone to nervous breakdowns and bouts of depression, was often gone from Mount Wilson in the 1910s. Sometimes his absences were due to war work, but often because he was recovering from his illnesses. Whenever that happened, Adams was in charge. A proper and dutiful man known for his frugal ways, Adams was so regular in his habits that staffers could "set their clocks by his comings and goings." An inveterate pipe smoker as well, Adams forged the "Lucky Strike" trail, a shortcut from the observatory to the cigarette stand of the rustic hotel then operating nearby on the mountain. Shapley often grumbled about Adams to his friends. "I feel very sure that if I should go away from here no opportunity would be given me to return so long as Adams has the deciding voice," Shapley once told a colleague. But the tension between them didn't seem to affect Shapley's innovative work while he was on staff.

The seed for Shapley's groundbreaking research was actually planted before he got to the mountain. While still a Princeton graduate student, Shapley had visited Harvard and there met veteran astronomer Solon Bailey, who suggested to the young man that he use Mount Wilson's new 60-inch telescope "to make measures of stars in globular clusters." When Bailey was stationed at Harvard's Peruvian outpost in the 1890s, serving as its head, he had begun discovering large numbers of variables, hundreds of them (including Cepheids), in some of the clusters and sensed it was terribly important. He knew that a large telescope, such as the one on Mount Wilson, would be immensely valuable in extending this work. It would have the power to resolve variables in the crowded inner regions of a globular cluster and peg their pulsations.

Shapley ultimately took Bailey's advice and, starting with the variables found by his wife, placed a firm stake in this domain. Shapley and globular clusters quickly "became synonyms" atop the mountain. Shapley's involvement became so intense that he eventually contacted Bailey, to make sure the Harvard astronomer didn't feel Shapley was trespassing on his celestial territory. "I have not intended to intrude upon your field, and I think that you do not feel that I am," wrote Shapley. "Very much of my work on clusters has been the direct result of my conversation with you in Cambridge three years ago when you suggested the advantages of the Mount Wilson instruments and weather." Bailey, a kind and gracious man, was in fact delighted by Shapley's joining in. "I hope you will appreciate the fact that I claim no proprietorship in these clusters," replied Bailey, "but . . . welcome other investigators in this field." It was fortunate that he was so affable. Bailey was primarily a data gatherer; Shapley by nature was a bold interpreter, a trait enhanced during his apprenticeship with Russell, who advocated problem-driven research. And that made all the difference in advancing the science.

A globular cluster appears through a telescope as an assembly of brilliant specks of light hovering around a dense and blazing core. With stars packed in like subway commuters at rush hour, the cluster offers a far more exotic celestial environment than our local stellar neighborhood. Alpha Centauri, the star closest to the Sun, is some 4 light-years away. But if the Sun were in the center of a packed globular cluster, it would have thousands of stars closer than that, covering Earth's sky like a sequined blanket visible both day and night. Near misses between stars would be commonplace.

That a globular cluster is a highly spherical collection of stars was not known until the 1600s, with the advent of the telescope. Before that, ancient astronomers simply noted the objects on their sky charts as a "lucid spot" or a lone "hazy star." Today, these clusters are known to be arranged as a globelike halo, surrounding the disk of the Milky Way somewhat like bees buzzing around a hive. But as late as the 1910s, when Shapley began his observations, astronomers didn't know that, nor exactly how big an individual globular cluster was. Some even pondered if they were island universes in their own right. Shapley himself believed that was true when he was starting out: "It is quite obvious that a globular cluster . . . is in itself a stellar system on a great scale—a stellar unit which without doubt must be comparable to our own galactic system in many ways," he wrote in the first paper of his study. Some dabbled with the idea that a spiral nebula was an early stage of a globular cluster about to form: Like an open flower closing at twilight, the spiral over time would fold up into a ball. Shapley's goal was to learn the globulars' true sizes, distances, and compositions and see if such ideas were valid.

Shapley's initial observations were fairly basic. Using the 60-inch

Globular Cluster M80
(*The Hubble Heritage Team* [*AURA/STScI/NASA*])

telescope, he simply surveyed the colors and magnitudes of the stars in the most prominent clusters. These included Omega Centauri (the biggest of them all), the Hercules cluster, and M3, a globular noted by Charles Messier in 1764. Shapley had no idea where this would lead, but that was standard practice in astronomy: Gather as much data as you can when faced with the unknown and keep your eye out for unusual trends. If anything, Shapley hoped his observations might help Hale in his quest to understand how stars aged and evolved, still quite a mystery to astronomers in the early twentieth century.

As his collection of photographs mushroomed, though, Shapley began to identify Cepheids, which he knew would serve as his measuring tape out to the globular clusters. He was quite aware of the paper that Henrietta Leavitt had published just a couple of years earlier and intended to apply it. "Her discovery . . . is destined to be one of the most significant results of stellar astronomy," Shapley later wrote to her boss, Pickering.

What was needed was a reliable distance to a Cepheid—any Cepheid, anywhere in the sky—that could serve as the calibration for determining the distances to all other Cepheids using Leavitt's period-luminosity law. That was the beauty of her discovery: Know the distance to just one Cepheid and you know the rest.

Distance measurements have long been a problem for astronomers. To our eye, the celestial sky resembles a dark bowl with pinpoints of light affixed to it—everything appears to be the same distance away. But in reality the stars we see reside at vastly different ranges. Bluish-white Sirius, the brightest star in the heavens, is located 8.6 light-years from Earth; Vega, the prominent summertime star in the constellation Lyra, lies 25 light-years away. How do astronomers arrive at these numbers? "Parallax" is one surveying technique. Parallax is the apparent change in a star's position on the sky when observed first at one end of Earth's orbit and then six months later at the other end (similar to the way an object close by will appear to shift when you view it first with one eye, then the other). By setting the radius of Earth's orbit as a baseline and knowing the angle of shift in the star's parallax, a bit of geometric triangulation determines the star's distance from the Sun directly. Astronomers devised the term *parsec* to describe the distance between Earth and a celestial object that displays a *par*allax of one arc*sec*ond of angular measurement on the sky. (One parsec equals 3.26 light-years.) The parallax method is useful out to several hundred light-years. After that, the change in a star's position is too small to be discernible by ground-based telescopes, which is why Leavitt's law was so treasured. It would enable astronomers to extend their distance surveys much farther outward. It would have been nice if a Cepheid resided fairly close to our Sun; then astronomers could have measured the star's parallax and gotten their calibration fairly easily. Unfortunately, there was no Cepheid within reach of a direct parallax measurement from Earth in Shapley's day. Nature was not so accommodating to astronomers. (The closest Cepheid to us is Polaris, the North Star, located about 430 light-years away. Polaris is actually a three-star system, one of which is a large yellow Cepheid that completes its dim/bright cycle every four days.)

The first person to try to confront the Cepheid distance problem was Ejnar Hertzsprung, who had initially recognized that Leavitt's twenty-five variables in the Small Magellanic Cloud were specifically Cepheid stars. He began to look at the Cepheids best studied within the Milky Way, thirteen in all. He couldn't measure their parallax (they were too far away), but he could consult a chronological sequence of astronomical atlases to see how far the stars had moved across the sky, at right angles to our line of sight, in their travels through the Milky Way. It was a matter of determining how their celestial coordinates had changed over the years. Astronomers refer to this advance as a star's "proper motion." From another type of catalog he looked up how fast they were moving either toward or away from Earth based on the stars' blueshifts or redshifts (a rough gauge of their overall velocity). In an imaginative leap, he then estimated the Cepheid's distance by comparing the star's measured velocity with how fast it *appears* to be moving across the sky from our faraway vantage point. The more distant the star, the slower it seems to journey across the sky. (His actual mathematical procedure, which also involved the Sun's motion through the galaxy, was more complex, but this provides the basic idea.) Hertzsprung's approach in the end provided a crude statistical calibration, one that he then applied to Leavitt's Cepheids in the Small Magellanic Cloud. He concluded that the cloud was 30,000 light-years distant, one of the greatest distances then measured for a celestial object. This demonstrated for the first time the potential power of Leavitt's discovery.

"I had not thought of making the very pretty use you make of Miss Leavitt's discovery," Henry Norris Russell wrote Hertzsprung when this result came out. Russell had employed a similar technique around the same time, but his aim was to determine the average magnitude of a Cepheid. In the process, he concluded that they were giant stars, far bigger than our Sun. Inspired by Hertzsprung, Russell proceeded to make his own distance calculation to the Small Magellanic Cloud, arriving at 80,000 light-years. Both estimates were highly uncertain and turned out to be far less than current distance measurements (210,000 light-years), but each figure was still astoundingly huge for its day.

Shapley soon adopted Hertzsprung's approach, although he used only eleven of Hertzsprung's thirteen Cepheids in his calibration, suspecting that two of them were peculiar. Just like Hertzsprung, he counted on a simple rule of perspective: The farther away a moving object is located from you, the slower it will appear to travel. A far-off plane seems to crawl along the sky, while a plane closer in going at the same speed would zoom right past you. After estimating an average velocity for a star, Shapley checked how his eleven Cepheids were journeying across the sky. The slower the apparent velocity, the more distant the Cepheid.

It was at this point, though, that Shapley parted company with Hertzsprung. He didn't use Leavitt's period-luminosity relationship, which was based solely on stars in the Small Magellanic Cloud, but instead constructed his own relationship based as well on the Cepheids in the Milky Way, in order to obtain an "improved and extended" period-luminosity law combining both sets of variable stars. He then applied his new rule to the Cepheids found in the globular clusters. He would monitor a Cepheid to peg its period and then calculate the star's distance from his graph.

This worked as long as Shapley could find Cepheids in his globular clusters. Some of the clusters had none at all, as far as he could observe. What they did harbor were variables that were not quite the same. These variables changed quite rapidly, in a matter of hours rather than days or months. There was no guarantee that they behaved in the same way as Leavitt's Cepheids.

Shapley tried mightily to check with Leavitt on this question, writing several times to her boss, Edward Pickering, on whether she had detected fast variables in the Magellanic Clouds and found them to obey her rule. Pickering assured him that photographs were being taken. But progress on the question was occurring at a glacial pace. Pickering was keeping Leavitt busy with work he considered more important. "Routine stuff," decried Russell to Shapley at one point. "I fear, however, that I am not the man who may justly raise my voice in criticism."

Eager to move forward, Shapley simply decided to treat his fast variables as if they did follow Leavitt's rule. He extended the Cepheids' period-luminosity relationship to include all these variables, both slow and fast. "This proposition scarcely needs proof," he had boldly asserted in one early paper, though it was a very controversial decision. But by doing this, Shapley was able to determine the distances to the nearest globular clusters—a formidable task, as the stars were very faint. For clusters farther out, too remote to spot any variables, he resorted to using the brightest stars as distance markers. He just assumed that the brightest stars in a distant globular cluster had a similar magnitude on average to the brightest stars in a nearby cluster. And when the stars themselves could no longer be adequately resolved, he judged distance by the apparent size of the globular cluster in the sky. "The whole line of reasoning . . . was brilliant," concluded astronomer Allan Sandage in a review of this technique decades later. Bailey, at Harvard, could have carried out this effort before Shapley, but he was overly cautious about the variables. To him there were too many uncertainties about their nature, so "definite conclusions from these data cannot be safely made," he reported. Shapley had no such qualms.

But it was yeoman's work, painstaking routines that took four years to complete. Shapley was securing the distance to every Milky Way globular cluster known at the time, sixty-nine in all. With the assistance of Edison Hoge, he took some three hundred photographs. Some exposures were only ten seconds in length, but others lasted up to two hours. Most took minutes. Afterward there was the brutal labor at the worktable analyzing what the images revealed. By 1917 he was writing a colleague that "the work on clusters goes on monotonously—monotonous as far as labor is concerned, but the results are continual pleasure. Give me time enough and I shall get something out of the problem yet." By then the war was on, but Shapley didn't sign up. He claimed that Hale had convinced him to stay at his job.

Some of the globular clusters (circled) surrounding the Milky Way
(*Harvard College Observatory, courtesy of AIP Emilio Segrè Visual Archives*)

Shapley didn't particularly enjoy his nights alone with the stars. What drove him back to the telescope month after month were his findings. With the first hint of dawn in the east, as the dome slit slowly closed with a noisome squeal, nearby coyotes would answer in kind with a serenade of high-pitched howls. At night's end, he and the other astronomers would walk back to the Monastery, sometimes whistling a merry tune if the viewing went well and forgetting that they might be disturbing the daytime observers—the solar astronomers—who were still fast asleep. Once in bed themselves, though, the nighttime observers could easily be wakened by the stirrings of the daytime crew. Both sides were together at noontime lunch, which offered the opportunity to settle any squabbles.

Shapley was curious about nearly everything that came his way when on the mountain. "The most unwarranted fun of all comes from *bugs*," he wrote a colleague while trapped in a snowstorm on Mount Wilson. "Not that I know much about them, but I am so interested that I would like to turn biologist." In a way, he did. He began to study the travels of ants around the observatory, noticing that the higher the temperature, the quicker their pace. One species ran fifteen times faster once the Sun heated the insects by an additional 30°C. As he put it, he had discovered the "thermokinetics of ants." Setting up "speed traps" to gauge the ants' pace precisely, he boasted he could estimate the day's temperature to within one degree by their perambulations. "Another method is to read your thermometer," he wryly added. His findings were published in scientific journals. For further rest and relaxation, he and his wife climbed all the nearby mountains—little and big—collected plants, and killed any rattlesnakes that came their way.

Between 1916 and 1919 Shapley published his growing body of data on the globular clusters in an extended series of papers, collectively titled "Studies Based on the Colors and Magnitudes in Stellar Clusters." Each article progressively added another piece to the puzzle. Shapley was taking his reporting skills to a new beat. And in carrying out this endeavor he was ultimately forced to alter his original mental picture of the universe. It began to dawn on Shapley that the Milky Way was far larger than anyone had previously conceived. The first hints arrived when he estimated that some well-known star clusters *within* the Milky Way were at least 50,000 light-years distant. Later he was finding that the distances to the globular clusters ranged anywhere from 20,000 to 200,000 light-years.

With the globulars acting in a way like surveyor posts, marking the boundaries of our galactic borders, the Milky Way was growing by leaps and bounds. As a result, the globular clusters could no longer be thought of as similar in size to the Milky Way, as Shapley once thought. The clusters were now far smaller by comparison. "This is a peculiar universe" was Shapley's reaction to this new cosmic landscape.

So what did this mean for the spiral nebulae, which Heber Curtis and V. M. Slipher were now enthusiastically hawking as separate galaxies? Around this time Shapley's Mount Wilson buddy van Maanen was claiming to see some spirals rotate, an impossible feat if they were lying at a great distance. To perceive a rotation from so far away over a short period of time would mean the spirals had to be spinning at close to the speed of light!

To understand why this would be so, imagine a kitchen clock sitting right by you on the wall. The second hand is sweeping around the dial at a speed of about 1 centimeter per second. But then imagine the face of that clock covering the entire surface of the Moon, its apparent size looking just like the clock on your wall. Yet the clock in reality is now much bigger, so the second hand has to travel at a faster clip, about 110 miles per second, to make a full circuit over a minute's time. Now if that clock were as big as the Milky Way, the second hand would be moving at a demonic pace. If van Maanen's spiral nebula was truly a distant galaxy and he was able to detect its arms shift over a matter of years, then he was seeing it rotate at light-defying speeds.

Not willing to tolerate such bizarre behavior, Shapley at first was doubtful of van Maanen's finding. In fact, he published an article in 1917 saying that "the minimum distance of the Andromeda Nebula must be of the order of a million light-years," based on some dim novae detected within the nebula and the faintness of its brightest stars. "The difficulty is obvious," he continued, "in reconciling van Maanen's measures of internal proper motion with the hypothesis of external galactic systems. We are not prepared to accept velocities of rotation of the order of the velocity of light." The issue in question was not whether spiral nebulae truly rotate. In the 1910s Vesto Slipher had already detected evidence that they spin around. The proof was found within the spectra of the spirals that he was examining. Like a Frisbee thrown outward and spinning in flight, the rotation on one side is directed forward, adding to the measured velocity, while on the other side the spin is aimed back, subtracting from the overall speed. This difference manifests itself as a slight inclination in the spiral's spectral lines. But this motion was certainly not rapid enough to notice by eye alone when comparing photos taken just a few years apart. Moreover, the spectral signatures indicated that a spiral was closing up, wrapping its arms tighter around the nebula's center, "like a winding spring," reported Slipher. But this clashed directly with van Maanen's claim that a spiral was opening up. Slipher, so modest and reticent, didn't make an issue of this contradiction. If he had made a clamor, loudly and persistently publicizing his proof, van Maanen's assertions might have been dismissed far earlier. But, as it turned out, Slipher's conflicting result was essentially neglected, occasionally discussed among astronomers privately but rarely singled out in print.

Shapley soon asked his Princeton adviser, Russell, whether he too questioned van Maanen's rotations. "V. M. does a little, Hale a little more, and I much," wrote Shapley. In time Russell replied that he was "inclined to believe in the reality of the [spirals'] internal proper motions, and hence to doubt the island universe theory. But if [the spirals] are not star clouds, what the Dickens are they?"

Considering what happened soon after the receipt of this letter, Shapley likely took Russell's counsel very, very seriously. As Russell's protégé, he highly respected his former professor's opinion and would have had difficulty ignoring Russell's astronomical advice. It was well known in the community that Russell's "word was law," which could "make or break a young scientist." Shapley eventually had a change of heart, a transformation that was triggered not only by Russell's advice but also by the further evaluation of his immense pool of data.

Starting in November 1917 Shapley fired off, with great rapidity, his next group of papers. In his ongoing series on the globular clusters, he completed articles six through twelve in just six months. It was as if he were back at his old newspaper job, pounding out an exclusive on his typewriter to meet a daily deadline. The first of these papers announced his grand goal straightaway: He intended to report on nothing less than "the general plan of the sidereal system . . . bearing on the structure of the universe." Shapley made this bold claim because, while trying to make sense of all his data, he had a revelation. He came to believe that his observations were not only refashioning the Milky Way but the universe as well. Unlike many of his fellow astronomers, he was fearless at making extravagant leaps in speculation.

At this stage Shapley had finished slogging through his many observations and calculations and had plotted the positions of the sixty-nine known globulars onto a graph. This provided him with a feel for how they were distributed through space in three dimensions. The result, he noted in paper number seven, was "striking." Most of the clusters resided in one particular direction, over by the constellation Sagittarius. Like moths lingering by a streetlamp, they were symmetrically arranged about a spot rich in stars and nebulae within our galaxy. It was said the star clouds in this region were so thick that it was "impossible to count every star shown; the images of the faintest stars . . . merged into one another forming a continuous gray background." The galactic coordinates for this spot did not match those for our solar system. The globular clusters were not arranged around the Sun at all (as might be expected). Good old Sol was situated off to the side—by Shapley's initial estimate around 20,000 parsecs, or 65,000 light-years away.

Other astronomers had noticed this peculiar distribution of the globular clusters before. In 1909 the Swedish astronomer Karl Bohlin even dared to suggest that the center of the galaxy was in that direction, with the clusters all huddled around it. But no one at the time, including Shapley, took this idea seriously. It was just assumed that the solar system resided in the heart of the galaxy (or close to it). Now Shapley was confirming what Bohlin had suspected all along. His observations forced him to radically alter his original opinion.

From this point on, Shapley's progress was swift. Papers eight through eleven, submitted for publication in December and January, provided the technical details on his methods, assumptions, and calibrations. Shapley knew his conclusion was going to be revolutionary, so he stacked his ammunition with orderly care. Page by page he was stepping toward his grand finale. The full-scale assault took place with paper number twelve, titled "Remarks on the Arrangement of the Sidereal Universe." This particular article was not fully ready for submission to the *Astrophysical Journal* until April, in the waning days of World War I, but Shapley couldn't wait that long to spread the news. On January 8, 1918, he wrote the noted Arthur Eddington in England that "now, with startling suddenness and definiteness, [the cluster studies] seem to have elucidated the whole sidereal structure"—in other words, the architecture of the Milky Way. Not only were the globular clusters uniformly scattered around the center of the galaxy, with the Sun shoved off to the hinterlands, but the Milky Way was far larger than anyone had formerly presumed. Shapley now gauged it was an astounding 300,000 light-years from one end of the galactic disk to the other, ten times greater than previous estimates. "You may have been completely prepared for the result," Shapley told Eddington, "but I was only partially successful as a prophet."

"While I cannot pretend to have anticipated the view of the stellar system that now seems to be emerging," responded Eddington, "I do not feel any objection to it either." This was a confidence booster for Shapley, who was still essentially a rookie in astronomical circles and assuredly grateful for support from such a renowned figure.

Shapley didn't forget to give his boss, George Ellery Hale, then out of town on stressful war business, advance notice as well. "May I impose upon your time for a little while, with an off-hand talk about my astronomical work—divert your attention from earthly troubles to heavenly affairs?" wrote Shapley. The young staff astronomer hardly knew where to begin and for brevity's sake he cautioned Hale that he was leaving out the "probablies, perhapses, maybes, apparentlies, and other such necessary weaknesses in scientific exposition. . . . So my assumed surety . . . is neither over-confidence nor a whistling in the dark, but an agreement between us."

True to character, Shapley fashioned one of his ubiquitous tall tales to present his case to Hale: "The first man, away back in the later Pliocene, who knocked out a hairy elephant with his club, or saw his pretty reflection, or received a compliment, became suddenly conceited (it was a mutation) and there immediately evolved the first reflective thought in the world. It was: 'I am the center of the Universe!' Whereupon he took himself a wife, transmitted this bigotry of his germ plasm, and through hundreds of thousands of years the same thought without much alteration has been our heritage." And now, he assured Hale, he would furnish the remainder of the story.

Shapley reminded Hale that he was determining the distances to all the globular clusters then known and was getting ready to publish a series of papers announcing the results: twenty pages of tables, nearly a dozen figures, and around a hundred pages of text in all. For Hale, Shapley summarized his results in three, single-spaced typed pages. The bottom line, he said, was this: The Milky Way is huge, some 300,000 light-years in width, and the Sun is far away from the hub. "Start a messenger on a light-wave down the main highway from the center," wrote Shapley, and he'd end up at Earth about sixty-five thousand years later. Moreover, he said, "there is no plurality of universes. . . . The galaxy is fundamental in what we call the universe." True to his impetuous nature, Shapley threw caution to the wind. The Milky Way was now so big, he figured, it had to be the dominant feature of the universe.

When the Milky Way was thought to span only 10,000 or 30,000 light-years, it was easier to think of the spiral nebulae as separate galaxies. But everything changed when Shapley claimed the Milky Way was far larger. If the Andromeda nebula was also a galaxy and with similar dimensions to the Milky Way, its distance would have to be farther out than anyone anticipated to appear as it did on the sky. And that meant that the novae that lighted up within Andromeda's disk were even more luminous than any known physical law could possibly explain. Within a matter of months, Shapley made a complete about-face regarding the spiral nebulae. Once a believer in island universes, Shapley now considered it more reasonable to assume that Andromeda and the other spirals were simply closer: either nestled cozily inside our galactic borders or situated just outside, as smaller outlying colonies. They were no longer the Milky Way's equal in grandeur and power but mere appendages. He even speculated at one point that they might be blobs of nebular material somehow being repelled by the Milky Way at high velocities, perhaps due to radiation pressure or electrostatic forces. As our galaxy travels through space, surmised Shapley, it might be driving "the nearby spirals to either side much as the prow of a moving boat cuts through the waves."

Formerly suspicious of van Maanen's findings, Shapley now came to like that his friend was detecting the spirals rotate, for it strongly backed his own, newly constructed model of the universe. It meant that the spirals had to be close by, merely secondary members of the Milky Way. Our galaxy reigned supreme. "I believe the evidence is quite against the island universe theory of spirals. I should *guess* the Andromeda nebula to be not further away than 20,000 light-years," Shapley told Hector MacPherson, a popular British writer on astronomy.

With all his advance notices and public declarations, the thirty-two-year-old Shapley was crowing and, like a little boy, wanted his elders to notice his cleverness at completely renovating the image of the universe. "The observational problems opened up are unlimited; the amount of stupid measuring ahead of me is almost discouraging," he told Hale. "But I am enjoying it all except for a considerable nervous strain at the last."

. . .

Shapley released his findings around the same time that the Mount Wilson Solar Observatory dropped the word *solar* from its name. Shapley most of all was broadening observations from the mountaintop post to questions far beyond the Sun and into the depths of space and time. What Shapley had done was to hugely extend the Copernican rule. Just as Copernicus in the sixteenth century had removed Earth from the center of the solar system, Shapley relocated the solar system from the heart of the Milky Way. "The solar system is off center and consequently man is too, which is a rather nice idea because it means that man is not such a big chicken. He is incidental—my favorite term is 'peripheral,'" Shapley bluntly wrote in a 1969 memoir. "If man had been found in the center, it would look sort of natural. We could say, 'Naturally we are in the center because we are God's children.' But here was an indication that we were perhaps incidental. We did not amount to so much."

Shapley had carried out a tour de force, and his findings hit the astronomical community like a lightning bolt. Praise for the work, from the most eminent corners of astronomy, was immediate. After reading Shapley's completed papers, Eddington wrote Shapley that "this marks an epoch in the history of astronomy, when the boundary of our knowledge of the universe is rolled back to a hundred times its former limit." In a *Scientific American* article, Russell described the results as "simply amazing." And British theorist James Jeans told Shapley that his newly published papers were "certainly changing our ideas of the universe at a great rate."

Mount Wilson astronomer Walter Baade later remarked that he "always admired the way in which Shapley finished this whole problem in a very short time, ending up with a picture of the Galaxy that just about smashed up all the old school's ideas about galactic dimensions. It was a very exciting time, for these distances seemed to be fantastically large, and the 'old boys' did not take them sitting down."

While the news spread quickly within the astronomical community, it took longer to reach the general public, likely due to the shadow of the war and its aftermath. Not until May 31, 1921, did the *New York Times* report on its front page that Shapley had multiplied the universe's size immensely. Our galaxy, noted the *Times* reporter, was now 300,000 light-years from end to end, a "super–Milky Way. . . . The young astronomer has proved to his satisfaction by various calculations that the sun, the little speck of light around which a tiny shadow called the earth revolves, is 60,000 light years from the centre of the universe."

"Personally I am glad to see man sink into such physical nothing-ness, and it is wholesome for human beings to realize of what small importance they are in comparison with the universe," says Shapley in the article. (If any readers were made queasy by Shapley's news, the story was conveniently set just above a tiny advertisement for Bell-Ans pills, a popular 1920s indigestion remedy.) The *Chicago Daily Tribune* published the same announcement on page 1 as if it were entertain-ment news: Earth, proclaimed the headline, was now a "Rube . . . Miles Off Sky Broadway."

Not everyone was convinced of this new cosmic scheme. Critics pointed to several weak points in Shapley's arguments, including Hale, who wondered whether the spirals were something other than Shapley imagined. But the Mount Wilson director still supported Shapley's bravado: "You have struck a trail of great promise. . . . I think you are right in making daring hypotheses, and in pushing the work ahead as you have done, as long as you . . . are prepared to substitute new hypotheses for old ones as rapidly as the evidence may demand," Hale responded to Shapley from his wartime post in Washington. Hale pre-ferred his astronomers to take chances. He didn't want them turning solely into unimaginative data gatherers, as Pickering, at Harvard, was wont to do.

But since Shapley had based his results using such novel methods as the Cepheid beacons and rashly ignored many uncertainties, accep-tance was hardly unanimous. Other astronomers had been slowly and methodically measuring the galaxy's dimensions over many years by essentially counting stars, tracing out their distributions and movements over the sky and deep into space. The leader in this endeavor was the highly respected astronomer Jacobus C. Kapteyn, at the University of Groningen in the Netherlands. Though not possessing a good telescope, he organized a massive effort to measure the positions of hundreds of thousands of stars on plates taken at other observatories, partially with the help of state prisoners placed at his disposal. He reached the pinnacle of his life's work when he introduced what became known as the "Kapteyn Universe." In this model, the largest portion of stars in our galaxy (there was a smaller fraction farther out) were gathered in a space roughly 30,000 light-years wide and 4,000 light-years thick, a sort of squashed football. Moreover, the Sun retained its plum position near the center. But Shapley was declaring that the Milky Way was ten times larger and the Sun pushed far off to the sidelines. It was extremely difficult for Kapteyn and his colleagues to imagine that their time-honored methods for tracking stellar distributions could be so flawed. Others thought so, too. Shapley had constructed a formidable distance ladder outward, but its calibration rested on a measly eleven Cepheids, whose motions were still highly uncertain. If those were wrong, the entire construct toward his fundamental verdict—what he called his "Big Galaxy"—would fall apart like a celestial house of cards. Kapteyn told Shapley that he was "building from above, while we are up from below.... When will the time come that we thoroughly mesh?"

Conservative astronomers were most disturbed by the many analytical leaps of faith made by Shapley, who tended to speak, it was said, with a "carnival barker's certainty of truth." Though brilliant and original, he was often quick to jump to conclusions based on meager observations. Accuracy seemed to be less important to him than developing a broad, grand picture. Walter Adams, for one, was sure that the fast and slow variable stars that Shapley lumped together in his computations were actually "two different breeds of cats." (He was right; they were later found to be RR Lyrae stars, variables that are less massive and fainter than Cepheids.) And then there was the issue of Shapley's borrowing ideas and techniques from other astronomers without proper acknowledgment. Adams complained to Hale that Shapley "has never given the credit where it belongs." In one paper published in the *Proceedings of the National Academy*, Shapley made no mention of either Hertzsprung or Leavitt, who had both certainly paved the way. This infuriated Adams. As one Harvard astronomer later put it, "I have never seen a quicker mind, a more agile sense of humor, or a more complete absence of what usually passes for humility."

Shapley's critics were right to be cautious. In hindsight, he did get certain things wrong. Astronomers, for example, would later reduce the Milky Way's girth from 300,000 to 100,000 light-years, once they better understood the difference between the fast and slow variable stars and affirmed the presence of interstellar dust, which made celestial objects appear dimmer and hence more distant than they actually were. This made Shapley mistakenly believe that the Milky Way was more extensive than it actually was. Yet even when the galaxy's width was reduced to 100,000 light-years, it was still bigger than Kapteyn and his supporters had been hawking. Shapley's discovery held up over time on the essentials: first of all, that the Milky Way was a far larger metropolis of stars than previously suspected and, second, that the Sun was situated in its suburbs.

Shapley's shift of the Sun's position was fully confirmed in the mid-1920s when Bertil Lindblad, a Swedish expert on stellar dynamics, and Jan Oort, at the Leiden Observatory, in the Netherlands, demonstrated that stars were circulating within the Milky Way around a point situated in Sagittarius, exactly where Shapley had pegged the galactic center. Once Lindblad worked out the theory, Oort rounded up the evidence to prove it. If anyone was still questioning Shapley's pushing the Sun off into the galactic boondocks, Lindblad and Oort swept away all doubts. Like the horse on a carousel, the solar system travels in a continual loop, completing one full circuit around the galactic disk roughly every 250 million years. The last time we were in this neck of the celestial woods, the Appalachian and Ural mountains were just being formed and the dinosaurs were getting ready to rule the Earth.

Shapley's new model of the Milky Way had broad repercussions, especially regarding the spiral nebulae. The idea of island universes, then on the verge of acceptance, was back on shaky ground. "With the plan of the sidereal system here outlined," reported Shapley, "it appears unlikely that the spiral nebulae can be considered separate galaxies of stars." There was still the problem of the exceptionally bright novae seen earlier in the spiral nebulae. How do you explain that? asked Shapley. And then there were van Maanen's rotations to take into account. Not everyone was swayed by Shapley's worries; the most ardent believers in external galaxies still held fast to their convictions—not only Curtis but also such major players as Arthur Eddington, W. W. Campbell, and V. M. Slipher. It was the undecideds who were most affected by Shapley's arguments and so remained huddled on the fence. What resulted were two completely different views of the universe, which were difficult to reconcile. The writer MacPherson poetically put it this way: "We may compare our galactic system to a continent surrounded on all sides by the ocean of space, and the globular clusters to small islands lying at varying distances from its shores; while the spiral nebulae would appear to be either smaller islands, or else independent 'continents' shining dimly out of Immensity." As the Roaring Twenties was about to make its appearance, Shapley voted for the "smaller islands," Curtis for the "continents."

[9]

He Surely Looks Like the Fourth Dimension!

Astronomy was not the only field in a tumult as the nineteenth century turned into the twentieth. Physics, too, was in upheaval.

Doctors across the globe were still reeling over their newfound ability to use X rays, discovered in 1895 by the German physicist Wilhelm Röntgen, to peer inside the human body. Soon after, in Paris, Henri Becquerel accidentally stumbled upon a phenomenon, what came to be called radioactivity, when he was investigating the properties of uranium salt crystals. And in England J. J. Thomson identified the first particle smaller than an atom—the electron. In the new, topsy-turvy world of quantum physics, light itself was soon imagined as either a wave or a particle, and physicists were realizing that their trusted laws of motion, dating back more than two hundred years to Isaac Newton, could not reliably gauge how light, whatever its form, whizzed through space. Using Newton's laws of gravity and motion, scientists arrived at one answer, but upon applying James Clerk Maxwell's laws of electromagnetism, they obtained a differing result. It took a rebel—a cocky kid who spurned rote learning throughout his schooling, always questioned conventional wisdom, and had an unshakable faith in his own abilities—to blaze a trail through this baffling territory, one that involved an entirely new take on space, time, gravity, and the behavior of the universe at large. Before anyone else, Albert Einstein discerned that a drastic change was needed, "the discovery of a universal formal principle," as he put it.

This was not the iconic Einstein—the sockless, rumpled character with baggy sweater and fright-wig coiffure—but a younger, more romantic figure with alluring brown eyes and wavy dark hair. While in his twenties and thirties, he was at the height of his prowess. Among his gifts was a powerful physical instinct, almost a sixth sense for knowing how nature should work. This often involved his thinking in images, such as one that began haunting him as a teenager: If a man could keep pace with a beam of light, what would he see? Would he see the electromagnetic wave frozen in place like some glacial swell, as Newton's laws were suggesting? "It does not seem that something like that can exist!" Einstein later recalled thinking.

After pondering this issue long and hard, Einstein came to realize in 1905 that since all the laws of physics remain the same—whether you're at rest or in steady motion, sitting quietly on a beach or reading on a train—then the speed of light has to stay constant as well in both situations. He had found the answer to his question. No one can catch up with a light beam, no matter how fast they are traveling. Whether your feet are firmly planted on Earth or aboard a spacecraft speeding toward a far planet, you'd measure the exact same pace to light's motion, 299,792 kilometers (186,282 miles) per second.

How is that possible? It seems to go against common sense. But Einstein ingeniously deduced that if the speed of light is identical for all observers, no matter what their state of motion, then something else has to give. And that something else was absolute time and space. With his special theory of relativity, Einstein completely altered the traditional perspective of classical physics that had been firmly established by his illustrious predecessor. "Newton, forgive me," said Einstein in his autobiographical notes. "You found the only way which, in your age, was just about possible for a man of highest thought—and creative power." In Newton's world, there was one universal clock and common reference frame, which made time and space the same for one and all throughout the cosmos. But that scheme no longer held. Instead, space and time were now "relative," flowing differently for each one of us depending on our motion. Einstein intuited that length and time are adjustable. If two observers are uniformly speeding either toward or away from each other, each will measure space shrinking and time proceeding slower for the other. Their clocks and yardsticks will not match up, as they did under Newton's laws. The only thing that they will agree on is the speed of light, a universal constant that remains unchanged for both travelers. The reason this seems counterintuitive is that we can't readily discern these differences in length and time in our rather humdrum surroundings. The changes are only apparent when the speeds between two objects are enormous, a sizable fraction of the velocity of light.

Soon Einstein was not satisfied with that adjustment alone. Special relativity was just that—special. It could only explain the properties of objects moving at an unvarying velocity. But that restricted its use to a great extent. Most events in nature don't behave so methodically. What if something were speeding up, slowing down, or changing direction? What if an object were accelerating under the force of gravity? Einstein knew that he had to develop a more *general* theory to deal with these situations, and he struggled with the problem for nearly a decade. It was a formidable job, as he had to do nothing less than recast Newton's venerable laws of gravity in the light of relativity.

For years success eluded him as he struggled to figure out how to make his equations truly universal and still reproduce Newton's law of gravity for the simplest cases, when gravity was weak and velocities were low. After all, Einstein couldn't just throw out a law that had been time-tested for more than two centuries. His new theory had to agree very closely with Newton's in the everyday realm where physicists had long been conducting experiments, a place where space-time distortions were too small to be overt. But then the theory would have to merge smoothly into either the intense gravity or high-velocity regimes in which the strange effects of relativity at last become obvious. "In all my life I have labored not nearly as hard," he wrote a colleague in the midst of his deliberations. ". . . Compared with this problem, the original relativity is child's play."

The breakthrough for the thirty-six-year-old physicist finally came in November 1915. Over that month Einstein reported weekly to the Prussian Academy of Sciences on his final progress toward a new theory of gravitation. A key moment arrived in midmonth, when he was able to successfully explain a small displacement in the orbit of Mercury, a nagging mystery to astronomers for decades. Einstein later remarked that he had palpitations of the heart upon seeing this result: "I was beside myself with ecstasy for days."

Complete success arrived on November 25, the day he presented his concluding paper. In this culminating talk, Einstein presented the decisive modifications that allowed him to secure a comprehensive theory. Written in the terse notation of tensor calculus—shorthand for a larger set of more complex functions—the general theory of relativity looks deceptively like a simple algebraic equation. It fits on one line and is the embodiment of mathematical elegance:

$$R_{uv} - \tfrac{1}{2}\, g_{uv} R = -\kappa T_{uv}$$

On the left side are quantities that describe the gravitational field as the geometry of space-time. In fact, the Rs denote how much space-time is curved. On the right side is a representation of mass-energy and how it is distributed. The equal sign sets up an intimate relationship between these two entities. As Princeton physicist John Archibald Wheeler liked to put it, "Spacetime tells mass how to move and mass tells spacetime how to curve."

Einstein showed that the three dimensions of space and the additional dimension of time join up to form a real, palpable object. While it's impossible for us to visualize these four dimensions, it can be pictured in three. Think of space-time as a boundless rubber sheet. Masses, such as a star or planet, then indent this flexible mat, curving space-time. The more massive the object, the deeper the depression. Planets thus circle the Sun not because they are held by invisible tendrils of force, as Newton had us think, but because they are caught in the natural hollow formed by the Sun in four-dimensional space-time, much as a rolling marble would circle around a bowling ball sitting in a trampoline. With this image in mind, the pull of gravity could now be easily explained; it's merely matter sliding like a downhill skier along the undulations of space-time. When Einstein's younger son, Eduard, later asked his father why he was so famous, Einstein singled out this elegant and lucid illustration of gravity as curving space-time. "When a blind beetle crawls over the surface of a curved branch, it doesn't notice that the track it has covered is indeed curved," he explained. "I was lucky enough to notice what the beetle didn't notice."

This realization was why Einstein was so excited by his successful result regarding the planet Mercury. It was clear evidence of this fantastic new image of gravity, its geometric representation. His insight centered on the fact that planets do not orbit the Sun in perfect circles but rather in ellipses, one end being slightly closer to the Sun than the other. And it was long known that the point of Mercury's orbit that is closest to the Sun—its perihelion—shifts around over time due to the combined gravitational tugs of the other planets. But there is an added shift—an extra 43 seconds of arc (or arcseconds) per century—that could never be adequately explained. Astronomers had even postulated an undiscovered planet called Vulcan—even closer to the Sun than Mercury—to explain the anomaly.

Here's where the relativistic geometry makes a difference: Because Mercury is situated so close to the Sun, whose mass has created a sizable space-time crater, it has more of a "dip" to contend with, more so than the other planets. Einstein declared that the added shift in Mercury's orbit was caused solely by Mercury's proximity to the Sun, not by some yet-to-be-observed inner planet. This wasn't just a vague prediction; the equations of general relativity accounted for Mercury's extra 43 arcseconds of shift per century with utmost precision.

Arthur Eddington, for one, was immediately smitten by Einstein's groundbreaking opus. "Whether the theory ultimately proves to be correct or not, it claims attention as being one of the most beautiful examples of the power of general mathematical reasoning," he wrote in his

Arthur Eddington (*AIP Emilio Segrè Visual Archives*)

account of general relativity, the first book on the subject to appear in English. With Eddington acting as Einstein's translator and champion, the two were often linked in people's minds. An accomplished popularizer of science, Eddington said that Einstein had taken "Newton's plant, which had outgrown its pot, and transplanted it to a more open field." Eddington was becoming so proficient at explaining relativity that "people seem to forget that I am an astronomer and that relativity is only a side issue," he lamented after one wearying interview with reporters.

For Eddington to serve as a spokesman for a radical new theory was somewhat out of character for him. He was usually reserved to the point of shyness, so shy, said physicist Hermann Bondi, that "he couldn't talk at all. . . . When anybody was with him . . . he played with his pipe, and emptied it and re-stuffed it, and occasionally said a word about the weather." A thin man of average height but with penetrating eyes, he lived with his sister, who served as homemaker and hostess at their Cambridge Observatory residence. A devout Quaker and pacifist, Eddington remained at Cambridge University in Great Britain during World War I, having been declared valuable to the "national interest" at his university post.

As both an astronomer and a theorist, Eddington divined early on the revolutionary significance of Einstein's ideas: that the general theory of relativity was offering a means to comprehend the workings of the cosmos within a rational and mathematical framework. While Newton's laws were fine for predicting the behavior of comets, planets, and stars, only general relativity could deal with the immensity of space-time as a whole. And at the moment Eddington was beginning to work on a translation of general relativity for his colleagues, Einstein was already at work applying his revolutionary new theory to the universe at large.

For Newton, space was eternally at rest, merely an inert and empty container, a three-dimensional stage through which objects moved about. But general relativity changed all that. Now the stage itself became an active player, since the matter within the cosmos sculpts its overall curvature. With this new insight into gravity, physicists could at last make predictions about the universe's behavior, an innovation that moved cosmology out of the realm of philosophy, its long-standing home, and transformed it into a working science.

Einstein was the first to do this. In 1917, just as Shapley in California was revamping the Milky Way, he published a paper in Germany titled "Cosmological Considerations Arising from the General Theory of Relativity." In it he explored how his new gravitational ideas could be used to determine the universe's behavior. Einstein had always been attracted to that age-old question: Is the universe infinite or finite in extension? "I compare space to a cloth . . . one can observe a certain portion," he mused. ". . . We speculate how to extrapolate the cloth, what holds its tangential tension in equilibrium . . . whether it is infinitely extended, or finite and closed." Einstein decided that the universe was closed, what is also referred to as a spherical universe, the four-dimensional equivalent of a spherical Earth. Though this shape has neither a beginning nor an end, its volume is finite. Travel forward through it long enough and you return right back to your starting point, just as you would circumnavigating our globe. In this scheme matter is so plentiful that space-time bends profoundly, so much that it literally wraps itself up into a hyperdimensional ball. Recognizing how frightfully odd this sounded, Einstein told a friend, "It exposes me to the danger of being confined to a madhouse." But he stuck with this strange notion, as it helped him get around other problems in applying general relativity to the cosmos. Einstein also preferred this model because, given his astronomical knowledge at the time, he assumed the universe was filled with matter and stable. In 1917 it certainly appeared to him to be steady and enduring. Truth be told, he liked the idea of an immutable cosmos, a large collection of stars fixed forever in the void.

From a theorist's perspective, this choice was mathematically beautiful, but it also presented a problem. Even Newton knew that matter distributed throughout a finite space would eventually coalesce into larger and larger lumps. Stellar objects would be gravitationally drawn to one another, closer and closer over time. Ultimately, the universe would collapse under the inescapable pull of gravity. So, to avoid this cosmic calamity and match his theory with then-accepted astronomical observations, Einstein altered his famous equation, adding the term λ (the Greek letter lambda), a fudge factor that came to be called the "cosmological constant." This new ingredient was an added energy that permeated empty space and exerted an outward "pressure" on it. This repulsive field—a kind of antigravity, actually—exactly balanced the inward gravitational attraction of all the matter in his closed universe,

Willem de Sitter
(*Courtesy of the Archives, California
Institute of Technology*)

keeping it from moving. As a result, the universe remained immobile, "as required by the fact of the small velocities of the stars," wrote Einstein in his classic 1917 paper.

Others soon followed up on Einstein's cosmological endeavor, most important Willem de Sitter. The esteemed Dutch astronomer, a tall and slender man with a neatly trimmed Vandyke beard, started keeping track of general relativity's development as early as 1911 and was one of the first to recognize its deep significance to astronomy. After meeting with Einstein in Leiden on several occasions in 1916, discussions in fact that inspired Einstein to conceive his spherical universe, de Sitter soon corresponded with Eddington on the subject. Intrigued by de Sitter's insights, Eddington asked him to write up his impressions of general relativity for the *Monthly Notices of the Royal Astronomical Society*, which resulted in three long papers on the topic, the first articles to make Einstein's accomplishment widely known to scientists outside Germany. De Sitter was obviously stimulated by the assignment, for in his third paper he offered up his own cosmological solution to the equations of general relativity, one that was very different from Einstein's.

When scientists originate an equation to describe some phenomenon, their job is far from done. They must still *solve* the equation—in the case of general relativity, figure out what values for those Rs and Ts make the equation come out right. This is a tall order. So, to progress, a researcher will often introduce a simplifying assumption about the equation that makes the problem easier. If a solution is found in this way—and there is no guarantee—the scientist trusts it will shed some light on the overall problem, leading them to a more complete understanding.

What de Sitter assumed was that the universe contained no matter. He discovered that Einstein's equation could be solved if he imagined that the universe was both stable and *empty*. On the face of it, this seemed like a ludicrous assumption, but de Sitter wondered if cosmic densities were so low that the universe could be considered essentially barren. By making this conjecture he was able to construct a model of space-time in which "the frequency of light-vibrations diminishes." That is, light waves get longer (more red) with increasing distance from their source. The unique properties of space-time that arose in his solution demanded it. Einstein was not up on the latest astronomical news, but de Sitter was. In fact, he would soon be director of the Leiden Observatory, in the Netherlands. He was very aware that V. M. Slipher, at the Lowell Observatory, had recently discovered some spiral nebulae seemingly racing away from the Milky Way—and at very high velocities as measured by their redshifts. De Sitter was one of the few at the time who was sure that the spiral nebulae being sighted by astronomers in ever greater numbers were probably "amongst the most distant objects we know," indisputably located beyond the Milky Way. And he surmised that their tendency to display appreciable redshifts could be proof of his model. In his paper, he suggested that the nebulae might only appear to be moving outward because their light waves were getting longer and longer (hence redder and redder) as the light traveled toward Earth. This set up the illusion of movement.

On the other hand, there was another way to interpret the effects in de Sitter's universe: Any bit of matter dropped inside its space-time would immediately fly off. That was another possible reason for the redshifts Slipher was noticing. Eddington liked to say that "Einstein's universe contains matter but no motion and de Sitter's contains motion but no matter."

Before the publication of his bizarre yet fascinating solution, de Sitter exchanged a number of letters with Einstein arguing over its details. Einstein was clearly flummoxed by de Sitter's quirky take on the universe. It "does not make sense to me," he wrote. Where was the "world material" in his cosmos, where were the stars? It didn't seem based in

Albert Einstein and Willem de Sitter working out a problem
at the Mount Wilson Observatory's Pasadena headquarters in 1932
(*Associated Press*)

reality. In Einstein's eyes, de Sitter's solution was physically impossible.
The properties of space could not be determined, he believed, without
the presence of matter.

De Sitter was certainly making a huge assumption by considering a
cosmic density so low that the universe could be regarded as devoid of
matter. But what was exciting about his model was that it was testable. If
distances to the spiral nebulae could be measured precisely, then
astronomers would be able to see if the redshifts truly increased "systematically," as de Sitter noted in his paper. That is, the more distant a spiral,
the larger its redshift. But in 1917 carrying out such rigorous measure-

ments was a pipe dream. At that time astronomers were still not settled on what a spiral was, much less able to figure out its exact distance.

Besides, few astronomers were paying attention to Einstein's theory as yet. World War I had kept Einstein's work from being widely circulated outside Germany, and when astronomers did hear of it, they weren't quite sure what to make of its unconventional and perplexing view of gravity. George Hale, like many astronomers at the time who were trained to observe rather than to tinker with mathematical equations, said he feared "it will always remain beyond my grasp." All of that changed, though, once the findings of a British solar-eclipse expedition in 1919 transformed the name of Einstein, the former Swiss patent clerk, into a synonym for genius.

At the time Einstein was working on general relativity, he had early on suggested a specific test that astronomers could perform to confirm his predicted curvatures in space-time: Photograph a field of stars at night, then for comparison photograph those same stars when they pass near the Sun's limb during a solar eclipse. A beam of starlight passing right by the Sun would be gravitationally attracted to the Sun and get bent, making it appear that the star has shifted its standard position on the celestial sky, the position it would have if the Sun were in another part of the heavens. In 1911 he computed a bending of 0.83 arcseconds, the same arising from Newton's laws alone. But a few years later, once his final theory was in place, Einstein doubled his predicted bending. The extra contribution, Einstein figured out, occurs due to the Sun's enormous mass warping space-time. He calculated that a stellar ray just grazing the Sun would get deflected by 1.7 arcseconds (a thousandth the width of the Moon).

Three solar-eclipse expeditions were launched prior to 1919 to detect this light bending but were unsuccessful due to either bad weather or the ongoing war. The results of a fourth effort, an American endeavor led by Lick astronomers W. W. Campbell and Heber Curtis, were plagued by data comparison problems and so were never published. That was a fortunate turn of events for Einstein. The shaky American results went against him, and some of the other expeditions were carried out when his theory, not yet fully developed, was predicting that smaller, incorrect deflection.

That's why scientists paid keen attention to British astronomers when they announced they would give it a try in 1919 during a favorable solar eclipse whose path crossed South America and continued over to central Africa. The eclipse was taking place against a particularly rich background of stars, the Hyades cluster, which offered excellent opportunities to detect a star's shift. Sir Frank Dyson, Great Britain's astronomer royal, first pointed out this fortunate occurrence more than two years earlier. "This should serve for an ample verification, or the contrary, of Einstein's theory," he noted at the time. And as the victors in World War I, the British had the necessary funding to organize and carry out the intricate venture.

On the evening before sailing, Eddington and his eclipse companion, E. T. Cottingham, joined Dyson in his study. The discussion turned to the amount of deflection expected from classical Newtonian theory compared to Einstein's predicted value, which was twice as great. "What will it mean," asked Cottingham playfully, "if we get double the Einstein deflection?" Dyson replied, "Then Eddington will go mad and you will have to come home alone!"

The next day Eddington and his assistant began their journey to the tiny isle of Principe, situated 140 miles off the coast of West Africa, a favorable site in the path of the eclipse. And to improve the venture's chances for a clear-weather view, two other astronomers traveled to the village of Sobral in the Amazon jungle of northern Brazil. On the day of the eclipse, May 29, a violent morning rainstorm almost doomed the Principe crew's operations. But by noon, the deluge ended and an hour and a half later they got their first glimpse of the Sun, already partially covered by the Moon. Too busy changing plates during totality, Eddington had only one chance, halfway through, to view the Sun's dark visage. "We are conscious only of the weird half-light of the landscape and the hush of nature, broken by the calls of the observers, and beat of the metronome ticking out the 302 seconds of totality," he later recalled of the adventure.

The astronomers in Sobral were more fortunate. There they had two instruments and better weather. With their astrographic telescope they clicked off sixteen photos, and eight more were taken with a 4-inch scope. On Principe, Eddington and Cottingham, too, took sixteen photographs, but most ultimately turned out useless because of the intervening clouds. For several days after the eclipse, Eddington spent the daytime hours taking a first stab at measuring the star images on the plates that did turn out well. Upon examining the preliminary results, he turned to his colleague and exclaimed, "Cottingham, you won't have to go home alone." He saw evidence that the streams of starlight had indeed bent around the darkened Sun according to Einstein's rules.

At a dinner soon after his return to England, Eddington entertained some fellow astronomers with a poem in the style of the Rubáiyát. "One thing is certain, and the rest debate—light-rays, when near the Sun, DO NOT GO STRAIGHT" was his rousing finale. In November the results of the expedition were officially reported at a special joint meeting in London of the Royal Society and the Royal Astronomical Society. Dyson spoke on behalf of the participants and behind him, almost like a stage setting, hung a picture of Isaac Newton, whose historic law of gravitation was undergoing its first modification. The best results supporting Einstein came from the Sobral 4-inch telescope. From its plates, the Britishers measured a starlight deflection of 1.98 arcseconds, a little on the high side of Einstein's prediction. The poorer images from Principe suggested a bending of 1.61 arcseconds, just below what Einstein calculated. Viewed together, though, Einstein was deemed a winner. These were the results that Eddington and Dyson stressed in their reports, which were widely hailed in newspaper headlines worldwide, turning Einstein into an overnight celebrity. "LIGHTS ALL ASKEW IN THE HEAVENS, Men of Science More or Less Agog Over Results of Eclipse Observations . . . Stars Not Where They Seemed or Were Calculated to be, but Nobody Need Worry," blared the *New York Times*. Suddenly the public's attention was riveted to all things relative. Awed by the contributions scientists had made in the war effort, the public was highly receptive to hear more from the physics frontier—at least until their attention was diverted by the tomb of a young Egyptian pharaoh named Tutankhamen ("King Tut") found a few years later almost completely intact.

Often neglected in the recountings of the famous 1919 solar-eclipse expedition was the team's largest data set from the Sobral astrographic telescope, which indicated a deflection of 0.93 arcseconds, in favor of Sir Isaac Newton. Because of various technical problems with the Sobral scope, including a blurring of its images, the British team decided to downplay that instrument's results. Eddington admitted he was unscientifically rooting for Einstein, but his instincts to reject the astrographic telescope results turned out to be good in the end. Campbell headed up another solar-eclipse expedition in 1922, which arrived at similar results and further confirmed Einstein's theory. When asked what he had been expecting, Campbell replied in all seriousness, "I hoped it would not be true." Relativity's thoroughly new vision of space and time, coupled with its complexity, made even several leading scientists reluctant to accept its predictions. Those primarily trained in classical physics were quite leery of general relativity's strange outlook on the force of gravity and wondered if the light-bending was actually a refraction effect in the Sun's atmosphere or perhaps was due to a physical distortion of the photographic plate from imaging the hot solar corona. Heber Curtis, who met Einstein during his first visit to the United States, was certainly no fan of relativity. "We met in quick succession Their Eminences, the Prince of Monaco, Dr. Einstein and President Harding, and were photographed in a group on the White House lawn," Curtis wrote Campbell soon after the meeting. Curtis, who was still convinced he had proven Einstein wrong with his flawed 1918 expedition results, would have been glad to see someone join him in overturning Einstein. "He surely looks like the fourth dimension!" joked Curtis about the German phenom. "Face is somewhat sallow and yellowish, redeemed by very keen bright eyes. But wears his hair a la [Polish pianist Ignacy] Paderewski in narrow greasy curls of small diameter and four or five inches long."

Even a full decade after general relativity was introduced, many scientists were still resisting Einstein's new view of the universe. At the National Academy of Sciences meeting in Washington in 1925, physicist Dayton Miller of the Case School of Applied Science in Cleveland announced he had seen evidence of an "ether drag," the speed of light changing with the motion of Earth. According to one conference attendee, this report was a "bombshell . . . which quite blew up the meeting of the Academy and *got more applause* than anything that happened . . . [disturbing] the relativists in general."

Einstein supporters were enraged that doubts over relativity still lingered at all. "I am really getting pretty tired of the fundamentalist's attitude of the opponents of relativity," said Henry Norris Russell in response to Miller's bolt from the blue. "Their psychology seems to me to be exactly similar to that of the most conservative theologians." In time, though, Miller's experiments proved faulty, and astronomers would eventually have to face up to the new cosmic order.

[10]

Go at Each Other "Hammer and Tongs"

The year 1920 was one of achievements—illustrious, infamous, resourceful, and humorous. American women got the vote, Joan of Arc was canonized by Pope Benedict XV, prohibition was initiated throughout the United States, an employee at the Johnson & Johnson company invented the Band-Aid, and the U.S. Post Office ruled that children may not be sent by parcel post. Moreover, astronomers didn't yet know exactly how the Sun generated its tremendous power or that the solar orb was largely composed of hydrogen, even though Einstein's newly introduced theory relating mass to energy, nicely summarized as $E = mc^2$, was offering a fresh clue.

What 1920 is best remembered for in the annals of astronomy is Harlow Shapley and Heber Curtis meeting in Washington, D.C., before members of the National Academy of Sciences to argue the arrangement of the universe. The sides were now clearly drawn, and it was time for a showdown. Shapley had, of course, recently pronounced that the Milky Way was far bigger than previously assumed, easily imagining the spirals as minor players hovering on the edge of our vast system of stars. Curtis, on the other hand, thought otherwise. This epochal encounter is commonly known as the "Great Debate," although in truth that's hardly an apt description at all. More like two lectures back to back, the event wasn't covered by even the science-oriented press. In astronomy circles, the venerable legend that surrounds that April session—the memory of it as the mighty clash of cosmic titans, astronomy's version of *High Noon*—developed gradually over time, the embroidery added so profusely over the years that it was eventually described as a "homeric fight," two opposing sides battling it out in the highest court of scientific opinion.

This odyssey, though, began quite simply. George Ellery Hale had suggested at a council meeting of the academy that the 1920 Hale lecture on a topic of interest to scientists, an annual event established in honor of his father in 1914, be held during the academy's upcoming spring conference. He himself was leaning toward a discussion of Einstein's general theory of relativity, the trendiest scientific topic of the era. But the academy's home secretary, solar physicist Charles Greeley Abbot, feared that the revolutionary new view of gravity would already be "done to death" by the time of the meeting. The triumphant British solar-eclipse expedition was the science story of the year, still garnering headlines around the world. More than that, Abbot was wary of relativity's radical and difficult-to-comprehend concepts: "I pray to God that the progress of science will send relativity to some region of space beyond the fourth dimension, from whence it may never return to plague us," he declared. Abbot wondered whether the "cause of glacial periods, or some zoological or biological subject" might be more appealing. The Prince of Monaco was even suggested as a lecturer, to speak on oceanography. But eventually Hale's second choice came to the forefront—the unresolved issue of the island-universe theory.

There was no question that Shapley, the thirty-five-year-old rising star, would defend his Big Galaxy idea. But who to pick for the other side? Lick Observatory director W. W. Campbell was briefly considered to champion the island universes, but Curtis, who had been devoting his professional life at Lick to this issue, was ultimately chosen, as by then he had become the leading spokesman for the claim. In terms of personalities, it was an interesting matchup. Shapley was acknowledged to be the "daring innovator, pressing the last bit of information from his observations, unafraid to extrapolate from the known to the unknown . . . occasionally depending upon intuition to supply connecting links." Curtis, on the other hand, was considered a "cautious, sometimes overcautious, conservative who weighed every observation and more often concluded 'not proven' than 'not so.'" Though not as prominent a figure as Shapley, Curtis was a respected astronomer nonetheless; by then forty-seven years old, bespectacled, and far less brash than his younger contender, he struck one as being a distinguished banker. Despite this stolid appearance, however, he proved to be the more venturesome one in regard to the upcoming talk.

With more professional experience under his belt, Curtis was quite comfortable at a podium and eager for a good tussle. But for Shapley, then ill at ease as a speaker, public exposure at this time was problematic. British historian Michael Hoskin first pointed out that Shapley had come to believe he was the front-runner for the directorship of the Harvard College Observatory, one of astronomy's most prestigious positions. Edward Pickering had recently died, having established a monumental legacy, and the search for his replacement was actively under way. Though young and completely untested in managing a world-class research institution, Shapley submitted his name for consideration, wanting to advance his career and strike while the iron was hot. Though he would be leaving the world's biggest telescopes, Shapley was enticed by Harvard's extensive collection of photographic plates, which offered a lush resource for the problems in which he was most interested. "Perhaps Harvard is amateurish, compared with Mount Wilson," he told Russell, "but you and I . . . realize the enormous possibilities of the place." More than that, it was an opportunity for Shapley to get away from his troubled relationship with Mount Wilson's deputy director, Walter Adams. Given this ambition, he worried how he would come across to certain members of the National Academy audience, who might have influence in the final decision. Curtis was known to be a dynamic lecturer; Shapley feared he would look bad by comparison. A letter from Curtis before the debate didn't calm his fears: "I am sure that we could be just as good friends if we did go at each other 'hammer and tongs.' . . . A good friendly 'scrap' is an excellent thing once in a while; sort of clears up the atmosphere."

There was a flurry of correspondence between the participants and the National Academy in the months before the event, aimed at establishing the rules of engagement. Curtis was eager to air the controversy in a no-holds-barred debate. He told Shapley he wanted to " 'take the lid off' and definitely attack each other's view-point." But Shapley had a different agenda altogether. He wanted to discuss solely his new supersized model of the Milky Way and even informed Russell a few weeks before the debate that he didn't intend to say much about the spiral nebulae at all. "I have neither time nor data nor very good arguments," he lamented. In fact, Shapley was relieved that the chosen title for the lecture, "The Scale of the Universe," was ambiguous enough to allow him to carry out his plan. Shapley was quite reluctant to dwell on the spiral nebulae, a subject with such uncertain evidence. He hated airing science's dirty laundry in public.

Ardently voicing these concerns, Shapley convinced Hale that the so-called debate should be more of a discussion, "two talks on the same subject." And instead of forty-five minutes for each speaker, as originally posed, Shapley asked for thirty-five. "My sympathies are with the audience, always," he argued. "Could it listen to or endure nearly two hours of nebulosity?" Curtis was dismayed by this suggestion; he firmly believed he needed more time to lay out his scientific arguments. "We could scarcely get warmed up in 35 minutes," he pleaded with Hale. After a while, they all compromised at forty minutes. And there would be no rebuttals. "If you or he wish to answer points made by the other, you can do so in the general discussion," Hale told Curtis.

Shapley and Curtis were each paid an honorarium of $150, out of which they paid their travel expenses to journey from California to the East Coast. For Curtis it was $2 for the stagecoach to San Jose, then another $100 for the round-trip railroad ticket. By chance both Shapley and Curtis took the same train out to Washington via the southern route, but they agreed not to hash out their ideas ahead of time in order to keep their arguments fresh. When the train broke down at one point in Alabama, they got out and walked around for a while, keeping their conversation focused on flowers and the classics. Shapley didn't forget to collect a few native ants. In all likelihood, they were also, silently and unobtrusively, sizing up their competition.

The annual meeting of the National Academy of Sciences that year extended over three days. During the daytime sessions, a number of outstanding scientists presented talks. Franz Boas, the father of American anthropology, spoke on "growth and development as determined by environmental issues," and rocket pioneer Robert Goddard advocated the use of rockets in weather forecasting. The "debate," however, took place on the cool and showery evening of April 26, 1920, at the end of the conclave's first day. The audience of around two hundred to three hundred gathered in the Baird Auditorium of what is now the Smithsonian Institution's Museum of Natural History, prominently positioned along Washington's national mall directly across from the

Smithsonian "castle." In a news report the day before, the *Washington Post* announced that "Dr. Harlow Shapley, of the Mount Wilson solar observatory, will discuss evidence which seems to indicate the scale of the [Milky Way] to be many times greater than is held.... Dr. Heber D. Curtis, of the Lick Observatory, will defend the old theory that there are possibly numerous universes similar to our own, each of which may have as many as three billion stars."

The proceeding started at 8:15 p.m., and Shapley was the first to speak. He had been right to be nervous; two friends of Harvard president A. Lawrence Lowell—George Agassiz, a member of the Harvard astronomy department's visiting committee, and Theodore Lyman, chairman of its physics department—were in the audience to size him up. But Shapley came prepared. He made sure that Russell, still a valuable supporter of his cosmic model, was in the audience to back him up during the discussion period.

What exactly happened that night—the tenor of the speakers, the reception of the audience—is largely guesswork, based on the limited evidence left behind. Recollections of the event are riddled with false memories. Shapley, for instance, recalled an interminable banquet beforehand with honored guest Albert Einstein whispering to his tablemate that he "just got a new theory of Eternity." But the conference dinner was the following night, and the noted theorist of relativity didn't make his first visit to America until the following year. However, Shapley did save the typescript of his talk, complete with last-minute scribbles (some in shorthand, a talent honed in his reporting days), which revealed his style and manner. Given the diversity of his audience, many not schooled in astronomy, Shapley chose to avoid technicalities and spent a good portion of his time just presenting basic astronomical facts: He carefully described the size of the Milky Way, its structure, and its constituent parts—the stars, gaseous nebulae, and clusters. He accompanied it with slides of the 100-inch telescope, the Moon, the Sun, the Pleiades cluster of stars, globular clusters. It was a visual tour of the known universe, with special attention paid to making the audience understand the meaning of a light-year. "You do not see the sun where it is, but where it was eight minutes ago," he instructed. "You do not see these stars as they are now, but more probably as they were when King Cheops was a little boy."

Instead of talking about the nature of the spiral nebulae, the very reason for the encounter, Shapley focused on his Big Galaxy model. He figured that if he proved the Milky Way was immense, the spiral nebulae would automatically be relegated to minor status in the cosmic scheme of things, mere hangers-on. Anticipating that Curtis would challenge his use of the Cepheids as standard candles in determining the globular cluster distances, Shapley simply ignored the technique in his remarks. "[Curtis] may question the sufficiency of the data or the accuracy of the methods of using it," he said. "But this fact remains: we could discard the Cepheids altogether, use instead the thousands of B-type stars upon which the most capable stellar astronomers have worked for years, and derive just the same distance [to the globular clusters] . . . and obtain consequently the same dimensions for the galactic system." But Shapley was being disingenuous. Two years earlier he had reported to the Astronomical Society of the Pacific that the Cepheids carried "so much greater weight" for his distance measurements and that the magnitudes of red giants and blue stars "can best be used as checks or as secondary standards."

Shapley went on to stress his finding that the Sun is not at the center of the Milky Way: "We have been victimized by the chance position of the sun near the center of a subordinate system [of stars], and misled by the consequent phenomena, to think that we are God's own appointed, right in the thick of things." As for the spirals? "I shall leave the description and discussion of this debatable question to Professor Curtis," he said. Shapley conceded that the possibility remained that they were comparable galactic systems, but only if the Milky Way were cut down to a tenth of his newly defined dimensions. He believed that unlikely and preferred to think of the spirals as nebulous objects. He maintained "that it is professionally and scientifically unwise to take any very positive view in the matter just now."

One can imagine Curtis's mounting dismay as his opponent was progressing through his talk. Shapley had spent most of his time on just the basics of astronomy, while Curtis had prepared a full-fledged analysis, laden with scientific detail. The Lick astronomer was about to address the audience on issues that Shapley had never brought up. While he anxiously awaited his turn at the podium, his mind raced, wondering whether he should change his approach on the fly, making his presentation more relaxed and general. But in the end he decided to stick to his original plan.

Unlike with Shapley, a copy of Curtis's script no longer exists, but some of his slides, displaying his essential points, do survive, and they provide a glimpse of the flow of his arguments that evening. Contrasting sharply with Shapley's popular approach to the topic, Curtis's talk was more technical, although by all accounts he spoke more spontaneously. At first he focused on one of his major disagreements with Shapley: the size of the Milky Way. He carefully outlined his reasons for believing that the Milky Way was a tenth the size that Shapley was hawking. Mainly, he had no confidence in Shapley's use of the Cepheids. Shapley himself knew that his momentous refashioning of the Milky Way's size stood on the foundation of eleven "miserable" Cepheids, as he had earlier described them in a letter.

From there Curtis went on to focus on the spiral nebulae, the subject that Shapley conveniently avoided. Curtis showcased his best evidence, echoing many of the points he had made to the Washington Academy of Sciences just the year before: He stressed that the spiral nebulae displayed the spectra typical for collections of *stars*—not gas; that not one spiral had ever been found within the Milky Way itself; that the spirals are primarily seen away from the Milky Way, because obscuring matter blocks the view through the plane of our galaxy. He paid special attention to the many novae being sighted within some spirals. He showed that if the flare-ups in Andromeda were half a million light-years distant, their luminosity would roughly match those seen in our own galaxy. Any closer and they would be far too bright. And then there was the movement of the spirals detected by Slipher; the spirals traveled speedily through space unlike any other celestial object in the Milky Way, which suggested that they had to be located outside our galaxy's borders.

All in all, the two men were simply talking at cross purposes. Shapley primarily defended his new vision of the Milky Way—its unexpected bigness—while Curtis hammered away on his contention that the spiral nebulae were far-off galaxies. In hindsight, each turned out to be partly right and partly wrong. Shapley argued for his larger Milky Way (true) but insisted that the spirals were local (wrong). Curtis still believed in a smaller home galaxy (wrong) but persevered in his belief that the spiral nebulae were situated far outside the Milky Way and rivaling it in size (true). At the end of the day, it was a wash.

Everyone in essence went home maintaining the beliefs they held at the start of the lecture. The data were so muddled that Curtis and Shapley could take the same facts and arrive at completely contradictory conclusions. At the time of the debate, there was no overwhelming evidence to settle the inconsistencies either way. Both men were traveling along a precarious road, each viewing his destination through an obscuring fog and interpreting the hazy view in different ways.

There was a winner, however, for best presentation. Curtis headed off that night feeling pretty good about his performance. He received assurances afterward that he "came out considerably in front." Shapley, on the other hand, was judged more poorly. Russell wrote Hale afterward that his former student sorely needed to enhance the "gift of the gab." Agassiz, the Harvard evaluator, was not impressed by Shapley's performance at all. "He has . . . a some what peculiar and nervous personality . . . lacks maturity and force, and does not give the impression of being a big enough personality for the position," he reported back to Harvard's president two days after the event. More attractive to Agassiz was Russell, who spoke quite eloquently that night in support of Shapley's arguments during the audience response period. He said that Russell had "more balance more force and a broader mental range."

The two opponents came to acknowledge what others sensed all along over the course of that April evening. "Yes, I guess mine was too technical," admitted Curtis to Shapley a couple of months after the debate. "I thought yours would be along the same line, but you surprised me by making it far more general in character than I had expected." As captivating scientific theater, the so-called Great Debate was ultimately a letdown.

A full year after the debate, however, the two astronomers battled it out once again within the *Bulletin of the National Research Council.* The original intent was to simply print the lectures they had given before the National Academy. But as the articles were being prepared, each man deepened and extended his arguments. It was not during that misty spring night in Washington that Shapley and Curtis had their great debate but rather within the pages of the *Bulletin.* It was the written version, vastly altered and amended, that ultimately established the legend handed down by succeeding generations of astronomers, many coming to believe it was the bona fide transcript of the April scrap.

At first Curtis wasn't keen on publishing his comments, but he indicated he would be willing if both he and Shapley delved more deeply on the technical issues. Shapley agreed. They were originally asked to keep to ten pages, which Curtis joked would force him to follow the laws of writing "generally observed in composing telegrams." Perhaps, he wrote Shapley, they could "shoot our arrows into the air, to let them fall we know not where." With his customary down-home wit, Shapley suggested that he would provide "ten pages of buncombe and flapdoodle," while Curtis could supply "ten more pages of wisdom."

Shapley also wondered if they should exchange their papers, providing the opportunity to rebut each other's arguments. "Should I go ahead, shoot my shot (or wad), then you use your shillelah (or hammer), then I sneak up behind you and apply my ole horn-handle." Curtis was game, and over the ensuing months a lively train of drafts and comments went back and forth between them. In the process Curtis pushed Shapley to devote more space to the spiral nebulae, "at least a brief statement of how you explain them if not island universes." Upon completion, their published remarks each expanded from ten pages to twenty-four, and though it wasn't mentioned until his penultimate paragraphs, Shapley's strongest and freshest ammunition against Curtis involved the spirals. It was there at the end that he played his definitive trump card: The spiral nebulae could not possibly be island universes, because the rotations measured by Adriaan van Maanen at Mount Wilson "appear fatal to such an interpretation." Shapley now seemed more at ease in dismissing the spirals as simply minor objects. "I see no reason for thinking them stellar *or* universes," he told Russell during the course of his writing the *Bulletin* article. "What monstrous assumptions that requires before you get done with it." From that point on, Shapley's strongest weapon against supporters of distant galaxies was van Maanen's twirling spirals.

Although in his heart of hearts he never believed it would happen, even Curtis had to grudgingly concede in his published response that if van Maanen's findings held up "the island universe theory must be definitely abandoned." Over the succeeding years, van Maanen and his observations stood like a giant wall before island-universe advocates. If the spiral nebulae were truly remote and massive galaxies, how could you possibly explain seeing them rotate over just a few years from so far away? The island-universe theory would not gain general acceptance until its supporters figured out how to breach this formidable rampart.

Van Maanen had begun his measurements of the spiral nebulae in 1915 and continued into the early 1920s. Astronomers took his results seriously because his reputation was exemplary. He was known to be a careful observer who followed intricate astronomical procedures to the letter. And it was easy to accept his conclusions, as they supported an idea of the universe that many readily believed at the time: The Milky Way defined the universe, and the spirals were mere appendages that from their swirling appearance had to be turning. Stars, planets, and moons rotated; planets revolved around the Sun; rotation was a natural feature of the universe. Given that, it was not surprising to hear that the spirals were rotating. In 1914 Slipher had already reported on a spiral rotation from his spectroscopic data, but simply viewing the curving

Adriaan van Maanen (left) with Bertil Lindblad (*Photograph by Dorothy Davis Locanthi, courtesy of AIP Emilio Segrè Visual Archives*)

lines of a spiral's misty arms, captured so vividly in photographs, made it impossible to think otherwise.

Van Maanen was the descendant of an aristocratic family in the Netherlands, whose ancestors were ministers, teachers, and noted jurists. Those who knew him attested to his meticulous integrity and high sense of personal honor, instilled by his family's esteemed heritage. After earning his doctorate in 1911, van Maanen had traveled to the United States to work as a volunteer assistant at the Yerkes Observatory. But after his mentor, the noted Dutch astronomer Jacobus Kapteyn, brought him to the attention of Hale, he was soon offered a permanent position at Mount Wilson. The observatory wanted him for his superb and proven skills in gauging the motions and distances of stars. In 1917, in carrying out such observations, he discovered the second known white-dwarf star, a rare find at the time.

Van Maanen was drawn as a student toward this line of work, an endeavor that other astronomers tried to avoid because of the tedium and difficulty in discerning the change in a star's position over time. The procedure involved comparing, with intense concentration, photographic plates taken over intervals of months or years. But to van Maanen the routine was heaven; he even went back to the pursuit two years before his death. "One always returns to one's first love," he scribbled on a copy of his 1944 paper on stellar parallaxes. To carry out the task, he superimposed the pictures of the stars taken at different times in his special stereocomparator (more often called the "Blink"). This machine allowed the viewer to quickly alternate between two photographic plates taken of the same field at different times. The blinking proceeded so rapidly that an object that had moved between pictures would immediately stand out, while those that remained fixed appeared still. Van Maanen could then slowly turn a micrometer screw to measure the star's exact advancement across the sky. The number of turns measured off the change in the star's position—the amount the star had moved over the years. It was his most cherished instrument at the observatory's Pasadena headquarters, and everyone knew it: The warning "Do not use this stereocomparator without consulting A. van Maanen" was blatantly posted on its front. The sign remained there for decades, long after he had left.

Sociable and well-liked, "Van," as everyone called him, played a good game of tennis and made sure newcomers to the mountain felt right at home. He was a lively storyteller and also a bit of a playboy.

Once Shapley arrived, he and van Maanen became fast friends, as they were nearly the same age. "He could go to a dinner and soon have the whole table laughing," recalled Shapley. An accomplished chef, van Maanen relished throwing parties where he could put his culinary skills into practice and prepare fine dishes for his cohorts. Shapley and van Maanen further bonded when they discovered they were both disliked by Adams, who was suspicious of them for their liberal outlooks and ambivalence toward the ongoing war in Europe, as well as their ambitions. "Van Maanen and I are in ill-favor because we do or try to do too much," confided Shapley to a friend.

One of van Maanen's first jobs at Mount Wilson was to measure photographs of spectra taken over the face of the Sun, an endeavor that helped Hale map the Sun's magnetic field. Early reports suggested that the strength of the magnetic field varied with solar latitude, and van Maanen always seemed to see this effect, even though it was later found to be a mistake. Van Maanen's persistence in finding the change was a harbinger of trials to come in succeeding years—not concerning the Sun but rather the spiral nebulae.

Van Maanen first got involved with the spirals in late 1915, when George Ritchey asked him a favor. Ritchey was then using Mount Wilson's 60-inch telescope to produce superb photographs of spiral nebulae. Everyone agreed the images were breathtakingly beautiful. Part of this success was due to Ritchey's inventiveness. He had developed a fast camera shutter, which allowed him to build up an image from a series of short exposures, each taken when the atmosphere was calm. The total exposure time could last anywhere from two hours to more than eight hours, sometimes stretched over two or three nights. This resulted in rich nebular details never before captured.

When he approached van Maanen, Ritchey had recently taken a picture of the spiral nebula M101, the Pinwheel, which he had also photographed in 1910. With both images in hand, he asked van Maanen to put the plates into his trusty stereocomparator and see if any changes could be detected in the nebula over those intervening years. Van Maanen at first measured no variation but got permission from Ritchey to keep the plates to study them further. Adapting methods he had used previously in other work, he chose thirty-two stars, equally bright and positioned uniformly around the nebula on each plate, and measured how dozens of points within the spiral nebula may have shifted in comparison to those stars. Extending his study, he borrowed additional plates of M101 from the Lick Observatory, photographs taken in 1899, 1908, and 1914. He didn't rush. Van Maanen was so meticulous that he even made sure the temperature in the room where he was making his measurements was tightly controlled. He wanted to take care that any thermal expansion, in either the glass photographic plate or the measuring machine, would be negligible.

In the end, he decided that the nebular material within M101 was moving after all, although exactly how was not immediately obvious. "If the results . . . could be taken at their face value, they would certainly seem to indicate a motion of rotation or possibly motion along the arms of the spiral," he reported. If turning at his measured rate, M101 was completing one full rotation every eighty-five thousand years. As noted earlier, that meant if the Pinwheel were truly the size of the Milky Way and located way off in distant space, the nebula's edge had to be traveling faster than the speed of light, an impossibility given Einstein's special theory of relativity, which said that no bit of matter can move fast enough to overtake a beam of light.

Given what was at stake, van Maanen followed all the precautions: He switched the plates in the holders to eliminate machine error, and he got a colleague to redo the measurements with a different machine, to make sure there wasn't an instrumental error or personal bias. He came to believe that the matter within the spiral was drifting away from the center—outward along the arms—and noted in his report that this agreed with the Chamberlin-Moulton model on the origin of spiral neb-ulae, which involved a collision between a star and a nebula. Thomas Chamberlin was elated to hear the news from Hale. "While the recent revival of the notion that spiral nebulae are mere distant constellations has not seemed to me to have any substantial basis, it is a satisfaction to feel that definite evidence is about to give it a quietus," he responded.

Van Maanen was aware that his work "might indicate that these bod-ies are not as distant as is usually supposed to be the case," but he kept that speculation out of his early reports. That's partly because in 1917 he measured a rotation for the Andromeda nebula with error bars larger than his result. "So that we do not know yet if this is an island universe!" he told Hale.

But that was the exception. Van Maanen primarily got the answer that many expected: Spiral nebulae exhibited internal motions and so must be relatively nearby. Moreover, the announcement was being made by a widely respected astronomer working at the world's premier observatory, whose expertise in stellar measurements was lauded. "His wide experience in astrometric work," Walter Adams later recalled, "gave his conclusions a high standing among astronomers." Other observers even confirmed that the spirals were changing; concurring reports came out of Mount Wilson, the Lowell Observatory, and observatories in both Russia and the Netherlands. It became the conventional wisdom among astronomers. And why not? It fit the general opinion of the time.

Only a few, such as Heber Curtis, openly disagreed. Curtis, with his wealth of spiral nebulae photographs at Lick, had earlier attempted to measure a change in the spirals over the years but could only conclude that "a much greater time interval will probably be necessary before nebular rotations can be definitely established." He knew that many of the older plates that van Maanen was perusing were very poor and useless for measuring anything. To Curtis comparing a photograph made at Lick in 1900 with a more modern picture taken with a completely different telescope at Mount Wilson was a fool's errand, which is why Curtis found it easier than Shapley and others to dismiss van Maanen's data right away. "The mean of five measures each of which is not worth a damn, has a maximum value of only five damns," he liked to say sarcastically.

But Curtis's warning was not heeded. With hindsight, it now seems easy to dismiss van Maanen's measurements. But at the time it was extremely difficult to assess. Van Maanen was no slouch at telescopic measurements and his finding a spiral rotating appeared quite reasonable. One of the era's leading theorists, the Britisher James Jeans, was especially eager to jump on van Maanen's bandwagon. Upon hearing the Dutch astronomer's results, Jeans speedily sent off a letter to the journal *Observatory*, saying they were "entirely in agreement with some speculations in which I have recently been indulging." In calculating the behavior of a blob of gas, rotating and condensing, he had determined that tidal forces would lead to the formation of spiraling arms. And now Van Maanen was providing the observational evidence to back him up. Jeans eventually wrote up his ideas in the book *Problems of Cosmogony and Stellar Dynamics*, which exerted a tremendous influence on astronomers at the time. Moreover, both van Maanen and Jeans began to calculate higher masses for the spirals. So, instead of a single solar system in the making, they began to think of a spiral nebula as the start of a dense (but still small) cluster of stars.

As more plates became available, van Maanen expanded his study to include other spirals. He measured a rotational period of 160,000 years for M33 (Triangulum), 45,000 years for M51 (Whirlpool), and 58,000 years for M81, a handsome spiral in the Ursa Major constellation. Other nebulae followed. All were rotating in such a way that the spirals appeared to be unwinding their arms, spreading them farther outward. He figured the spirals were no more than several hundred light-years wide and ranged in distance from one hundred to a few thousand light-years away.

Soon van Maanen was running out of spiral nebulae to measure, as few had been regularly photographed for comparison over the years. As a double check on his dexterity with the Blink, he measured a simple globular star cluster, M13, which was known not to rotate. If there were

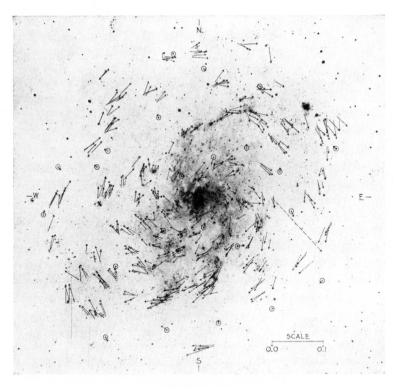

INTERNAL MOTIONS IN MESSIER 33

Adriaan van Maanen's markings on a photo of M33 indicating the rotation he measured (*From* Astrophysical Journal 57 [*1923*]: *264–78, Plate XIX, courtesy of the American Astronomical Society*)

any instrumental error, he should have mistakenly measured a motion, but he didn't, which seemed to imply his methods were valid. A British astronomer independently checked his methods as well and concluded that no one "would be so bold as to question the authenticity of the internal motions. . . . In fact, the more one studies [van Maanen's] measures, the greater is the admiration which they evoke."

"I finished . . . my measures of M51," van Maanen wrote Shapley in the spring of 1921. "The results look more convincing than M101. . . . Motion outwards along the spirals + some motion away from the center. . . . By this time Curtis and [Swedish astronomer Knut] Lundmark must be the only strong? defenders of the island-universe theory."

"Congratulations on the nebulous results!" responded Shapley. "Between us we have put a crimp in the island universes, it seems, — you by bringing the spirals in and I by pushing the Galaxy out. We are indeed clever, we are." Shapley reported on his friend's latest results at that summer's American Astronomical Society meeting in Connecticut. "I think that your nebular motions are taken seriously now," he told van Maanen afterward, "and nobody . . . dared raise his head after I explained how dead the island universes are if your measures are accepted."

The two were feeling quite cocky. At this stage, van Maanen at last made it publicly known, in the *Proceedings of the National Academy of Sciences*, that his observations "raise a strong objection to the 'island-universe' hypothesis." If M33, the prominent spiral in the Triangulum constellation, for example, were several million light-years distant, he pointed out, the motions he detected would represent velocities near the speed of light, "which, obviously, are extremely improbable . . . [and] afford a most important argument against the view that these nebulae are systems comparable with our galaxy."

But this declaration was hardly a resolution to the Great Debate. While Shapley and van Maanen were smugly celebrating, Knut Lundmark was visiting the Lick Observatory, using the Crossley reflector to gather the extremely faint light of M33. It was a difficult task, requiring extremely long exposures, one totaling thirty hours collected over four nights. Lundmark eventually saw that the light from the nebula's spiral arms resembled nothing less than the light of ordinary stars. Where other astronomers had seen a fuzzy patch in a spiral arm and called it a "nebulous star," Lundmark pondered whether each mistlike spot was instead "a great number of very distant stars . . . crowded together [to] give the impression of nebulous objects." That led to his cogent conclusion: that his observations of the spiral arms "speak for a large distance." The respected Swedish astronomer soon became one of the loudest voices championing the existence of other galaxies, and Shapley began to feel sizable pressure on his beloved model of the universe under Lundmark's onslaught.

Meanwhile, Slipher, in Arizona, had dispatched a story to the *New York Times* revealing that he had found a new "celestial speed champion," a faint spiral nebula that he judged had to be "enormously large" and "many millions of light years" away. And in the following year, 1922, Ernst Öpik, at the Dorpat Observatory, in Estonia, carried out an elegant calculation demonstrating how the Andromeda nebula must be some 1.5 million light-years distant. He did this by assuming that its mass and luminosity were comparable to those of the Milky Way. This "increases the probability," reported Öpik in the *Astrophysical Journal*, "that [Andromeda] is a stellar universe, comparable with our Galaxy." Thrust and parry. Thrust and parry. The duel over the island universes continued. Nothing would be settled until astronomers obtained a clear and unequivocal distance measurement to a spiral nebula—an observation so clear, so decisive, so comprehensive, that it immediately quelled all doubts.

Poor Shapley, it turns out, did put himself in jeopardy with his performance before the National Academy of Sciences gathering. Still in his thirties, Shapley was judged as too impetuous and immature to be the head of the Harvard College Observatory. Instead, his Princeton mentor, Russell, was offered the position. "Shapley couldn't swing the thing alone," Russell confided to Hale two months after the conference. "I am convinced of this after . . . observing Shapley at Washington. But he would make a bully second . . . if he grew intellectually he would be a prodigy!"

Russell gave the Harvard directorship intense consideration, with the understanding that Shapley would be his assistant. "At this point," continued Russell to Hale, "I would like to see your expression! I know I have my nerve with me: but,—and here I am very serious indeed,— consider what Shapley and I could do at Harvard! Between us, we cover the field of sidereal astrophysics pretty fully . . . and I might keep Shapley from too riotous an imagination,—in print."

But Russell, after nerve-racking deliberation and an attractive counteroffer from Princeton, ultimately declined the job ("I would rather do astronomy," he confided to Shapley). Harvard came back to Shapley, but not for the top position. Harvard officials brought up the title of "Chief Observer or something of the sort." He, a bit miffed, curtly turned it down. A month later, however, Shapley reversed his decision when Harvard (spurred by a suggestion from George Hale) agreed to try him out for a year as chief of staff, starting in the spring of 1921. He obviously passed muster, for he was soon named full director and served at the post for thirty more years, working at his unique desk that turned like a wheel—"a kind of rotating galaxy for ideas," noted a friend.

Shapley breathed new life into the sclerotic institution, bounding up the stairs two steps at a time and greeting everyone with a sporty cheerfulness. "He cast spells over people," said one staff member. Pickering had run the observatory like an absolute monarch. Under the youthful and energetic Shapley, it became a band of enthusiastic workers. Leo Goldberg, a student at Harvard in the 1930s, compared him to a benevolent Mafia Godfather. On the one hand, "he inspired us all," said Goldberg. "He pepped us up, he raised us out of the depths of discouragement many times." But a darker side lurked within Shapley as well. Adopting a "divide and rule" principle, he could be a father figure to some, while a tyrant to others. He also stubbornly ignored new scientific data at times, if it conflicted with his personal vision of how the universe should work.

Even as Shapley settled into Harvard, his former employer requested one more task from him. He was asked to contribute to Mount Wilson's annual report, to recount the final work he carried out there in 1920. "I thought I told you that I left Mount Wilson just to avoid this ordeal," he replied playfully. "Suppose I had lived wickedly and unrepenting died—would you even then haggle with His Majestic Nibs for your annual tithe of Blood-and-Brain?" Shapley was again being Shapley. It was his last hurrah for the California observatory. Mount Wilson got its notes.

Meanwhile, Curtis, who could have done much toward solving the mystery of the spiral nebulae, stepped out of the race entirely. Just a few months after the debate, he left the Lick Observatory to become director of the Allegheny Observatory, the same post that James Keeler once held. He had actually tendered his resignation ten days before the Washington debate took place. Being in charge of an observatory, a more highly paid position with increased prestige, was an opportunity

Harlow Shapley at his wheel-like desk at the Harvard Observatory
(*Harvard College Observatory, courtesy of AIP Emilio Segrè
Visual Archives*)

hard to pass up, especially for a family man. But, as with Keeler, the
urban setting, cloudy weather, and poorly equipped telescope at the
Pennsylvania observatory ultimately prevented Curtis from making any
further cutting-edge discoveries. Some considered it "the biggest mistake he ever made." Even Curtis later confessed to his former boss
Campbell that "the California combination of instruments PLUS climate is a hard one to beat. . . . There is no place like the hill [Mount
Hamilton] for astronomical work and . . . any man who leaves these
opportunities is bound to be sorry for it." A visiting colleague found him
at Allegheny one day puttering with an instrument and chided him for
turning into a toolmaker. "You play golf don't you? Well, this is my golf,"
he responded.

Despite their differences in cosmic outlooks, Curtis and Shapley
remained cordial over the years and kept in touch through correspondence. More than two years after the debate, Curtis looked back on the
event—what he called their "memorable set-to"—with good humor. "I

have always thought that the clubs we wielded at each other were all the more effective because politely padded," he told Shapley, "and regard with approbation the view-point of the old lady who warmed the water in which she drowned the kittens. . . . I fancy we both are as stiff-necked as ever; am sure that I am; cant [sic] see that my views have changed in the slightest." With his new responsibilities at Allegheny, though, Curtis had to remain on the sidelines, resigned to simply "watching the strife with interest," as he put it.

A few years later, a friend from Lick asked Curtis what he would have done with the Crossley if he had stayed on Mount Hamilton in 1920. Curtis replied that he would have just kept "photographing, photographing, and yet more photographing." He had in mind a "program of about 30 min. exposures of all the larger spirals at frequent intervals, to hunt for novae and variables." In a nutshell, he would have done everything that Edwin Hubble later carried out at Mount Wilson using its 60- and 100-inch telescopes, but with a few years' head start. Was the Crossley up to the task? Curtis had total faith in his beloved telescope: "I am copying that instrument in my design far more than any other," he said. "Could a 'race' be run between the 60" and the Crossley, would bet on the Crossley every time." Others, too, later judged the Crossley as having had a fighting chance at clinching the distance to Andromeda. But once Curtis left for Pennsylvania, no other Lick astronomer was interested in photographing the spiral nebulae. In effect, once Curtis left the Lick Observatory, it handed the baton over to Mount Wilson.

[11]

Adonis

From the mile-high summit of Mount Wilson, you can look a dozen miles to the southwest, across a wide valley, and catch sight of Hollywood and its lower-lying hills. The movie studios situated there in the 1920s were rapidly growing in allure and generating their mythic aura. This enchanted atmosphere must have somehow wafted over to the San Gabriel Mountains, for the man who eventually solved the mystery of the spiral nebulae looked as if he had come straight out of central casting.

In the eyes of his friends Edwin Hubble was an "Adonis," a tall and robust figure with compelling hazel eyes, a cleft chin, and wavy brown hair that glinted of reddish gold. Pronounced cheekbones cast attractive shadows in his photographs, lending his face a movie-star look. A woman screenwriter considered him too handsome for his occupation, comparing him to box-office idol Clark Gable. "Had we been casting [the role of a scientist] at M.G.M., Edwin Hubble would have been turned down as 'unrealistic,'" said Anita Loos, author of *Gentlemen Prefer Blondes.*

Raised in a solid middle-class household, Hubble somewhere along the line acquired a profound yearning to be singular and distinct. Fiercely determined to rise in the ranks, he reinvented himself upon reaching adulthood—adopting a British accent, dressing like a dandy, and adding dubious credentials to his curriculum vitae. The young man was seemingly intent on burying the most boring aspects of his midwestern family heritage and over time crafted a persona as big as the silver screen. By marrying into a wealthy southern California family, Hubble attained many of his lofty social and financial goals, and his wife, Grace, became his accomplice. She idolized her husband and, long after his death, propagated the legend he established, of which numerous details were highly edited or demonstrably wrong. She put him on a pedestal. And the longer time went on, said astronomer Nicholas Mayall, who had once worked with Hubble, the higher the pedestal got. Hubble's discovering the modern universe didn't seem to be glory enough.

Born on November 20, 1889, in Marshfield, Missouri, Hubble was the third of seven surviving children and christened Edwin Powell, although he generally avoided using his middle name or initial. His father, John, who grew up in Missouri, was trained in the law but earned a living working in his family's insurance business. When not traveling, he ruled his domestic realm with a firm puritanical hand, a strictness that was balanced by the more forgiving and accessible mother, Virginia Lee ("Jennie") James, daughter of a local physician.

It was in Missouri, the "Show Me" state, that Hubble began his love affair with the heavens. His maternal grandfather, William James (a distant relation to the famous outlaw Jesse James), had built a telescope, and as a present on his eighth birthday young Edwin was permitted to stay up past his bedtime and use it to peruse the pinpoints of light, sparkling like brilliant gems, in the nighttime sky. The impression made on him that pitch-black winter evening, the starry wonders he beheld, lasted a lifetime. Two years later his family moved to the Chicago area, eventually settling in the village of Wheaton, Illinois, just west of the city. In high school Ed, as he was known to his friends, blossomed, regularly maintaining an A average and excelling in track, football, and basketball. The two areas in which he was downgraded a few times came in "application" and "deportment," as he wasn't afraid to argue with his teachers in class. With his peers, he remained aloof and at times arrogant—both a dreamer and schemer. "He always seemed to be looking for an audience to which he could expound some theory or other," recalled a childhood friend. Two years younger than most of his classmates, he may have been putting on a knowing front to appear older and more self-assured.

Graduating in 1906 at the age of sixteen, Hubble was awarded a scholarship to the University of Chicago, partly due to his superb athletic skills. But what he would major in became a contentious issue. Never forgetting his childhood experience with his grandfather's telescope, Hubble earnestly desired to study astronomy, but his father, a practical man, wanted his son to take up the law. According to one of Hubble's sisters, John Hubble considered being an astronomer an "outlandish" career choice. Hubble compromised by taking science classes—mathematics, astronomy, physics, chemistry, geology—as well as the prerequisite courses in the classics, including heavy doses of Greek and Latin, that would prepare him for a legal career.

In regard to learning science, the timing for Hubble was perfect. Though a relatively new institution, the University of Chicago had

Edwin Hubble (left) in 1909 with a teammate on the
University of Chicago track team (*Reproduced by permission
of the Huntington Library, San Marino, California*)

already attracted two top physicists, Albert Michelson and Robert Millikan, who would go on to receive Nobel Prizes for their seminal work. And the Yerkes Observatory, affiliated with the university, offered one of the best telescopes then in existence. The early 1900s was a time, Hubble later recalled, when the world was astir: "Motor cars, at last, were successfully competing with horses. Airplanes were trying their wings. Bleriot had just flown the English channel, and . . . the wireless was groping its way over the map. Marconi . . . transmitted a message from Ireland to Buenos Aires, 6000 miles away. . . . Technology strides across the modern stage like some gigantic, streamlined god."

Hubble inhaled the charged air of this exhilarating era deeply. A classmate described him as being a "whiz" at calculus, who "often utterly dumfounded" the professor. By the end of his sophomore year he was singled out as the best physics student. He also participated in track (though seldom winning) but did better in basketball, as his exceptional height for the day (six feet two inches) gave him an advantage playing center. He and his teammates were national champions in 1909. Moreover, Hubble did some boxing at an off-campus gym, becoming so good as an amateur heavyweight that Chicago promoters were eager for him to turn professional (or so he claimed). Such diverse activities and coursework may have been all part of a plan, for early on he had set his sights on obtaining a Rhodes Scholarship. Cecil Rhodes, the British imperialist who made his fortune mining South African diamonds, had set up the program to strengthen the relationship between Great Britain and the United States. Every year, in each state, a young man was chosen to attend Oxford University in England for postgraduate studies. In his will Rhodes stipulated that Rhodes scholars should be bachelors between nineteen and twenty-five, good in academics but not "mere bookworms." Each was to be a manly chap, exhibiting a "moral force of character" and proficient in both athletics and leadership. Hubble made sure that his accomplishments in college covered all the bases. In his senior year, he even served as vice president of his class, a position he acquired with ease as he shrewdly knew he would be running unopposed.

After passing the initial Rhodes examination, Hubble became one of the two finalists in Illinois. He may have won the slot once the judging committee saw the glowing letter of recommendation written by Millikan. Hubble had served as a laboratory assistant in Millikan's elementary physics course at the University of Chicago. To Millikan, Hubble was a "man of magnificent physique, admirable scholarship, and worthy and lovable character. . . . I have seldom known a man who seemed to be better qualified to meet the conditions imposed by the founder of the Rhodes scholarship than is Mr. Hubble."

Hubble arrived at Oxford in October 1910, living for the next three years on an annual stipend of fifteen hundred dollars. There he walked the very halls where Edmond Halley once strode and joined a cozy club of privileged young men from England's wealthiest families, who were training for select positions in the military, banking, industry, government, and diplomatic services. With continued pressure from both his father and grandfather, Hubble dutifully studied the law and completed the jurisprudence coursework in two years instead of the usual three. He received second-class honors. But, always in the background, astronomy beckoned. He couldn't let it go, so deep was his passion for the celestial specialty. Sensing it would create a ruckus, he didn't let his parents know that he was cozying up to Oxford's top astronomer, Herbert Turner, visiting his home several times.

A Rhodes official jotted in Hubble's record that he showed "considerable ability. Manly. Did quite well here. I didn't care v[ery] much for his manner—but he was better than his manner. Will get A." His "manner" had become decidedly British, but in an exaggerated, almost cartoonish way. It was during his Oxford sojourn when Hubble underwent his bewildering metamorphosis, adopting a distinct style that he maintained for the rest of his life. Becoming a full-fledged (some might say rabid) Anglophile, Hubble began to regularly speak with an upper-crust accent, smoke a pipe, brew a proper cup of tea, and wear a black cape with great flourish. Some were not impressed, including Rhodes scholar Warren Ault, who believed that Oxford "had transformed [Hubble], seemingly, into a phony Englishman, as phony as his accent."

This theatrical transformation clearly signaled that Hubble was desperately in search of an identity, as well as a profession in which he could make a lasting mark. In his third year at Oxford he chose to specialize in Spanish, a respite from his grueling law curriculum. "I sometimes feel that there is within me, to do what the average man would not do," Hubble had earlier written his mother, "if only I find some principle, for whose sake I could leave everything else and devote my life." His ambition was unmistakable. When a classmate declared he'd rather be first in the provinces than second in Rome, Hubble snappily replied, "Why not be first in Rome?"

In January 1913, Hubble's father died, after years of fighting nephritis, a disease of the kidney. When first hearing of his father's declining health, Hubble had wanted to return home, but his father ordered him to stay. Though a devastating event to any child, his father's passing was in many ways a liberation for the Rhodes scholar. He was no longer shackled by the career path preordained by his stern father, although a full emancipation took some time. Upon finishing his studies in Great Britain at the end of May, Hubble first returned to Louisville, Kentucky, where his widowed mother and siblings were now settled, to help out his family and figure out what he would do next.

When writing about this point of his life, right after his return from Oxford, Hubble's early biographers uniformly noted that he soon passed the Kentucky bar examination and briefly practiced law in Louisville. That was the story that Hubble told everyone, and it became the standard line for all of his life and several decades afterward. But in truth he did neither. According to a later biographer, Gale Christianson, the closest Hubble came to a legal career was translating what may have been legal correspondence for a Louisville import company conducting business in South America. There is no evidence whatsoever that Hubble handled a professional legal case, despite what he wrote his chums back in England. Hubble was adding more mythic gloss to his résumé. All this time he was actually teaching at the high school in New Albany, Indiana, across the Ohio River from Louisville. For a year he taught physics, mathematics, and Spanish—an experience he never openly discussed later, even though his students were obviously fond of him. After Hubble coached their basketball team to an undefeated season and a third-place finish in the state tournament, they lovingly dedicated the school's 1914 yearbook to him.

High school teaching, though, hardly satisfied Hubble's steadfast hunger for a more illustrious career. He would come to see the fellow Rhodes scholars in his group become respected journalists, authors, poets, and congressmen. He yearned to match their potential in the field of science. "So I chucked the law," Hubble later reminisced, maintaining the fiction of having had a legal career, "and went back to astronomy and the test was this—I knew it was astronomy that mattered and that I would be happy in astronomy if I turned out to be second-rate or third-rate."

Earliest known photo of Edwin Hubble with a telescope. Taken in 1914
in New Albany, Indiana, upon his return from Oxford. (*Reproduced by
permission of the Huntington Library, San Marino, California*)

With his father's daunting presence no longer an obstacle to his
long-standing aspiration, Hubble contacted his favorite astronomy professor at Chicago, Forest Ray Moulton, to inquire about returning for
graduate studies. Moulton wrote a glowing letter of recommendation to
Edwin Frost, then director of the Yerkes Observatory. Hubble, said
Moulton, was a "splendid specimen," who showed "exceptional ability"
in science. Frost promptly took him on, offering a scholarship that covered his $120 tuition and provided $30 a month for basic living
expenses.

Frost, who hailed from New England, had joined the staff of Yerkes
just months after it first opened in 1897, chosen by Hale to be its first
professor of astrophysics. He was best known for his measurements of

the radial velocities of the stars (how fast they were moving either toward or away from the Earth) and also served as managing editor of the *Astrophysical Journal*. One day he received a telegraph from a reporter with a plea: "Send us three hundred words expressing your ideas on the habitability of Mars." A man of good humor, Frost replied, "Three hundred words unnecessary—three enough—no one knows."

As director, Frost divided the nights at Yerkes, with astronomers working only the first or second half, so they could get some sleep. Certain hours were given over to spectroscopic work, other hours for determining the distances to the stars. In the remaining hours, the observers would carry out such tasks as photometric studies—determining the brightness of stars—or visually observing interesting objects, such as double stars.

In the winter temperatures at Yerkes could reach 15° to 20° F below zero, yet the dome couldn't be warmed as the temperature had to closely match that of the outside. Otherwise, currents of warm air rising in front of the lens would spoil the resolution of the celestial objects in the telescope's sight. "Those who have visited a large observatory on such a night," Frost recalled, "say that they will never forget that cold eerie place, silent except for the persistent ghostly ticking of the driving clock and the wind howling around the slit in the dome. But there the astronomer sits in his Eskimo suit or fur coat and cap with his eye glued to the eyepiece of the telescope, watching closely to see that his star does not drift away from the crossed spider-threads which mark the center of his field while a plate is being exposed."

Occasionally visitors were invited to look through the telescope. A favorite target was a dazzling cluster of stars in the constellation Hercules. One man, upon viewing the great cluster, remarked to Frost right before the 1908 presidential election, "So you say that each of those points of light is a sun and each one is larger than ours. And you allege that this cluster is so far away that the light requires thirty thousand to forty thousand years to reach us? Well"—with a sigh—"if this is so, I guess that it doesn't really matter whether Bryan or Taft is elected."

Graduate work in astronomy at the University of Chicago was primarily carried out right on the premises, at the observatory itself. When Yerkes first opened, it was one of the foremost observatories of its time. But when Hale, its founder, moved to California to build his even greater astronomical establishment atop Mount Wilson, taking with him the cream of Yerkes' observers, only the older astronomers whose creative years were long over or the second tier stayed behind. Frost himself was slowly losing his eyesight due to cataracts and could no longer observe, the ultimate tragedy for an astronomer. With a few exceptions, most students who completed their PhD at Yerkes around this time made no major contributions to astronomy. Hubble, though, did not let Yerkes' declining fortunes deter him.

Right before starting at Yerkes in the fall of 1914, Hubble attended a meeting of the American Astronomical Society held on the campus of Northwestern University in Evanston, Illinois. It was the eventful meeting when Vesto Slipher presented his awe-inspiring results on the speedy velocities of the spiral nebulae, which created such a commotion. Mingling with the great astronomers of the day, Hubble must surely have sensed the importance of Slipher's announcement. Hubble was after astronomical fame (though just inducted into the AAS, he managed to get in the front row for the meeting's group photo) and here was a compelling mystery garnering everyone's attention. It could be that his decision to focus on the nebulae was made that very week, as he joined in the ovation, standing up with the audience, clapping his hands in honor of Slipher's achievement.

As a graduate student, the lowest rung in the observatory hierarchy, Hubble didn't have regular access to Yerkes' grand 40-inch telescope. But, driven and self-reliant, he took advantage of the equipment that was available to him and took over the observatory's 24-inch reflector, then standing idle, a curious situation since it was the same two-footer that George Ritchey had built years earlier to compete with Lick's productive Crossley reflector. Hubble attached a camera to the telescope and proceeded to take pictures of various nebulae. Soon these images became the topic of his doctoral thesis, "Photographic Investigations of Faint Nebulae." His first discovery was finding that certain faint nebulae could change. He compared his photographic plate of a nebula called NGC 2261, a comet-shaped cloud of gas, with ones taken earlier at other observatories. His latest photo displayed distinct differences, indicating that the nebula had to be relatively small and close by. (This object, located within the Milky Way, is now known as Hubble's variable nebula.)

In many ways, this endeavor became a trial run for his later work on galaxies. Although Hubble's telescope was small by modern standards, he was able to discern that the faint white nebulae were not all spiral disks (as many then believed); some were also bulbous, what later came to be known as elliptical galaxies. He could also see that many of these nebulae crowded together on the sky. Much as astronomers did in earlier centuries with stars, Hubble was hoping to learn something about nebulae from their distribution over the sky. "Suppose them to be extrasidereal [outside the Milky Way] and perhaps we see clusters of galaxies," wrote Hubble about his findings. "Suppose them within our system, their nature becomes a mystery." He even estimated that if they were separate galaxies, each the size of the Milky Way, they would have to be millions of light-years distant to appear so small.

While seemingly prescient in his speculations, his findings at this point were not terribly revolutionary. Others, like Curtis and Slipher, had already made similar statements. Today astronomers judge Hubble's thesis as not very good technically, as it contains few references to earlier work and offers confusing theoretical ideas. "But it shows clearly the hand of a great scientist groping toward the solution of great problems," Donald Osterbrock, Ronald Brashear, and Joel Gwinn emphasized in an evaluation of Hubble's work. "Hubble was never an outstanding technical observationalist . . . but he always had the drive, energy, and enough skill to use available instruments so as to get the most out of them. . . . He recognized the right questions to ask, and he had the self-confidence to see what was on his plates, and describe it, where others who had perhaps seen it before had ignored it, or worse, tried to ignore it, because it did not fit the current pictures of the universe that they had in their minds."

Hubble was certainly astute enough to realize that his initial research on the nebulae merely scratched the surface. In his thesis he made sure to note that his "questions await their answers for instruments more powerful than those we now possess." He was already thinking ahead, keenly aware that another place, the Mount Wilson Observatory, in southern California, was swiftly becoming the world's premier astronomical institution, with a record-breaking 100-inch telescope under construction. In turn, Mount Wilson's director, Hale, was similarly aware of Hubble. He had been hearing reports of an exceptional young man at Yerkes who was looking into faint nebulae and, after consulting with University of Chicago professors, he offered Hubble a job, contingent on the successful completion of his doctoral degree.

"I have offered Hubbell [*sic*] a position with us at $1200. per year," wrote Hale to Adams, his second in command. "He will talk the question over with Frost in the near future." Frost had no problem with Hubble leaving Yerkes. In fact, the Yerkes director was probably relieved, for he didn't have the money to offer his graduating student a well-paid position, as he had hoped, and was glad to hear that Hubble had another prospect in hand, and an excellent one at that.

In the course of working toward his degree, Hubble had spent hundreds of hours at his scope, photographing a vast array of nebulae and classifying them. Yet, as published, Hubble's thesis consisted of just nine pages of text, eight pages of tables, and two photographic plates. That it turned out a bit thin was largely due to the unusual circumstances of its final preparation. Hubble had planned to finish up in June 1917, but on April 6 of that year Congress approved President Woodrow Wilson's plea for the United States to enter World War I. Within days Hubble asked Frost for a letter of recommendation to obtain a commission in the army. Upon hearing that officers' training camp was starting in mid-May, Hubble hurriedly submitted the latest draft of his thesis, which he knew was decidedly "scimpy." On the advice of Frost, he plumped it up a bit by attaching his paper on NGC 2261, his variable nebula. Even then, Frost did not find it suitable for publication in the prestigious *Astrophysical Journal* and eventually sent it to the lesser *Publications of the Yerkes Observatory*. War fever obviously allowed Hubble's shakily composed thesis to pass without major rewrites. The young recruit handled himself superbly during his final oral examination, though, and the six-man committee awarded him his doctoral degree magna cum laude. Three days later, on May 15, he reported for duty at Fort Sheridan, a military reservation on Lake Michigan, north of Chicago.

As for his promised position at Mount Wilson, Hubble had already sent a letter to Hale a month earlier telling him of his desire to enter the reserve officers corps and asking how it would affect his job offer. Hale replied it was "natural" to apply for a commission and said he hoped "to renew as soon as you are able to accept it." Hale even supplied one of Hubble's needed recommendations to enter officer training.

For a year, Hubble's army division stayed in the United States, largely relegated to teaching new recruits. At one point his astronomy background came in handy. His commanding officer requested that he instruct his fellow trainees how to use the stars to guide their nighttime marches. While others joined the artillery and were commissioned as lieutenants, Hubble chose the infantry, where he could enter at a higher rank, as a captain. By September he was put in charge of the 2nd

Battalion, 343rd Infantry Regiment of the 86th Division at an Illinois base. "Stirring times," Hubble wrote a friend from his new camp. "I can't picture myself missing the gathering, as it were, of the clans."

Commended for his contributions, Hubble was promoted to major, just eight months after he joined. He finally made it to Europe in September 1918, his men reassigned to various divisions to serve as replacements. Hubble was sent to a combat training camp in France, but what exactly happened afterward is debatable (his full military record was destroyed in a fire). Hubble always claimed he saw some action in the trenches and later told his wife that he had been rendered unconscious at one point by a shell exploding nearby and awoke in a field hospital, whereupon he quickly dressed and departed. Nowhere in his discharge papers, though, is there a record of his participation in any battles, engagements, or skirmishes. Beside each listed category, only the word *none* appears. Furthermore, no "wound chevrons" were authorized for him to put on his uniform. Perhaps his exploits were never precisely documented in the fog of war or possibly Hubble was fastening more adornments to his reinvented self, tall tales that Grace proceeded to faithfully record, with unquestioning belief, in a memoir after his death. What seems most honest and unadorned is what Hubble wrote to Frost right after the war ended: "I barely got under fire."

With his skill in languages and his expertise in the law, Hubble purportedly took on postwar assignments at the U.S. Army of Occupation headquarters in Germany, the Combat Officers Depot in France, and the American Peace Commission in Paris. Along the way, he learned of a U.S. Army program for officers to study in British universities while awaiting shipment back home. He quickly arranged to be assigned and arrived at Cambridge University in March 1919, along with two hundred other American officers and enlisted men. Like James Keeler, Hubble was a skilled networker and made sure to hobnob with the noted astronomers who were there in Cambridge. Soon he was being proposed for membership in the Royal Astronomical Society. Upon arriving at a posh dinner hosted by the best and the brightest of British astronomy at this time, visitors from Mount Wilson were surprised to see their prospective staff member, junior at that, seated in a place of honor between a noted British physicist and Great Britain's astronomer royal.

By May 1919, worried that his promised job at Mount Wilson might have evaporated given the added delay of his postwar activities, Hubble dashed off a brief note to Hale for reassurance. He reminded Hale, "My interest has for the most part been with nebulae especially photographic study of the fainter ones." Hale soon replied. "I had been hoping to hear from you," he wrote, "and am pleased to find that you still wish to come to the Observatory." Hubble's salary offer rose to $1,500, and Hale promised him rapid advancement, should his work prove worthy. But he urged Hubble to come as soon as possible, "as we expect to get the 100-inch telescope into commission very soon, and there should be abundant opportunity for work by the time you arrive."

Hubble arrived in New York on August 10. After a one-day stop in Chicago to meet with his mother and sister, who had specially traveled down from their new home in Wisconsin for the brief reunion, he quickly journeyed to California, resplendently attired in uniform and introducing himself around as Major Hubble, a moniker that many people continued to use from that time on. But before showing up at Mount Wilson, just after being discharged in San Francisco, Hubble sent Hale a telegram: "Just demobilized. Will proceed Pasadena at once unless you advise to contrary." He was either being obsequious or still incredulous that Hale had held the position open for him so long.

The job was assuredly his, and Hubble couldn't have turned up at the Mount Wilson Observatory at a more perfect time. On September 11, 1919, just about a week after his arrival in Pasadena, the great 100-inch telescope came into full use for the staff. It was a moment that observatory director Hale had been anticipating since 1906.

[12]

On the Brink of a Big Discovery– or Maybe a Big Paradox

George Ellery Hale could never rest on his laurels. He was a man of endless enthusiasms. British theorist James Jeans said he possessed "a driving power which was given no rest until it had brought his plans and schemes to fruition." After being awarded nearly every major scientific honor before the age of forty—from election to the National Academy of Sciences in the United States to the gold medal of Great Britain's Royal Astronomical Society—Hale craved additional triumphs. "He has reached a place where scientific work and honors are not enough," George Ritchey suggested darkly after a conflict with Hale. "He must have vast *power* also; power to dictate the welfare, the making or unmaking, the *positions* even, of scientific men both in the observatory and outside of it—as far as his influences can possibly reach."

Even before Mount Wilson's 60-inch telescope went into operation in 1908, Hale was thinking ahead to a new adventure. In the summer of 1906 he spent a weekend at the home of John Hooker, a wealthy Los Angeles businessman and a founder of the Southern California Academy of Sciences, and excitedly discussed his latest dream. Again, it was to be an even bigger telescope. Like a compulsive climber, Hale was always looking ahead to the next challenging mountain. He captivated Hooker, an amateur astronomer, with his description of a mirror one hundred inches in width that would gather nearly three times more light—the very lifeblood of astronomy—than the 60-inch. Hale and Ritchey followed up with a letter to Hooker, outlining the usefulness of such a large mirror, including the tens of thousands of nebulae that would likely be revealed, unlocking the secret of their mysterious nature.

Hale's charismatic personality, coupled with Ritchey's technical expertise, worked their magic. Within weeks Hooker, who had made his fortune in hardware, pledged the money to construct the mirror, even though no one (neither Hale nor Ritchey) knew at the time whether such a disk—four and a half tons of pristine glass—could even be cast, polished, or mounted. No glass that large had ever been made before. Hale's younger brother, Will, once called George the greatest gambler in the world. Ordering up a 100-inch mirror was his biggest bet ever. And he almost lost.

In December 1908, the giant glass arrived from France, where it had been manufactured, but as soon as the crate was unpacked at the observatory's headquarters on Santa Barbara Street in Pasadena, everyone could see the blank was seriously flawed—bubbles were dispersed throughout the disk and the glass incompletely fused. From the side it looked like a three-layered cake. Such defects jeopardized the mirror's ability to expand and contract uniformly and so maintain stable images as temperatures in the telescope dome changed over the nighttime hours. "We don't pay for this!" declared Hale.

A new disk was ordered, but the best candidate broke as it was cooling. With his funds exhausted, Hale decided to have the first disk ground and polished, despite its imperfections. Both Hooker and Ritchey opposed this decision with intense vehemence. To compound Hale's trials, Hooker grew increasingly jealous and antagonistic over Hale's friendship with Mrs. Hooker. Previously an ally, Hooker was now Hale's demoralizing opponent who balked at any new request. Faced with these multiple struggles, Hale snapped. Having inherited the high-strung and anxious temperament of his reclusive mother, he experienced the first of many nervous breakdowns that plagued him for the rest of his life, attacks that included horrendous nightmares and blinding headaches. His exuberance, once deemed inexhaustible, finally flamed out. In a poignant letter to Walter Adams, Hale's wife wrote that she now wished "that glass was in the bottom of the ocean."

Hale's recurrent psychiatric episodes gave rise to the popular myth that he sometimes hallucinated during these breakdowns, literally seeing a little "elf" who would advise him on the conduct of his life. Helen Wright first recounted this tale in her noted biography of Hale, referring to the specter as Hale's "little man." Expanding on Wright's account, other authors began to use the word *elf*. The legend is rooted in a letter that Hale wrote to a friend, in which he refers to a "little demon" plaguing him. Psychiatrist William Sheehan and astronomer Donald Osterbrock have made a good case that Hale only intended the demon to be taken figuratively, not literally, as the personification of his depressions, much the way Winston Churchill referred to his "black dog" when facing a bout of melancholy.

In the end, Ritchey carried out Hale's orders concerning the imperfect glass. Gritting his teeth and complaining all the way, he initiated the grinding and polishing of the flawed disk in 1910, an arduous task that was finally completed in 1916. Over those six years, the disk was figured to exquisite perfection. The curved glass surface was subsequently coated with silver, transforming it, at last, into a true astronomical mirror. All the while, the materials for the mounting and dome—every bolt, rivet, and steel beam—were laboriously transported up the mountain by truck. The nine-thousand-pound mirror went up on July 1, 1917. To Walter Adams "there was more publicity . . . than was desir-

Full view of the 100-inch Hooker telescope on Mount Wilson
(*AIP Emilio Segrè Visual Archives*)

able" during the event. The Pasadena police had received word that there might be trouble on the road. As a result, the bridges were guarded, and deputies accompanied the mirror to the top.

The Hooker telescope got its first trial exactly four months later in the midst of wartime, which resulted in the 100-inch's assuming the nickname of a famous German howitzer—the "Big Bertha" of light. Among those present on that first evening of November were Hale, Adams, and the British poet Alfred Noyes, then visiting Pasadena as a university lecturer. Hale, as director, was the first to climb up the black iron steps to the observing platform and look through the eyepiece at the chosen target, Jupiter, then brilliantly shining in the nighttime sky. To his horror, he saw six overlapping images of the planet rather than one. The mirror was somehow distorted. Was it a physical defect—the numerous bubbles indeed wrecking havoc, as Ritchey had warned—or merely a temporary warping, caused by workmen having left the dome open that day and heating the mirror? "To add to the gloom," recalled Adams many years later, "news of the great disaster to the Italian army at Caporetto had just arrived, and I remember our sitting around on the floor of the dome speculating on whether Italy was completely out of the war."

After waiting many excruciating hours for the mirror to cool in the nighttime air, trying but finding it impossible to sleep at one point back at the Monastery in their spare rooms, furnished only with bed and desk, first Hale then Adams returned to their cathedral of brass and steel at around 2:30 in the morning. Jupiter was now out of reach, so the night assistant swung the telescope around, its massive weight smoothly rotating with little friction because its bottom supports floated in tanks of mercury. The scope's new target was the bright blue star Vega. Hale once again peered into the eyepiece, and this time let out a joyful yell. The stellar image was exquisite. To everyone's relief, the mirror was not permanently damaged after all. Noyes later paid homage to this historic launch in his poem "Watchers of the Sky," fairly bursting with metaphors inspired by the ongoing conflict in Europe:

> High in heaven it shone,
> Alive with all the thoughts, and hopes, and
>   dreams
> Of man's adventurous mind.
>   Up there, I knew

The explorers of the sky, the pioneers
Of science, now made ready to attack
That darkness once again, and win new
    worlds.
            . . . they hoped to crown the toil
Of twenty years, and turn upon the sky
The noblest weapon ever made by man.
War had delayed them. They had been
    drawn away
Designing darker weapons. But no gun
Could outrange this. . . .
We creep to power by inches. Europe
    trusts
Her "giant forty" still. Even to-night
Our own old sixty has its work to do;
And now our hundred-inch . . . I hardly
    dare
To think what this new muzzle of ours
    may find. . . .

But there were delays in the final preparation of the long and imposing telescopic "muzzle," keeping it from full operation. "The truth is
the war work here has completely stopped work on the 100-inch," said
Shapley to a colleague a year later. "Very little has been done with
it . . . because of the war contracts in the shop." Ritchey, for example,
had to turn his attention to making lenses and prisms for such military
items as binoculars, range finders, and periscopes. Once the United
States officially entered the war on the side of the Allies, the Mount
Wilson optical shop was quickly engaged in the effort.

Observations with the 100-inch did not really get going until the war
was over and necessary personnel had finally returned from their military duties. Its first images—of the Moon, of nebulae—surpassed the
promises that Hale had made to Hooker years earlier when a telescope
of such a tremendous size was only a far-off aspiration. "In such an
embarrassment of riches the chief difficulty is to withstand the tempta-
tion toward scattering of effort, and to form an observing programme
directed toward the solution of crucial problems rather than the accumulation of vast stores of miscellaneous data," said Hale.

High on Hale's list of priorities was determining once and for all the true size and nature of the universe, a job that Hubble took on with single-minded devotion.

Hubble's first night of observing on the mountain was on October 18, 1919. It took about an hour then to make the journey in a motorcar. Via the single stretch of telephone wire that ran to the top of Mount Wilson, the tollhouse keeper at the bottom alerted the observatory that a car was on its way, as the road was only wide enough for one car. Away from active observing for more than two years due to the war, Hubble initially freshened up his telescopic skills that autumn evening by using a 10-inch refractor called the Cooke lens. Though small, the telescope's wide-angle view enabled him to explore the sky quite handily. He took photos of the North America nebula (a diffuse cloud in the Cygnus constellation) and then directed the telescope to a nebulous loop of gas near the "belt" of Orion. He was getting back his sea legs, perusing familiar celestial territory, and mulling over his observing strategy for the coming months.

Seven days later Hubble tried out the 60-inch telescope. He took a photograph of the nebula NGC 1333, a rich star-forming region in Perseus, and later checked out how his beloved variable nebula, the one he first noticed as a graduate student at Yerkes, was doing. He noted that "striking changes have happened [in it] since 1916," which was the last time he had taken a look.

Milton Humason, who became Hubble's devoted observing partner a decade later, first met the young astronomer during these opening runs at the observatory. "He was photographing at the Newtonian focus of the 60-inch, standing while he did his guiding," recalled Humason, many years later. "His tall, vigorous figure, pipe in mouth, was clearly outlined against the sky. A brisk wind whipped his military trench coat around his body and occasionally blew sparks from his pipe into the darkness of the dome. 'Seeing' that night was rated extremely poor on our Mount Wilson scale, but when Hubble came back from developing his plate in the dark room he was jubilant. 'If this is a sample of poor seeing conditions,' he said, 'I shall always be able to get usable photographs with the Mount Wilson instruments.' . . . He was sure of himself—of what he wanted to do, and of how to do it."

Hubble got his first crack at the 100-inch telescope, what he called his "magic mirror," on Christmas Eve. So immense was its light-gathering power that it could spot a candle from five thousand miles away. Hubble couldn't have asked for a more fitting holiday present; the atmosphere was almost at its best at the start of the evening, and it was also dark-sky time, a waxing crescent Moon having just set in the west. That was the prime opportunity to seek out the sky's faintest objects. He first photographed a hazy star near the Pleiades cluster. With a sixty-minute exposure, its nebulosity showed up fairly well. Afterward, he perused two more objects, a wispy planetary nebula and (again) his variable nebula NGC 2261. After Hubble aimed the giant scope at this target, he was able to obtain his best photo of the night. The variable nebula soon became his observational "mascot."

At the end of his observing runs, if he was particularly eager to see his results, Hubble would go right to the dark room and develop his plates. Once dry, each was entered into his official Observing Book and put away in a numbered envelope. For marking his plates, Hubble used a special code: H 31 H, for example, stood for the hundred-inch telescope, plate number 31, taken by Hubble.

One of Hubble's first tasks on Mount Wilson was working with Frederick Seares to determine the color of "nebulous stars," stars surrounded by diffuse clouds of luminous matter, such as the ones in the Pleiades. For that project he primarily worked with the 60-inch and got a paper published fairly quickly in the *Astrophysical Journal*. It was a warm-up session for Hubble's main purpose for being at Mount Wilson. He was going to finish what he started in his doctoral dissertation—figure out exactly what those faint spiral nebulae truly were. As he later told Slipher, he was committed to one issue and one issue only: "to determine the relation of nebulae to the universe."

Henry Norris Russell was getting nervous about the spirals around this time. There were so many conflicting observations. The novae occasionally discovered within the spiraling clouds suggested they were far-off stellar systems. But then van Maanen was seeing them rotate, at an impossible rate if they were truly distant. "We are on the brink of a big discovery—or maybe a big paradox, until someone gets the right clue," ventured Russell.

The dawn of the 1920s seemed the right time to break the impasse. With the war over, pent-up energies were fueling a plethora of inventions and clever ideas. Heber Curtis, now settled at the Allegheny Observatory, was particularly enamored of a newfangled entertainment medium. "I have just gone into the lecture room, pressed a button, and heard records by Galli-Curci and Rachmaninoff sent out by wireless telephony from East Pittsburgh, ten or twelve miles away," he wrote his former Lick boss, Campbell. "As soon as the Westinghouse people start a broadcasting station at San Francisco, the mountain would enjoy one of these receiving and amplifying sets. They send out music, stock market reports, news bulletins, speeches, etc. . . . We have one of their experimental models here on loan (hope they will eventually give it to us). It is called 'the Aeriola Grand'; is the size of a small phonograph such as you have; is simplicity itself; has only one button and one dial,— no adjustments; about 75 feet of a single wire forms our aerial. Sermons on Sunday, with no collection possible!"

At Mount Wilson Albert A. Michelson and Francis Pease mounted a special instrument called an interferometer on the front of the 100-inch telescope and made the first successful measurement of a star's diameter. Betelgeuse, in the constellation Orion, was their target. They learned that if the red giant star on Orion's right shoulder were placed inside our solar system, it would engulf the planets out to Jupiter. And, of course, Harlow Shapley at this time was also on the mountain resizing the Milky Way.

Hubble's and Shapley's employment at Mount Wilson overlapped for about a year and a half, until Shapley moved to Harvard. Their relationship over that brief period, though, could hardly be called collegial. Both were from the heartland of America, but they might as well have been born continents apart. Hubble cultivated an air of sophistication and restraint around his colleagues. The cold and standoffish persona of his youth never went away. Hubble kept his distance and maintained a regal air. With his ever-present pipe, he would occasionally blow smoke rings out into the room or flip his lighted match and catch it, still alight, as it came down. As other astronomers put it, he was a "stuffed shirt," who couldn't "write an inter-office memo without it sounding like the Preamble to the Constitution." Shapley, on the other hand, retained his brassy and chummy country ways. Hubble's affectation for wearing jodhpurs, leather puttees, and a beret while observing or going around and saying "Bah Jove" was simply too much for Shapley to bear. An unadorned "Missourian tongue" was good enough for him. The fact that Shapley was a close friend of Adriaan van Maanen's made it even more difficult for the two midwesterners to cozy up. "Hubble disliked van Maanen from the time he himself arrived on Mount Wilson; he scorned him," claimed Shapley years later. It may have been because

Edwin Hubble wearing his knickers on Mount Wilson
(*Courtesy of the Archives, California Institute of Technology*)

van Maanen, more senior than Hubble, openly displayed his jealousy at having to share time on the 100-inch. To Shapley, though, "Hubble just didn't like people. He didn't associate with them, didn't care to work with them."

Part of the coolness and tension between Hubble and Shapley had to do with their differing experiences during the war. Hubble had immediately volunteered, putting his professional life on hold and taking the risk that his research would be taken up by others. Shapley, who hated war, remained at Mount Wilson—the "conscientious slacker"—

weakly suggesting that Hale convinced him to stay and taking on work that Hubble had hoped to tackle, such as the globular clusters. But, fortunately for Hubble, analyzing the mysterious nebulae was still a wide-open field when he returned from overseas. And once Shapley left for Harvard, Hubble at last had the chance to step out of the formidable shadow Shapley, then the golden boy of astronomy, had been casting on Mount Wilson.

Hubble first carried out an extensive study of the diffuse nebulae within the Milky Way, identifying the various types and describing the sources of their luminosity. But he also kept track of the "non-galactic nebulae" that he came across as he carried out this research. Hubble's sympathies certainly leaned toward the island-universe theory. When he was a graduate student at Yerkes he especially noted that the high velocities of the spiral nebulae "lend some color to the hypothesis that the spirals are stellar systems at distances to be measured often in millions of light-years." But he became more circumspect once he became a staff member at Mount Wilson, at least in print. Caution became his byword. He emphasized in a 1922 *Astrophysical Journal* paper that the term *non-galactic* didn't mean the spirals were necessarily "outside our galaxy" but that these nebulae tended to avoid the galactic plane. At this point, Hubble's publications no longer contained grand references to island universes or other galaxies, as those of Heber Curtis and Vesto Slipher were doing. Hubble started to keep his words fairly neutral, adopting the guarded language that came to be a trademark of his research reporting. He was now consciously hiding his biases to avoid criticism.

Hubble was far more vocal and forthright, though, about his observational plans. In February 1922 he sent a lengthy, typewritten letter to Slipher, a member of the Committee on Nebulae for the International Astronomical Union, on his long-term strategy for studying the nebulae. It was going to be an all-out attack. Hubble planned to determine their structure, peg their distribution across the heavens, and measure their dimensions. And as a stealth advocate of the island-universe theory, Hubble wanted to obtain undeniable proof that stars—vast collections of stars—resided in the spiral nebulae. He knew that finding novae were crucial in doing this and urged the IAU that "half a dozen of the largest spirals in addition to Andromeda should be followed carefully for novae." Major Hubble was now applying his lessons on military tactics to conquering his astronomical targets.

"I must confess that I am rather dazed by [Hubble's] letter," said Lick astronomer William H. Wright, who had also received a copy of Hubble's agenda. "One can see that the nebulae will have no private life when he has his way. Hubble is a great lad, and I only hope that he will have the strength and energy to carry out a fraction of the work he would like to see done."

Hubble, who had just gained a seat on the committee, was particularly fired up about a nebula classification scheme he wanted the IAU to adopt. To Hubble, properly categorizing the nebulae was an essential first step in determining their physical nature. By 1923 he had divided the nongalactic nebulae into two categories: the ellipticals and the spirals. An elliptical was an amorphous blob shaped somewhat like an egg. The spirals, of course, were the stunning pinwheels. If the bright center of the spiraling disk was a round bulge, he called it a "normal spiral"; if elongated, a "barred spiral." The nongalactic nebulae that didn't fit either class, like those resembling the chaotic Magellanic Clouds, were tagged "irregulars." But the IAU committee dragged its feet on Hubble's naming system and desired some changes, a rebuke that may have had long-term effects. At one point in the long wait, Knut Lundmark published a similar scheme, which enraged Hubble. He accused the Swedish astronomer of plagiarism. Afterward, Hubble was never keen to work on committees, attend general astronomy meetings, or share in collaborations. With a few exceptions, he tended to work alone. There might have been another reason for this as well. Though displaying a commanding public presence, Hubble was actually "pathologically shy around colleagues with whom he had little . . . contact," contends Allan Sandage, who knew Hubble in his later years. Hubble proceeded to classify the nebulae in his own way and over time his arrangement was eventually accepted by the astronomical community.

Throughout 1923, over a total of forty-seven nights on the mountain, Hubble used both the 60-inch and 100-inch telescopes to survey a variety of nebulae around the celestial sky. He was on a reconnaissance mission. Though scarcely any nebulae were repeated, he did pay special attention to NGC 6822, a nebula in Sagittarius first discovered in 1884 by his former Yerkes colleague E. E. Barnard. The nebula stood out

The 100-inch and 60-inch telescopes (left, right) side by side on Mount Wilson (*Courtesy of the Archives, California Institute of Technology*)

from the pack because it looked strikingly similar to the Magellanic Clouds in the southern celestial hemisphere.

By July Hubble found five variable stars in NGC 6822 and informed Shapley at Harvard, suggesting that Shapley investigate the object on the plates stored away at the Harvard observatory. "What a powerful instrument the 100-inch is in bringing out those desperately faint nebulae," responded Shapley. "As for N.G.C. 6822, I think there is no doubt but that it is another star cloud like the Magellanic Cloud." Although there was no love lost between Shapley and Hubble, the two astronomers maintained a courteous correspondence, perhaps adhering to that old adage, "Keep your friends close but your enemies closer." In actuality, they needed each other. Shapley oversaw the world's foremost collection of astronomical photographs, while Hubble had ready access to its largest telescope.

Shapley proceeded to estimate the distance of NGC 6822 by comparing its size and the observed magnitudes of its brightest stars to that of the Large Magellanic Cloud. Interestingly, he arrived at a distance of about a million light-years. "It appears to be a great star cloud that is at least three or four times as far away as the most distant of known globular clusters and probably quite beyond the limits of the galactic system," reported Shapley in his observatory's December 1923 *Bulletin*. A news report by *Science Service* promptly called it "the most distant object seen by man, another universe of stars." NGC 6822 wasn't a spiral nebula, but it certainly offered Shapley proof that large stellar systems existed beyond the Milky Way. Yet, for Shapley, his distance calculation for this stellar cloud simply had no bearing on the question of the spirals. For those he tenaciously held fast to his convictions and continued to spread the word in various publications that spiral nebulae were "neither galactic in size nor stellar in composition."

Hubble obtained more than fifty photographs of NGC 6822 over time and found fifteen variable stars. "Eleven . . . are clearly Cepheids," he eventually reported two years later. Using them as standard candles, Hubble calculated a distance of some 700,000 light-years, which was undoubtedly beyond the borders of Shapley's newly supersized Milky Way. "N.G.C. 6822 lies far outside the limits of the galactic system," stated Hubble, "and hence may serve as a stepping-stone for speculations concerning habitants of space beyond." His early work on NGC 6822 likely gave Hubble the confidence that he could pursue Cepheids as distance markers in spiral nebulae, observations he was carrying out at the same time and would actually report on first.

Observing with the 100-inch was a choreographed dance within the monumental dome a hundred feet high and nearly as wide. Sometimes Hubble could just lean back in a bentwood chair, his favorite, and serenely smoke his pipe in the darkness while taking a photograph. But other times he was perched high in the air on a platform that could adjust to any height via rails set on either side of the dome opening. With the telescope's clock drive shifting the telescope as the nighttime sky slowly moved overhead, he and his assistant made sure the advance stayed in synchrony with Earth's rotation. At the same time, they had to keep the dome rotated and the platform height adjusted, so that the telescope kept spying on the cosmos and not an inside wall. "This was the astronomical observing experience at its best," noted Mount Wilson astronomer Allan Sandage, "a dark, quiet dome, a silently moving monster telescope, and mastery of the dangerous . . . platform, all in the interest of collecting data on a problem of transcendental significance." Night after night, the cosmic waltz went on. If Hubble got clouded out,

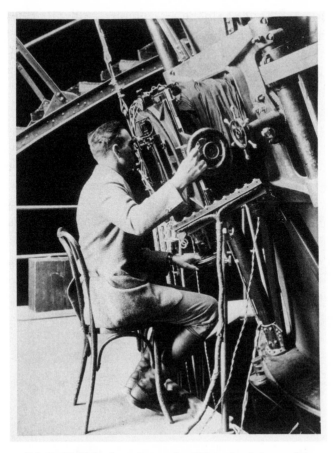

Edwin Hubble observing at the 100-inch, sitting on his
favorite bentwood chair (*Reproduced by permission of the
Huntington Library, San Marino, California*)

he had a backup: "You begin with deskwork, later you turn to heavy
reading, and later, to a detective story," he said.

The only scheduled break was "lunch," provided at midnight. In the
early years, it was simply hardtack and cocoa (Hale considered coffee
"unwholesome"), served in a concrete bunker beneath the 60-inch.
Later on, at a shack built halfway between the 60- and 100-inch,
astronomers were offered two pieces of bread, two eggs, butter and jam,
and a single cup of coffee or tea, a repast purposefully kept skimpy by
the observatory's notoriously frugal administrator, Walter Adams. Hubble gained the respect of the night assistants when he washed his own
dish afterward, giving them a break from their cleaning responsibilities.

The assistants also liked Hubble's no-nonsense attitude. Unlike some other Mount Wilson astronomers, he always arrived for his scheduled runs with a well-thought-out observing plan in hand. He delegated authority and expected a professional performance in return. "You knew where you stood with him," said Humason.

Hubble's observations at this time were fairly routine as he methodically went from target to target. He was noted for carrying a map of the heavens in his head; the hundred-odd Messier objects were as familiar to him as the alphabet. On July 17 he stopped to confirm a new and wispy nebula that Shapley had reported seeing on two occasions earlier that year in the Boötes constellation. But even after a 150-minute exposure, Hubble came up empty-handed. He saw nothing in that area of the sky. "Shapley object is probably an accident," wrote Hubble in his logbook. From a photo he took on August 15, he spotted the track of an asteroid passing by. Week by week the routine continued.

And then came the October surprise.

Discovery

[13]

Countless Whole Worlds...
Strewn All Over the Sky

October 4, 1923. The seeing was poor, but it was good enough (just barely) to stalk some celestial quarry that autumn evening. Hubble first pointed the 100-inch at NGC 6822, the far-off, Magellanic-like cloud of stars that he had long been studying. As the giant scope swung around, there was a whine, a series of loud clicks, and then a final *clang* as the instrument was secured into place. After taking an hour-long photographic exposure, Hubble went on to examine M32, a small and roundish nebula, for a spell. He then maneuvered the telescope just a fraction of a degree to photograph M31—the famous Andromeda nebula, the target of choice in the island-universe debate. By then the seeing had deteriorated to a point that other astronomers might have closed up shop. But Hubble persevered, and despite the mediocre viewing, soon noticed a new speck of light within Androm-eda's cloudy veil. It was exactly what he was hoping to find one day as he conducted his extensive survey of the nebulae. Novae had been seen before in Andromeda; that wasn't startling. But Hubble was sure that additional sightings would help reveal Andromeda's secret. "Nova suspected," Hubble neatly wrote in black ink in his logbook for Plate H 331 H. After photographing Andromeda for forty minutes, he went on to observe another nebula, a barred spiral, before ending his run.

The very next night Hubble returned to the 100-inch to follow up. This time the atmosphere was better—clear and steady, at least for a while. When the sky was at its best, he aimed the telescope at Androm-eda and again saw the new pinpoint of light. "Confirms nova suspected on H 331 H," Hubble noted in his logbook.

Everything, though, cannot be readily seen through the telescopic eyepiece or by a quick peek at a newly developed photograph. Plate H 335 H, the forty-five-minute exposure taken on October 5 to verify the nova, was analyzed in more detail later, back in Hubble's Pasadena office. There he confirmed not just one but rather *three* new pinpoints of light within Andromeda. He figured he was seeing two additional novae and wrote "N" beside each one on his plate to mark their location.

From his earlier work on the Magellanic cloud–like NGC 6822, Hubble knew that he had to make sure his newly spied objects were truly novae and not some other phenomenon. For a further check, he turned to the massive collection of plates archived in a quakeproof vault at the observatory headquarters. He began perusing previous photos of the Andromeda nebula, taken by observatory astronomers as far back as 1909. By comparing his latest photographic plate with those from the past, he could easily see that two of his spots of light were indeed novae — never-before-seen stellar flares. But one spot, the one farthest out from the center of the nebula, had been around before. Going from plate to plate, Hubble could see that this tiny dot of light was brightening and dimming over time. It was not a nova at all, but instead some kind of variable star. At this point Hubble went back to plate H 335 H, crossed out the N beside this particular dot, and beneath it wrote "VAR!" instead. His exclamation point emphasized the significance of this discovery: He had struck celestial gold. Once he had this stellar nugget in his hand, he didn't let it go.

Hubble more carefully tracked the ups and down of his variable's luminosity from the archival photographic plates. He also continued his survey of the heavens, making sure to check back on Andromeda again and again, as this was the time of year when Andromeda was in full view. He found more novae and another variable. He kept track of his finds, numbering each nova and variable and marking their positions in the spiral with a tiny red dot or circle on photos of Andromeda.

Three nights in February 1924 proved especially crucial. Over the fifth, sixth, and seventh of that month he directly observed his first variable in Andromeda brightening by more than a magnitude, doubling its luminosity, a tremendous break. From the data he had on hand he could now sketch a reliable light curve. The variable star went through its complete cycle — from bright to dim and back to bright again — in a matter of 31.415 days. From the length of this period and the shape of

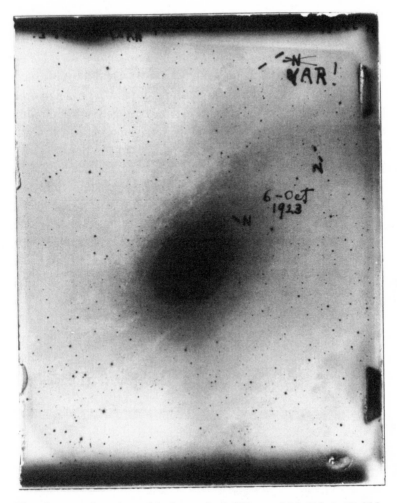

The photographic plate of Andromeda (M31) on which Edwin Hubble
identified a Cepheid variable star, mistaken at first for a nova, in a spiral
nebula—the first step in Hubble's opening up the universe (*Courtesy
of the Observatories of the Carnegie Institution of Washington*)

the curve (sharp rise and slow decline), Hubble now comprehended
that he had captured that elusive and rare celestial beast—a Cepheid
variable, a star seven thousand times brighter than our Sun. But it
appeared so dim—the barest smudge on his photographic plate—that
Hubble knew it had to reside at a great distance. It was on average more
than one hundred thousand times dimmer than the faintest stars visible
to the unaided eye.

At some point during these deliberations, Hubble went back to his logbook, page 157, and quickly scrawled an added note on the side of the page to amend the report of his October 5 observing run. Customarily reserved, Hubble at this moment is unmistakably restive. He didn't write his message in black ink, which he regularly did for his records, but instead in pencil. And his handwriting, usually so fluid and precise, was more hurried and askew. He was obviously elated: "On this plate (H 335 H), three stars were found, 2 of which were novae, and 1 proved to be a variable, later identified as a Cepheid—the 1st to be recognized in M31." To highlight the addition, he drew a big arrow, pointing directly downward at his historic news. In its broad stroke, the arrow makes his excitement visible upon the page. For once Hubble dropped his guard and figuratively clicked his heels at this moment of discovery.

Hubble couldn't help but notify his nemesis. On February 19 he wrote Harlow Shapley about his efforts over the previous months. Hubble didn't open with polite niceties or inquiries of health. He got straight to the point. "Dear Shapley:—You will be interested to hear that I have found a Cepheid variable in the Andromeda Nebula (M31). I have followed the nebula this season as closely as the weather permitted and in the last five months have netted nine novae and two variables." His glee in communicating this news jumped off the page as he then provided

Pages 156 and 157 of Hubble's 100-inch telescope logbook
(*Reproduced by permission of the Huntington Library,*
San Marino, California)

Shapley with all the technical details on color index corrections and magnitude estimations. Shapley was, after all, the world's reigning Cepheid expert—not only in using them as standard candles but figuring out early on, soon after he arrived at Mount Wilson, that they were pulsating stars, their atmospheres repeatedly ballooning in and out.

Accompanying this legendary letter was a graph that Hubble had fastidiously drawn in pencil on paper torn from a notebook. It displayed the light curve for his "Variable No. 1" in M31—a roller-coaster ride that peaked at eighteenth magnitude, dipped a bit below nineteenth magnitude, and then rose once again to its maximum brightness over a period of thirty-one days, "which, rough as it is," he told Shapley, "shows the Cepheid characteristics in an unmistakable fashion." And here was the kicker. Hubble used the exact same technique for gauging a distance to the spiral that Shapley had devised for mapping the arrangement of globular clusters around the Milky Way. Applying the Cepheid period-luminosity formula that Shapley had derived, Hubble calculated a distance to Andromeda of around 1 million light-years ("subject to reduction if star is dimmed by intervening nebulosity," he carefully noted). No more oblique evidence or convoluted reasoning, such as

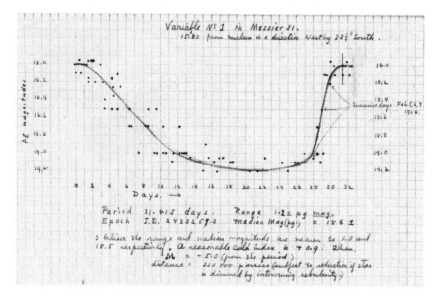

Edwin Hubble's graph of the periodicity of Variable No. 1
in Andromeda, included in his letter to Harlow Shapley that
destroyed Shapley's universe (*Harvard University Archives, UAV 630.22,
1921–1930, Box 9, Folder 71*)

Heber Curtis was forced to use. The Cepheid provided a direct and indisputable yardstick out to the nebula. Andromeda was indeed an island universe.

The second Andromeda variable, which Hubble had later found at the very edge of a spiral arm, was too faint for him to make a reliable distance measurement as yet. But no matter. "I have a feeling that more variables will be found by careful examination of long exposures. Altogether the next season should be a merry one and will be met with due form and ceremony," said Hubble at the close. He was having a fine time at Shapley's expense.

Shapley, upon reading the letter, immediately grasped that Hubble's finding spelled doom for his cherished vision of the cosmos. Harvard astronomer Cecilia Payne (later Payne-Gaposchkin) happened to be in Shapley's Harvard office when Hubble's message arrived. He held out the two pages to her and exclaimed, "Here is the letter that has destroyed my universe." Hubble was at last confirming the speculation that had been circulating through the astronomical community since the days of Thomas Wright, Immanuel Kant, and William Herschel. The Milky Way was not alone, but merely one starry isle in an assembly of galactic islands that stretches outward for millions of light-years.

Though Shapley assuredly sensed this sea change, he continued for a while to put up a good front. He mischievously wrote back that the news of "the crop of novae and of the two variable stars in the direction of the Andromeda nebula is the most entertaining piece of literature I have seen for a long time." He wouldn't even concede that the variables were *in* the nebula, only "in the direction of." He admitted that the second variable is a "highly important object" but went on to caution Hubble that his first variable star might not be a Cepheid after all, which meant it would be unreliable as a distance marker. And even if it were, he went on, Cepheids with periods greater than twenty days are "generally not dependable . . . [and] are likely to fall off of the period-luminosity curve."

Hubble was undeterred by Shapley's caveats and continued his searches at a brisk clip. His discovery spurred him to find even more Cepheid variables, in both Andromeda and other spiral nebulae. But cautious as ever, he made no public announcement. Not yet.

Just a week after sending off his triumphant communiqué to Shapley, in the very midst of these cosmos-altering observations, Hubble married, a surprise to many. His bride was Grace Burke Leib, thirty-five years old and the daughter of a wealthy Los Angeles banker. A smart and petite woman, Grace had graduated Phi Beta Kappa from Stanford University with a degree in English. She had compelling dark eyes and lustrous brown hair, but a stern mouth. She was more handsome than beautiful. Grace had been previously married to geologist Earl Leib, who specialized in assaying coal deposits and was tragically killed in a mining accident in 1921. Leib's sister was the wife of Lick Observatory astronomer William Wright, a connection that first put Grace in contact with Mount Wilson's most eligible bachelor while she was still married. When Wright visited Mount Wilson to carry out some observations in the summer of 1920, he took along his wife and sister-in-law, who stayed in a visitors' cottage on top of the mountain. Going over to a small library tucked away in the laboratory building one day to borrow some books, the two women came across Hubble. Years after Hubble's death, swept up in the nostalgic haze that colored most of her writings about her husband, Grace recalled that moment: "He was standing at the laboratory window, looking at a plate of Orion. This should not have seemed unusual, an astronomer examining a plate against the light. But if the astronomer looked like an Olympian, tall, strong, and beautiful, with the shoulders of the Hermes of Praxiteles, and the benign serenity, it became unusual. There was a sense of power, channeled and directed in an adventure that had nothing to do with personal ambition and its anxieties and lack of peace. There was a hard concentrated effort and yet detachment. The power was controlled."

By 1922 Hubble and Grace, who was now widowed, renewed their acquaintance and the couple, soon smitten, began a discreet courtship. She, more than anyone else, came to see Hubble's gentler side, his spontaneous and hearty laugh whenever someone surprised him or made an original remark. A reserved man not prone to idle chatter, he could still display a dry wit at moments. After Hubble had made the rounds of New York nightclubs one evening with a friend, his companion finally collapsed and said, "I've got to turn in. How can you stay up this way?" To which Hubble replied, "Do you think you can stay up later than an astronomer?"

Hubble wooed Grace with gifts of books and by reading to her and her parents when visiting the family's Los Angeles home. On February 26, 1924, they were married in a private Catholic ceremony

Edwin and Grace Hubble on their wedding day in 1924
(*Reproduced by permission of the Huntington Library,*
San Marino, California)

(Grace's faith), with none of Hubble's family members in attendance. After honeymooning at her family's cottage, set on six scenic acres near Pebble Beach, in Carmel, they toured Europe.

With their fondness for outdoor pursuits—riding, hiking, and fishing—and their stylish outfits, the Hubbles would have felt right at home in the countryside of aristocratic England. In California, they liked to mingle with the elite of Hollywood society rather than astronomers: writers, directors, and actors, such as Helen Hayes, George Arliss, and Charlie Chaplin. Given Hubble's fervent Anglophilia, they also hung out with members of Hollywood's long-established British colony, which at one point included the noted authors Aldous Huxley and H. G. Wells.

The Hubbles were a highly compatible match, as they both enjoyed the ways of high society (Grace grew up being chauffeured about in one of her family's two Cadillacs; Edwin got his suits and shirts custom-made in London) and always maintained a polite reserve; as one acquaintance noted, "A stranger could drop raspberry soufflé on the rug without hearing a murmur." Those who observed their interactions called the couple's relationship "quite out of the common." Given Edwin's astute powers of observation—he had a remarkable eye for detail—Grace said she "was Watson to his Sherlock Holmes."

As soon as Hubble returned in May from his three-month honey-moon—the very evening of his arrival, in fact—he was back on the mountain applying those Sherlockian skills to his study of the spiral nebulae. Throughout the remaining months of 1924 he found even more variables, tracking the ups and downs of each luminosity with care. It was plodding work. A dozen of the thirty-six variables he ulti-mately found in Andromeda turned out to be Cepheids, their cycles ranging from eighteen to fifty days. He did even better when he started studying M33, a striking face-on spiral in the Triangulum constellation, situated right next door to Andromeda toward the east. There Hubble found a total of twenty-two Cepheids with a similar range of periods, which provided him with a rich sample for calculating the nebula's distance.

In these days long before computers or handheld calculators, Hub-ble's computations for assessing the magnitudes of his Cepheids and determining their periods were scribbled on pieces of flimsy yellow paper or heavy graph paper—hundreds of pages now filed away in an archive. Points were carefully plotted on a graph to indicate a Cepheid's changing luminosity. As if playing connect the dots, Hubble then drew a crude line through the points, which displayed across the page the steady rising and falling of the Cepheid's light.

Hubble was not the best astronomer when it came to equipment. Anxious to see his results, he sometimes cut corners in the darkroom, not always using fresh developer or trimming the time for fixing and washing. The photographs and spectra he handled himself were often scratched up and required retouching before publication. But as a celestial accountant he was superb. Hubble patiently carried out his computations for variable after variable. Novae, as well, were studied

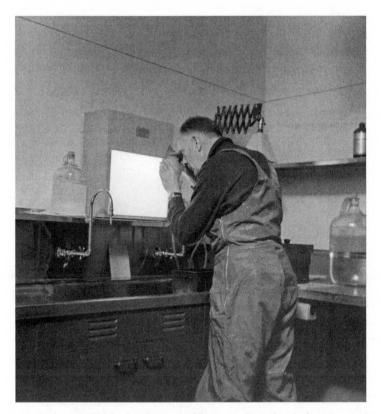

Edwin Hubble developing a photograph in the darkroom
(*Reproduced by permission of the Huntington Library,
San Marino, California*)

and tabulated. It's the very core of astronomical work, the endeavor that is never glorified, carried out as the astronomer is hunched over a desk far away from the telescope. It was there in his quiet, book-lined office on Santa Barbara Street in Pasadena—a spacious but spartanly furnished room once occupied by Hale—that Hubble truly discovered the universe. As Caltech astronomer Jesse Greenstein once said, astronomical observing "is one lump of beauty mixed with lots of incredible boredom and discomfort. . . . A single fact involves a tedious, incredibly long, difficult process."

And yet, despite his myriad pages of data—proof upon proof of a universe beyond the boundaries of the Milky Way—Hubble still did not publish. Given the relentless reconstruction he performed on his personal life story over the decades, it is obvious that Hubble's ego was frag-ile. But these boastful embellishments were attached to his life, never to his scientific achievements. Highly conservative when it came to ce-lestial speculation, Hubble never stuck his neck out in the arena of science, unlike Shapley, who readily (and loudly) broadcast his conjec-tures. Hubble's legal training might well have taught him to restrain his musings until the facts were firmly in hand, or perhaps he couldn't stand the thought of the disgrace if he had to retract his discovery, one that was going to remake the universe.

It was easier for Hubble at this stage to discuss his new findings infor-mally. In July, he wrote Vesto Slipher on routine astronomy committee matters and at the very end of his letter casually mentioned his latest work: "You . . . may be interested to hear that variable stars are now being found in the outer regions of Messier 31. Already a half dozen are definitely established and several others are under suspicion. . . . You can realize how eager I am to get curves for the others, and how bashful to discuss prematurely the Period-Luminosity relations." Hubble didn't know that Slipher had already heard about the intriguing finds. The news was rapidly spreading on the astronomical grapevine. Curtis became aware of Hubble's discovery the previous March; Shapley, of course, even earlier. And Princeton's Henry Norris Russell first heard it from James Jeans in England! The tendrils of the grapevine had a long and convoluted reach.

Besides Hubble, no one had more at stake on the outcome than Adriaan van Maanen. If Hubble's discovery held up, it meant he was wrong about his rotating spirals. So, van Maanen made sure to keep tabs on this new development at his observatory and glean all the latest gossip. "What do you think of Hubble's Cepheids," he wrote Shapley.

Shapley, meanwhile, was receiving updates from Hubble, hearing about the latest variables he was finding, including some in other spiral nebulae. "I feel it is still premature to base conclusions on these vari-ables in spirals," Hubble wrote him in August, "but the straws are all pointing in one direction and it will do no harm to begin considering the various possibilities involved."

Hubble was gaining more confidence in his findings. And Shapley, in response to the growing body of evidence, at last saw the scientific handwriting on the wall. He cried uncle, acquiescing speedily and gra-ciously. While visiting Wood's Hole in Massachusetts with his family for a summer holiday, helping dredge starfish at one point off Martha's Vineyard, Shapley briefly paused in his frolicking to respond to Hubble's August letter. He described the new results as "exciting."

"What tremendous luck you are having," he wrote. "I do not know whether I am sorry or glad to see this break in the nebular problem. Perhaps both." Shapley knew his change of heart now meant abandoning his Big Galaxy model of the universe and questioning the spiral rotation measurements of van Maanen, his good friend. He regretted that this had to happen, but Shapley was also relieved to have something definite about the spirals at last come to light. Once proven wrong, the Harvard Observatory director didn't look back and quickly adjusted to the new cosmic landscape, soon becoming its most boisterous promoter.

By the end of 1924 Hubble was finally starting to write a preliminary draft of his findings for the *Proceedings of the National Academy of Sciences.* He was dipping his toe into the proverbial water, but he was hardly leaping into the drink. As Hubble wrote Slipher on December 20, he was still hugely frustrated by van Maanen's contradictory observations on the spiral rotations. If the spiral nebulae truly resided in distant space, at least a million light-years away, no astronomer could possibly see them rotate in a matter of years. How could he make that conflict go away? "I am wasting a good deal of time investigating the possibilities of magnitude effect in van Maanen's measures. The suggestion is very strong among the comparison stars of M33 and M81 but I can not carry it through some of the others," he told Slipher. Had he truly discerned the source of van Maanen's error? Were the apparent magnitudes of the spiral stars that van Maanen picked out to make his measurements differing from plate to plate because observing conditions were dissimilar or the star was imaged on a different part of the plate? That could make it tricky to pinpoint each star's exact center, which would lead him to mistakenly measure the stars as moving, making it seem as if the entire spiral were rotating. Or was it something else? Before publishing anything, Hubble wanted to confront and overturn each and every result in van Maanen's work that was at odds with his discovery. He closed his letter to Slipher saying that he would not be attending the latest meeting of the astronomical society, starting in ten days in Washington, D.C.

Word of Hubble's discovery was still spreading like wildfire through the astronomical community. Though not yet official, the news even made it into the *New York Times.* Readers turning to page 6 on November 23, 1924, saw this headline (complete with misspelling): "Finds Spiral Nebulae Are Stellar Systems—Dr. Hubbell Confirms View That They Are 'Island Universes' Similar to Our Own." With Hubble revealing that the Andromeda and other nebulae were at least a million light-years distant, reported the newspaper, then "we are observing them by light which left them in the Pliocene age upon the earth."

Yet Hubble continued to stall, unwilling to rush his finding into the scientific literature. Though the island-universe theory had been gaining supporters, others persisted in regarding the spiral nebulae as minor entities. But the scent of resolution was in the air. At the December 1924 meeting of the British Astronomical Association, Peter Doig, a prominent figure in British amateur astronomy, presented a paper on the spiral nebulae that cautioned that "the rapid progress of knowledge, and the changing state of speculative theories of the nature and origin of these objects, perhaps make the compilation of . . . a paper [on the topic] . . . rather a risky procedure." Doig didn't realize how fast his prophecy would come true. The mountain of doubts and reservations concerning the spirals came tumbling down in less than a month.

Russell was so impressed by Hubble's accomplishment that he nominated the young Mount Wilson astronomer for membership in the National Academy of Sciences, quite an honor for someone still junior in his profession. Formerly a solid supporter of Shapley's cosmic model, the Princeton astronomer had now done a quick about-face. Just ten months earlier he had been lecturing that spirals were nearby, supported by van Maanen's evidence, but now Russell was telling the managing editor of *Science Service* that Hubble's find was "undoubtedly among the most notable scientific advances of the year." He contacted Hubble and encouraged him to publish his results as soon as possible, wanting him to present a paper at the thirty-third meeting of the American Astronomical Society, which was going to be held jointly that year with the annual conference of the American Association for the Advancement of Science.

"Heartiest congratulations on your Cepheids in spiral nebulae!" wrote Russell on December 12. "They are certainly quite convincing. I heard something about them from Jeans a month or two ago, and was wondering when you would be ready to announce the discovery. It is a beautiful piece of work, and you deserve all the credit that it will bring you, which will undoubtedly be great. When are you going to announce the thing in detail? I hope you are sending it to the Washington meeting, both, because we all want to know all about it, and because you ought, incidentally, to bag that $1000 prize." The Council of the American Astronomical Society was ready to nominate Hubble's paper for the prestigious $1,000 AAAS prize (a substantial sum of money in its day) given to the best paper read at the gathering. It was only the second year for the competition, and the *Washington Post* was reporting "considerable interest" in the outcome.

But Hubble was hesitant to change his plans. As he later related to Russell, "The real reason for my reluctance in hurrying to press was, as you may have guessed, the flat contradiction to van Maanen's rotations." Van Maanen was a more senior member of the Mount Wilson staff, and Hubble was hoping to avoid a public conflict, even fantasizing that there might be a way to reconcile the two contradictory sets of data. "But in spite of this," he admitted, "I believe the measured rotations must be abandoned. . . . Rotation appears to be a forced interpretation."

Russell assumed his letter (and the lure of the prize) would finally persuade Hubble to put aside his concerns and make the discovery official once and for all. As soon as Russell arrived at the Washington conference, he had dinner with University of Wisconsin astronomer Joel Stebbins, then secretary of the astronomical society, and eagerly asked Stebbins whether Hubble had as yet sent in his paper. When Stebbins replied no, Russell was flabbergasted and declared that Hubble was "an ass!! With a perfectly good thousand dollars available he refuses to take it."

A telegram was quickly drafted, urging Hubble to send his principal results by overnight letter. Both Russell and Shapley stood ready to take Hubble's data, whatever he chose to convey, and turn it into a proper paper for the meeting. But just as Stebbins and Russell were about to go over to the telegraph office, Russell noticed on the floor behind the hotel desk a sizeable envelope addressed to him. Stebbins spied Hubble's name in the return address. Hubble had mailed his paper after all, and in the nick of time. "We walked back to the group in the lobby, saying that we had got quick service," Stebbins later told Hubble. "That coincidence seemed a miracle."

In Hubble's absence, Russell stepped in and read the paper to the assembled conferees on the snowy morning of January 1, 1925. Hubble relayed that he had found twelve Cepheids in Andromeda and twenty-two in Triangulum, their telltale blinks indicating a distance for each of nearly a million light-years, confirming what others had gleaned with shakier methods. More than that, the 100-inch telescope had allowed Hubble to resolve the outer regions of the two nebulae into vast collections of stars. Astronomers could now be certain that the spirals were not simple nebulosities, not just clouds of dust and gas. At the end of his paper, Hubble hinted at more results to come, having by then sighted variable stars in M81, M101, and NGC 2403, some of the most commanding spirals in the celestial sky.

Astronomers in the audience could practically feel the universe changing as they listened to Russell—except for one. Curtis, who was briefly at the Washington meeting, took the announcement in stride. "As you know," he wrote a former Lick colleague the next day, "I have always believed that the spirals are island universes, and Hubble's recent results appear to clinch this, though I myself did not need the confirmation." You can almost hear him yawn between the lines.

Soon after Russell's presentation, the American Astronomical Society Council sent in its petition to the AAAS, nominating Hubble's paper (one of seventeen hundred presented at the conference that year) for the coveted prize. "Dr. Hubble," the council stated, "has found that the outer parts of the two most conspicuous nebulae, in Andromeda and in Triangular [*sic*], are resolved upon his best photographs into 'dense swarms of actual stars.' This has been suspected as a possibility for a century, but has never previously been unequivocally proved. . . . This paper is the product of a young man of conspicuous and recognized ability in a field which he has made peculiarly his own. It opens up depths of space previously inaccessible to investigation and gives promise of still greater advances in the near future. Meanwhile, it has already expanded one hundred fold the known volume of the material universe and has apparently settled the long-mooted question of the nature of the spirals, showing them to be gigantic agglomerations of stars almost comparable in extent with our own galaxy."

Although the great distances to the two nebulae flagrantly disagreed with van Maanen's data, most astronomers quickly rallied around Hubble's figures. The Cepheids were fast becoming the gold standard for measuring distances to the more remote starry regions of the universe. Nearly everyone came to assume that van Maanen was mistaken. "The great distances recently derived have made rapid rotation impossible," said Harvard astronomer Willem Luyten, "and the quick internal motion measured some years ago is now universally regarded as an optical illusion." James Jeans confirmed Hubble's distance results with an

Edwin Hubble and James Jeans at the 100-inch telescope
(*Reproduced by permission of the Huntington Library,
San Marino, California*)

alternate technique and wrote Hubble that "van Maanen's measurements have to go." The long and convoluted squabble on the nature of the spiral nebulae—centuries of debate—was finally over. The spirals were not adjuncts of the Milky Way at all but instead galaxies in their own right. The universe officially became far larger—and far more intriguing.

Russell's instincts, it turns out, were very good. Hubble in the end won the AAAS award for extending the boundaries of the known universe. He was informed by telegram on February 7, but the amount he received was cut by half. The young astronomer was told he was sharing the award with another scientist. Parasitologist Lemuel Cleveland of the School of Hygiene and Public Health at Johns Hopkins was also honored for his study of microscopic protozoa found inside the digestive tracts of termites. He showed that the tiny organisms were essential for a termite to digest cellulose. "To scientists," reported the *Los Angeles Times*, "the infinite and the infinitesimal are merely relative terms, alike in importance." The Hubbles had just bought an acre lot in San Marino to build their charming new home (designed in the style of a small Tuscan villa) and used the funds to help pay for pruning the live oaks on their property and clearing out the deadwood—a needed renovation for appreciating the fine view of Mount Wilson and the San Gabriel Mountains from their backyard. There at his home Hubble began collecting old books, specifically on the Renaissance years, when the old Aristotelian representations of the universe were crumbling. "If an old scrap of paper, published within the sacred period, contains the names of Copernicus, or Tycho Brahe, or Kepler, or Galileo, [Hubble] hankers after that paper more than a debutante hankers after orchids," wrote a local reporter. It was an apt hobby for the man who erected his own new-and-improved model of the cosmos.

Why was Hubble able to accomplish this magnificent feat while others were not? In actuality, there were several opportunities to resolve the island-universe controversy earlier. The Cepheids could have been hunted down and observed without the 100-inch telescope. It's somewhat surprising that more astronomers didn't sense the celestial riches to be found in distant space, just ready for mining. Having access to the world's largest telescope was not the essential key to Hubble's success (although it certainly helped). Mount Wilson's 60-inch telescope, erected in 1908, could have done the job just fine. Even the Crossley telescope at Lick had an outside chance. But few were interested in this area of endeavor, and those who did had bad luck. For example, the first person to find a variable star in a spiral nebula was not Hubble at all, but rather Wellesley College astronomer John Duncan. In 1920, while using Mount Wilson's 60- and 100-inch telescopes to search for novae, Duncan found three variable stars within the Triangulum nebula, M33. Over the next two years, he took additional images and also checked other photographs of the region made at the Yerkes, Lick, Lowell, and Mount Wilson observatories from 1899 to 1922 in an attempt to track the variables' periods but was unsuccessful. The data were simply too sparse at the time. And in his report of the find, Duncan refrained from directly linking the variables to the nebula. If he had followed up, the prize would have been his: The faintest variable he saw was later found to be a Cepheid and could have been used to peg the nebula's distance.

Why didn't Shapley himself, the world's Cepheid guru, search for these special stars in the spiral nebulae and garner one of the choicest discoveries in astronomical history? It seemed like it would have been a natural progression for him. But around 1910 his colleague at Mount Wilson, George Ritchey, had photographed thousands of "soft star-like condensations" in Andromeda and other spirals, which he figured were nebulous stars in the process of formation. This interpretation suggested that a spiral nebula was simply the early stage of a modest star cluster forming, rather than an entire galaxy. Shapley admitted he was deeply influenced by Ritchey's images at the time, as were many others. Years later, in his autobiography, Shapley also suggested that strict divisions were in place in Mount Wilson as well: Shapley was relegated to the globular clusters and Hubble to the spiraling nebulae. Moreover, his taking the Harvard Observatory directorship distanced him from the thick of the battle.

But in truth, Shapley had basically taken off his scientist's hat and become too wedded to his vision of the Milky Way as the defining feature of the universe. He ignored conflicting data longer than he should have, which kept him from extending his work to the spiral nebulae and beating Hubble to the punch. He saw no reason to search for Cepheids in spiral nebulae, since he had already convinced himself that they were not separate galaxies. He was enormously attached to his Big Galaxy concept and had built his career on it. It's not surprising that he would be reluctant to let his vision be supplanted.

Shapley, in the end, was simply human. He didn't view his ignoring the early doubts about van Maanen's work as a scientific lapse but rather a personal one. "I faithfully went along with my friend van Maanen and *he* was wrong on the . . . motions of galaxies. . . . [People] wonder why Shapley made this blunder. The reason he made it was that van Maanen was his friend and he believed in friends!" declared Shapley (oddly in the third person). He was also a man who was far too confident for his own good, if a popular tale often recounted at Mount Wilson is true. Around 1920 Shapley allegedly asked staffer Milton Humason to examine with the Blink some photographic plates of the Andromeda nebula Shapley had taken over the preceding three years. After comparing the plates for several weeks, Humason came to notice what appeared to be some variable stars in the nebula, possibly the same Cepheids that Hubble found a few years later. Humason, still in training, used a pen to mark off the suspects on the glass plates and went back to Shapley to show him the results. Shapley, not impressed, patiently explained to

Humason why his spots couldn't possibly be Cepheids. Shapley was so certain of his position that he proceeded to take a handkerchief out of his pocket and rub out the marks, wiping the plates clean—not to mention wiping out his chances for further astronomical glory. While Shapley was waiting in 1920 to hear from Harvard about the director-ship, he confided to one of his former Missouri professors, Oliver D. Kellogg, that he was frustrated by the university's indecision, since it disturbed his ability to prepare his research program for the next few years. Belying what he later said about there being "strict divisions" at Mount Wilson, Shapley noted that "spiral nebulae" were on his agenda and that "cosmogony" would be his future field. Had Shapley not gone to Harvard and instead stayed at Mount Wilson, he would surely have continued to look for novae in Andromeda, the purpose of his photo-graphic survey, and perhaps come to recognize the Cepheids after all. He might have scooped Hubble. That he didn't only added to the on-going rivalry between the two Missouri men.

Even decades later, when writing his memoir in the late 1960s, Shapley couldn't let go of the beefs with his rival. "The work that Hub-ble did on galaxies was very largely using my methods," he recalled sulk-ingly. "He never acknowledged my priority, but there are people like that." But then he grudgingly conceded that Hubble had "made himself very famous, and properly so. He was an excellent observer, better than I." Hubble was patient.

It was that patience that enabled Hubble to methodically carry out the measurements that eluded earlier astronomers. Others had approached the nebular mystery yet gathered only tantalizing and incomplete hints; Hubble performed the painstaking tasks that closed the deal. That meant searching for stars and novae at the very limit of his telescope's resolving power and using them to measure a distance. Curtis had removed himself from big telescope access; Shapley refused to consider that spirals could be huge stellar systems. Only Hubble pur-sued the question with dogged effort and even he had been looking for novae at first, not Cepheids in particular. Luck certainly played a small role, but as Louis Pasteur once put it, "In the fields of observation, chance favors only the prepared mind."

Once the news was out, reporters couldn't get enough of the tall and broad-shouldered Major Hubble, as they often addressed him. He was turning into an accomplished popular communicator. "There is just not one universe," Hubble told a local journalist about his discovery. "Countless whole worlds, each of them a mighty universe, are strewn all over the sky. Like the proverbial grains of sand on the beach are the universes, each of them peopled with billions of stars or solar systems. Science has already taken a census of nearly ten million galactic systems or individual universes of stars."

Newspapers vied with one another to come up with the catchiest headline to describe the new cosmic order: "Ten Million Worlds in Sky Census," "Gigantic Telescope Finds New Wonders in Heavens," "Mount Wilson Observers Now Study Stars on Newly Found Horizons," "Distances of Star Systems So Great 18 Ciphers Are Needed to Express Them in Miles," "Light Registered on Photographic Plates Started Million Years Ago." A London magazine ranked Hubble's achievement "with the greatest in the history of astronomy. Columbus discovered half of a known world, but Dr. Hubble discovered a host of new universes." Another droll scribe said Hubble had found "more systems of stars than there are hairs in the whiskers of Santa Claus."

Hubble had even gained enough fame to be joked about. "Professor Edwin Hubble announces that he has found another universe. Some people never seem to know when they have enough," said the caption of a *Nation* cartoon.

At lectures Hubble drew record crowds. At one Los Angeles talk, the room was filled to capacity while hundreds more jammed the doorways and windows to listen in. An additional five hundred were turned away. "Astronomy, as a matter of popular interest," reported the *Los Angeles Examiner*, "joined rank with football and prize fights" that night. When standing on the balcony of Mount Wilson's observatory laboratory one night with a reporter, the two gazing at the lights of the towns below, Hubble was asked how he carried out his work. "It is like looking at those lights," he replied, "and from them alone trying to tell what manner of people live there."

From that point on, the nebulae beyond the Milky Way became the sole subject of Hubble's professional life; he scarcely studied anything else—although he did by chance discover "Comet Hubble" in August 1937 while photographing a spiral nebula. When a friend asked him to name Jupiter's moons one day, he could recall three or four but no more. "I am commuting to a spiral nebula, and I forget the suburban stations," he responded apologetically.

Astronomers had been proceeding outward into space and time on stepping-stones. The first stops were at the globular clusters, followed by a giant leap to the spiral nebulae. The conventional understanding of the universe was changing and very swiftly. Just a few years after Hubble confirmed the existence of other galaxies, Jeans wrote that "astronomy is a science in which exact truth is ever stranger than fiction, in which the imagination ever labours panting and breathless behind the reality, and about which one could hardly be prosaic if one tried."

The English poet Edith Sitwell, upon a visit to Hubble's home, was ushered into the study, where she was shown slides depicting the myriad galaxies that cannot be seen with the naked eye, galaxies millions of light-years away. "How terrifying!" exclaimed Sitwell, to which Hubble replied, "Only at first. When you are not used to them. Afterwards, they give one comfort. For then you know that there is nothing to worry about—nothing at all!"

Except, perhaps, what to call them. There was much confusion at first on how to identify the newfound stellar systems. Everyone seemed to have a pet name, including anagalactic nebulae, nongalactic nebu-

Edwin Hubble in his office with a picture of the Andromeda galaxy
(*Hale Observatories, courtesy of AIP Emilio Segrè Visual Archives*)

lae, star clouds, cosmic nebulae, and island universes. Hubble preferred "extragalactic nebulae," using it in his lectures and publications rather than the term *galaxies*, the name regularly employed by Shapley at Harvard. "I want to get away from both the words universe and nebula in reference to these objects, as frequently as possible," argued Shapley. "Therefore I am adopting . . . the term galaxy, and from that the term inter-galactic space follows naturally."

But Hubble didn't see any pressing need to abolish the "venerable precedent" of preserving the word *galaxy* for the Milky Way alone. The term originated from *galakt*, the Greek word for milk. As a purist, Hubble chose the Oxford English Dictionary as his final arbiter. At the time its pages said the term *galaxy* was "chiefly applied to a brilliant assemblage . . . of beautiful women or distinguished persons." "The term *nebulae* offers the values of tradition; the term *galaxies*, the glamour of romance," concluded Hubble. According to historian Robert Smith, which code word an American astronomer used quickly pinpointed whether they came from the East or West Coast. The intense Hubble-Shapley rivalry had extended into a surprising new sphere. It wasn't until Hubble's death in 1953 that the term *galaxy* became the universally accepted moniker.

Van Maanen was obviously panicky once Hubble's findings were officially out. He soon wrote Shapley asking if there was a list somewhere in the literature of every observation of a nova. "I want to compare them with the novae in spirals," he said. "After Hubble's discovery of Cepheids I have been playing again with my motions and how I look at the measures." He was clearly baffled. "I cannot find a flaw in [my measurements of] M33, for which I have the best material. They seem to be as consistent as possibly can be." He understood that there were two sets of observations in circulation—his and Hubble's—that arrived at "radically different conclusions." He planned to take more plates for a reassessment.

But Shapley by now had completely switched sides and in response at last lowered the boom on his good friend. "I am completely at a loss to know what to believe concerning those angular motions; but there seems to be no way of doubting the Cepheids, providing Hubble's period-luminosity curves are as definite as we hear they are," he replied. When van Maanen a few years later again tried to defend his spiral work to Shapley, the Harvard Observatory director replied that he didn't "know what to think of your confounded spirals. . . . There is little chance that we can get the universe out of this mess." He avoided the topic with van Maanen from that point on.

Considering himself a gentleman at heart, Hubble didn't openly argue with van Maanen either, and hardly anyone else in the astronomical community appeared particularly concerned. But behind closed doors, it was another matter altogether. Personally Hubble felt that van Maanen's paradoxical findings lingered as a stubborn stain on his great accomplishment, a blemish that tarnished his otherwise sterling reputation. In her memoirs, Grace Hubble cheerily declared that the van Maanen episode hardly affected her husband at all, but she told others privately that "van Maanen's contradiction disturbed her husband so greatly from the late 1920s into the 1930s that he sometimes came home from the office and lay on his bed until his anguish abated." Hubble had been aiming a critical eye at van Maanen's findings for quite a while and had begun preparing a series of private manuscripts, even before he announced that the Milky Way was not the only galaxy in the universe. His sole objective: to find out where van Maanen had gone wrong.

For several years, Hubble kept his doubts to himself and his covert manuscripts stashed away in his office drawer. It appeared that the Hubble–van Maanen conflict would just wither away, likely remembered, if at all, as a minor episode in the history of the island-universe debate. That would have been the case, except that van Maanen was perversely unwilling to admit defeat. He began remeasuring some of his spirals and in Mount Wilson's 1931 annual report it was announced he had found in M101 "a decided internal motion in the same direction as was found in his original measures of this nebula." With this surprising new strike, the battle was reignited. "They asked me to give him time. Well, I gave him time, I gave him ten years," responded Hubble to the latest assault. Now faced with van Maanen's implicit slap in the face, the former boxer put his gloves back on and rushed headlong into one of Mount Wilson's most fabled tempests. It had already been simmering in regard to telescope use. Van Maanen was sure that Hubble had been heading up a cabal to deny him a fair share of time on the 100-inch. That's when van Maanen slapped his sign on the front of the Blink, warning others not to use the machine without his permission.

The skirmish even extended into the dining room atop Mount Wilson. Seating arrangements for lunch at the Monastery followed a strict protocol: The observer scheduled to use the 100-inch telescope always sat at the head of the table, the 60-inch-telescope observer to his right, and the solar-tower observer to the left. Down the table it went in order of diminishing telescopic prominence. But one day Hubble arrived on the mountain for a run on the 60-inch and slyly switched the napkin rings, each specially marked with a staff member's name. When the dinner bell rang van Maanen, then working on the 100-inch, proceeded into the dining room and found himself placed lower down, with Hubble victoriously positioned at the table's prime spot. It was the ultimate insult one could receive on the mountain.

Drawing on his former legal training, "Hubble skillfully employed trial tactics to attain a favorable verdict from the court of science," contends Hubble scholar Norriss Hetherington. First Hubble got his observing partner, Milton Humason, to photograph the Triangulum spiral over two nights in September 1931. He then compared this latest image with a photograph of the same galaxy taken in 1910. This was followed by new photographs of other prominent spirals long studied by van Maanen, such as the Whirlpool and Pinwheel galaxies. Hubble spent hours and hours comparing the old and new plates—picking out comparison stars, just as van Maanen did, and looking for telltale signs of rotation over the years. In the end, he concluded that "no evidence of motion" could be found. In a strategic coup de grâce, Hubble commandeered Seth Nicholson, who had assisted van Maanen in his earlier measurements, to examine the plates as well. This time Nicholson saw no changes whatsoever, at least within the range of probable error. The clever prosecutor had gotten a key witness to reverse his opinion on the courtroom stand. It appeared that van Maanen had made a personal error in regard to spiral rotation, simply finding what he expected to find.

Hubble wrote up his findings for publication, but his bosses were not pleased at all with his first draft. Breaking all the rules of dispassionate scientific discourse, Hubble's grudge with van Maanen was starkly visible upon the page. "Its language was intemperate in many places and the attitude of animosity was marked. He objected to any material change in the wording and a deadlock seemed to be indicated," confided Mount Wilson director Walter Adams to the president of the Carnegie Institution of Washington, John Merriam. Like the preparations for a treaty between two warring nations, resolution involved delicate diplomacy, although in this case the principals involved worked at the same place. Frederick Seares, who served as the editor for papers written at Mount Wilson, did not want the battle to go public. If he solely published Hubble's criticism of van Maanen's work, it would be as if he were taking sides. A serious man known for his courtly manner, Seares wanted to maintain a certain decorum. Otherwise, morale at the observatory could plummet.

Seares decided that it would be best to prepare a joint statement, to be published under all the names of the people involved in reviewing the case—Hubble, van Maanen, Nicholson, as well as Walter Baade, a new staff member who had also assisted. All the parties agreed to this cooperative effort—except for Hubble, who opposed it violently. He declared "no compromise, no compromise" as the truce was worked on, insisting on no watering down of his views of the evidence. Hubble was sure he was right and van Maanen wrong. Adams was appalled by this response. "I do not feel that Hubble's attitude in this matter was in any way justified. . . . This is not the first case in which Hubble has seriously injured himself in the opinion of scientific men by the intemperate and intolerant way in which he has expressed himself," Adams reported to Merriam. Seares was so exasperated by Hubble's pigheaded attitude that he was almost ready to tell him, "Print what you like, but print it elsewhere."

It was a moment when Hubble's discretion and judgment completely failed him. Although all the facts were assuredly in his favor, his obstinate manner in this episode deeply hurt his relations at the observatory. "The attitude of van Maanen in the matter was much superior to that of Hubble," concluded Adams. "Hubble, who had much the better of the general weight of evidence, showed a distinctly ungenerous and almost vindictive spirit." Hubble had become the big man in astronomy and could tolerate no lesser colleagues. He had begun to blithely ignore his duties on international committees when the chores didn't suit his schedule and was also less willing to join cooperative projects at the observatory, acting more as an individual driven by personal ambition than as a member of a larger staff. Adams lamented that he "recognized this curious 'blind spot' in almost every important dealing" he had with Hubble.

Hubble's increasing worldwide fame was inflating his ego, already outsized as it was. Never great pals with his astronomical colleagues, he widened the breach with his boorish behavior. He broke promises, ignored vital correspondence, took more travel than the norm (with pay), and failed to show up at meetings that he said he would attend.

Adams's remarks were a reflection of the growing irritation at Mount Wilson with this loutish conduct, but it was hard to rein in the observatory's most famous staff member. Hubble was, after all, the discoverer of the modern universe. Hubble's family, too, was deeply affected by his self-centered concerns. When his mother died in 1934, Hubble did not try to return from England, where he was then traveling, once he was cabled the news. By then he hardly interacted with his family or helped them much financially. Never once did Grace meet her in-laws. "Great men have to go their own way," his youngest sister, Betsy, said with resigned acceptance many years later. "There is bound to be some trampling. We never minded. . . . With Edwin, it was out of sight, out of mind. When he was with you, you were the only person in the world, but if you were away, he would forget you. His head was in the stars."

In the end, Hubble and van Maanen grudgingly arrived at a gentleman's agreement. After much discussion with Adams (and a lot of arm-twisting), Hubble at last consented to publish a brief statement on his own, which was to be accompanied by a paper by van Maanen in which he acknowledged the existence of possible errors in his research. Hubble's brief note came out in the May 1935 issue of the *Astrophysical Journal*. It was a mere four paragraphs plus a table, summarizing his measurements of M81, M51, M33, and M101. All arrived at the same conclusion: no "rotations of the order expected." In an orchestrated move, the *Astrophysical Journal* had van Maanen's paper immediately follow. After including new plates taken with the 100-inch telescope in his reevaluation, van Maanen conceded that his measured motions were now smaller. "[My] results, together with the measures of Hubble, Baade, and Nicholson . . . make it desirable to view the motions with reserve," he stated. Van Maanen promised a "most searching investigation in the future," but as the years progressed he never followed up.

The one nagging discrepancy keeping Hubble from his full triumph—the unquestioned discovery that the spiral nebulae were truly separate galaxies—was at last resolved. In print, the two adversaries symbolically shook hands and went their separate ways. But, from that point on, whenever the two passed each other in the observatory hallways, they exchanged not a word.

[14]

Using the 100-Inch Telescope the Way It Should Be Used

W hile it appeared that Hubble had clinched astronomy's brass ring, solving the mystery of the spiral nebulae once and for all, a nagging problem remained: how to explain the galaxies' astounding velocities, first spotted by Vesto Slipher in the 1910s. Why were the spiraling disks speeding away from us? They "shun us like a plague," exclaimed Eddington. It was a puzzle whose solution would prove to be even more momentous than Hubble's settling the island-universe controversy.

Hubble began to focus his full attention on the cosmic exodus in 1928. That summer the International Astronomical Union was holding its triennial general assembly in the picturesque city of Leiden, set along the Old Rhine in southern Holland. With fine weather to entice them, more than three hundred delegates attended the gathering, where they were entertained with boat excursions down the city's noted canals, gliding past scenery painted by Rembrandt three centuries earlier. It was the height of the Roaring Twenties, and Europe was overflowing with tourists. "Most of the Americans appear to be over here this summer, always on the run scooping up culture with both hands, buying walking sticks and spats, post cards," noted Lowell astronomer Carl Lampland, who attended the meeting.

Hubble had been appointed acting chairman of the IAU Nebulae Commission, and in and around its July session, he took the opportunity to sit down with Willem de Sitter to discuss relativity and its application to cosmology. Hubble was undoubtedly familiar with (though hardly an expert on) the Einstein and de Sitter solutions to the universe's struc-

ture. At the very end of his magisterial 1926 paper "Extra-Galactic Nebulae," Hubble had included a brief section titled "The Finite Universe
of General Relativity," in which he mentions both of them. Moreover,
the following year at Hubble's direction, Milton Humason had remeasured the redshifts of two nearby galaxies. In his brief report, likely
ghostwritten by Hubble, Humason specially noted that the galaxy
speeds computed from those redshifts were unusually low, "consistent
with the marked tendency already observed" for the closest galaxies to
have the smaller pace.

Hubble was thus certainly aware of the general trend of galaxy velocities outward, but at Leiden he seemed to have finally grasped the
tremendous hubbub the galaxy redshifts were generating among cosmologists and received some further lessons from one of the world's few
experts on general relativity. Eager to have his model of the universe put
to the test, de Sitter encouraged Hubble at this time to extend the redshift measurements of the spiral nebulae begun by Slipher at the Lowell
Observatory. With only a puny 24-inch refractor with which to work,
Slipher had essentially come to the end of his search. He had been able
to acquire the redshifts of the brightest spiral nebulae, over forty of
them, but trying to obtain a reading from ever fainter and smaller galaxies was impossible. Slipher had exhausted the power of his telescope
and could simply not gather enough photons. "The Flagstaff assault on
these objects stopped just short of some great excitement," Shapley later
pointed out. Most figured that to reliably establish whether a galaxy's
redshift was related to its distance in a predictable way would require a
far bigger telescope, like the 100-inch reflector available to Hubble at
Mount Wilson. It was the perfect match of problem to instrument. De
Sitter knew this, and Hubble was obviously convinced as well.

Upon returning to California, Hubble immediately made this pursuit his top observational priority. Having conquered the mystery of the
spiral nebulae, he was now commencing his next great challenge—to
see if there truly was a definitive trend to the redshifts of the galaxies as
they rushed headlong into distant space. It was at this time that Hubble
forged his industrious partnership with Humason, each taking on a specific task to get the overall job done. While Hubble searched for
Cepheid variables to determine the distances to a sample of galaxies, his
colleague focused on getting the redshift data to figure out the galaxies'
velocities (if that indeed was how the redshifts were to be interpreted).
Hubble's plan was to put these two pieces of information together and

determine if there was a law—a specific formula—that linked a galaxy's distance to its measured redshift.

Humason was not too happy at first upon hearing of his new assignment. Hubble had come home from Leiden quite excited and quickly suggested to Humason that he try to obtain a galaxy redshift that was not yet known. But the prism on the 100-inch spectrograph had started to yellow, and the photographic plates then on hand were considerably slow-acting for such work. Humason knew it would take several nights to get a decent spectrum. "I didn't feel much enthusiasm about these long exposures," he later recalled. "But [Hubble] kept at me and encouraged me." Hubble was after fainter and fainter objects, the ones too distant for Slipher to have studied with his smaller telescope, and some were low on the southern sky. "To get these," said Humason, "you had to climb onto the 100 inch and sit on the iron frame during the long winter nights, which was extremely cold and uncomfortable." For hours on end, through the freezing night, he would have to keep his guide star on the center of the cross wires, to make sure his image remained sure and steady. "The eye-strain, the monotony, the constant awareness—it was a test of endurance," he said. But Humason's unusual entrée into astronomy offered him superb preparation for this arduous undertaking. He was accustomed to hard work.

Born in Minnesota in 1891, Humason as a boy moved to the West Coast with his family and one summer as a teenager enjoyed a camping holiday on Mount Wilson. He fell in love with the mountain and soon dropped out of grammar school to work as a bellboy and handyman at the newly opened Mount Wilson hotel, a popular resort spot for local residents. He washed dishes, corralled the horses, and shingled the cottages. Once the 60-inch telescope was under construction, he drove the mule trains that took the equipment, piece by piece, up the rugged path to the top of the peak. When a mountain lion was found feasting on a prized goat in the area, Humason tracked the animal down and shot him between the eyes with a .22-caliber rifle. Several years after Humason married the daughter of the observatory's chief engineer, his father-in-law arranged for him to work as the observatory's janitor. Gradually he was allowed to help the astronomers as a night assistant and over time won their respect and trust in making observations on his own, despite his eighth-grade education and lack of formal training in astronomy. Seth Nicholson took the young man under his wing and taught him some mathematics; Shapley mentored him as well. With his round face and round eyeglasses, the quiet and self-effacing Humason came to look like an academic. In 1920 he was promoted to a staff position in the photography department and two years later moved up to assistant astronomer. Known for his patience and conscientious attention to detail, he became especially skilled at taking long photographic or spectroscopic exposures of the most faint celestial objects. A likable fellow and an inveterate gambler, he relieved the pent-up tension from this grinding work by playing poker with the other night assistants and shop workers. If his schedule allowed, he'd catch the late-afternoon horse races at the nearby Santa Anita racetrack, taking any astronomer who wanted to go with him.

Over time, Humason was even put in charge of arranging telescope time. Sharing Hubble's strong loyalty for the Republican Party, Humason tried to get as many Democrat observers as possible on the mountain, away from the polls, on election days. Solar astronomer Nicholson, a staunch Democrat, evened the score by making sure only Republicans were scheduled on the solar telescopes at the same time. Hubble, whose status could never be threatened by Humason's humbler origins, got along fine with his devoted junior partner.

Milton Humason at Mount Wilson
(*Courtesy of AIP Emilio Segrè Visual Archives*)

. . .

By 1929 Hubble had determined the distances to twenty-four galaxies (including the Small and Large Magellanic Clouds), the most remote then judged to reside some 6 million light-years away. He accomplished this feat by establishing a ladder of measurements, one rung leading to the next. First he used Cepheid variables, his most reliable yardstick, to directly obtain the distances to six relatively nearby galaxies; then he judged the magnitude of the brightest stars in those galaxies. Figuring such stars were similarly bright in other, more distant galaxies, he proceeded to use them as standard candles. He sought out these radiant stars in more far-off galaxies—fourteen in all—and estimated each galaxy's distance based on the stars' apparent luminosities. Then, taking all twenty of these galaxies into account (the first six and the subsequent fourteen), he estimated the brightness for an average galaxy and used that value for judging the distance to four more remote galaxies. Hubble's moving outward like this, rung by rung, was similar to Shapley's strategy for his globular cluster distance measurements, but here Hubble was making an even braver leap into distant space.

Hubble then paired each galaxy's distance with its measured velocity to see if there was a connection, some sort of organized flow in which the galaxies flew outward into the depths of space. Humason by then had redone a number of the redshifts, but when Hubble prepared his first paper on the findings ("A Relation Between Distance and Radial Velocity Among Extra-Galactic Nebulae"), he primarily used Slipher's original measurements.

Hubble was more vigilant than usual in preparing this landmark 1929 publication, chiefly because of the checkered history of the subject. An earlier and rather clumsy attempt by the Polish-American mathematical physicist Ludwik Silberstein to find a relationship between a galaxy's distance and its redshift had been met with derision, especially by noted astronomers Knut Lundmark and Gustaf Strömberg, who was Hubble's colleague at Mount Wilson. Silberstein had lumped globular clusters in with spiral nebulae, which led to a meaningless result. He was ridiculed for both his inept analysis and leaving out data that went against his prediction, which tainted everyone's outlook on the problem. To make sure this didn't happen to him, Hubble shrewdly sought out the advice of Silberstein's two harshest critics and specifically highlighted their contributions in his paper. "Mr. Strömberg has very kindly checked the general order of these values. . . . Solutions of this sort have been published by Lundmark," he wrote fawningly. Hubble knew he was dealing with a controversial finding, so he was taking every precaution. He was wooing potential enemies to his side. He didn't even like Lundmark, having earlier accused the man of plagiarizing his system for classifying the galaxies, and the point he had Strömberg verify was so simple it scarcely needed checking. As it was, Hubble held up publication of his data to make sure he had nailed down every argument, as well as gathered data on even fainter galaxies so he and Humason could quickly publish a follow-up and prevent others from jumping into what Hubble considered *his* field. He was being both careful and cunning; he was not just introducing an idea but selling it hard. Hubble knew he had to make an airtight case in order to convince his more skeptical colleagues. "There is more to the advance of science than new observations and new theories," historian Norriss Hetherington has noted. "Ultimately, people must be persuaded."

According to Hetherington, Hubble presented his first data on the problem as if he were standing before a judge and jury—again, not surprising given his legal training. Hubble even had witnesses. With Hubble citing their assistance, Strömberg and Lundmark were brought forward to serve as objective bystanders to verify his competence. What Hubble saw was a definite pattern to the galaxies' retreat, a rule that was simple and yet so elegant. The velocity of the galaxies was found to steadily increase—rise in a linear fashion, as scientists say—as astronomers peered ever deeper into space. At double the distance, a galaxy's speed doubles as well. A galaxy 10 million light-years away travels twice as fast as a galaxy 5 million light-years distant. Hubble also calculated the rate of that increase. This number has since been amended (as better and better measurements were made over the years), but at first Hubble found that for every million parsecs outward (around 3 million light-years), the velocity of a galaxy increased by 500 kilometers per second. He referred to this factor as K, the same term introduced by others in earlier analyses. By the late 1930s, though, astronomers were regularly referring to it as H, "Hubble's constant," later shortened to *the* Hubble constant.

Hubble did not really "discover" this relation but rather demonstrated an effect already suspected, with data that at last convinced his fellow astronomers. In previous attempts the plotted measurements looked like scattershot across the page. But on Hubble's graph, even though there was still some scatter, the galaxies lined up far more tightly. The sure straight line he was able to draw through his points, a diagram that is now an honored icon in cosmology textbooks, gave everyone confidence in the results.

Hubble actually carried out two separate computations. In one, he calculated his rate of recession using all twenty-four of his galaxies. In the second approach, he figured out a rate of recession when he combined the galaxies into nine groups, according to their distance and direction on the sky. Both methods led to similar outcomes. "For such scanty material, so poorly distributed, the results are fairly definite," concluded Hubble, almost with surprise.

In a clever move, Hubble didn't include on his historic graph the strongest bit of evidence then on hand. He left that for Humason to convey in a separate paper opportunely placed right before his in the *Proceedings of the National Academy of Sciences*. Humason's result was the attention grabber, setting up Hubble's paper, published under his name alone, for the fait accompli. After cutting his teeth on a few of Slipher's galaxies, Humason had gone after a fainter, previously unmeasured target, as Hubble directed. It was the galaxy NGC 7619 in the Pegasus con-

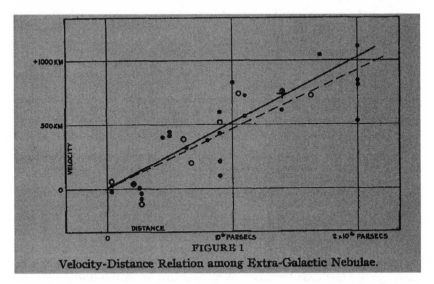

FIGURE 1
Velocity-Distance Relation among Extra-Galactic Nebulae.

Edwin Hubble's famous 1929 velocity-distance graph for a sample
of galaxies out to 2 million parsecs (~6 million light-years),
evidence for what later came to be understood as an expanding
universe (*From* Proceedings of the National Academy of Sciences *15*
[1929]: 172, *Figure 1*)

stellation. "I agreed to try one exposure," recalled Humason, who wanted to see whether it was even possible to venture farther out than Slipher. That exposure extended over a few nights, the sparse photons from the dim galaxy hitting the plate for a total of thirty-three hours. It was a lonely enterprise. Humason usually worked within the 100-inch dome with only the light of a tiny red bulb as company. For hours he would keep two crossed hairs—the guiding wires—smack-dab on his galaxy, a smudge of light barely visible through the barrel of the telescope. All this despite the fact, as one observer put it, "that the mountain itself is rolling eastward with the earth at ten times an express train's speed." For assurance, Humason went back and did the measurement again, this time for forty-five hours.

But making these observations was only the beginning. Back in the office, the spectrum on each photographic plate had to be put under a microscope and the shift of the spectral lines carefully measured. Humason often used the lines generated by calcium atoms glowing within the galaxy, which stood out sharply. Given Humason's lack of training in mathematics, though, he'd then take his measurements to an observatory "computer" (a person with a slide rule or adding machine), who used a formula to convert the physical measurement into a galaxy velocity. In the case of NGC 7619, Elizabeth Mac-Cormack calculated a final velocity of 3,779 kilometers per second, more than twice Slipher's highest velocity. The success spurred Mount Wilson officials to get Humason a better and faster spectrograph, appreciably reducing his bone-wearying exposure times, and the Kodak company was inspired by his problem to invent faster photographic plates, which was fortunate. Exhausted by his initial spectral observations, Humason was ready to quit Hubble's project if such improvements were not in the works.

With NGC 7619's speed record in hand, Humason was able to put a further gold star in Hubble's rising constellation by noting that his measurement meshed exactly with the relation that Hubble had just found between a galaxy's distance and its velocity. "The high velocity for N. G. C. 7619 derived from these plates," Humason was able to report, "falls on the extrapolated line." With that triumphant pronouncement, Humason was extending Hubble's findings even farther into space, out to 20 million light-years.

It seemed inevitable that the only astronomer to voice immediate doubts about Hubble's new law was his long-standing adversary, Harlow

Shapley, who was concerned that distances could only be certain for the nearest galaxies. He was perhaps envious, "in part regretting a lost opportunity to pursue such a relation himself," suggests historian Robert Smith. Shapley wrote a quick and pointed response to Hubble's paper, where he rightly argued that at a great distance a cluster of stars would be mistaken for a single star, making it a bad "standard candle." But on the opening page of this article, he didn't miss the opportunity to steal away a bit of Hubble's thunder and proclaim that ten years earlier he had published a notice that "the speed of spiral nebulae is dependent to some extent upon apparent brightness, indicating the relation of speed to distance." Shapley failed to disclose that in 1919 he still believed that the spiral nebulae were minor members of the Milky Way. So his claim, in the end, was meaningless.

Within two years, Hubble and Humason examined forty more galaxies beyond the several dozen that Slipher had earlier measured. They proceeded outward from what was then measured as 6 million light-years to as far out as 100 million light-years, a tremendous vault into the cosmos over such a short time. "Humason's adventures were spectacular," recalled Hubble many years later. "When he was sure of his techniques, and confident of his results, he set forth. From cluster to cluster he marched with great strides right out to the limit of the 100-inch." At one point Humason spent an entire week, night after night, gathering the light from just one faint galaxy in the Leo cluster to determine its redshift. Nicholas Mayall, a graduate student from Berkeley who was then assisting Hubble on a galaxy-counting project, was there as Humason developed the photograph at the end of the run. Holding the small plate up in front of the light box, Humason declared, "My God, Nick, this is a big shift!" The spectral lines, recalled Mayall, "were shifted way over to hell-and-gone from where they should have been. This proved to be a red shift of 20,000 kilometers per second, and it was probably more than twice the biggest one he had ever obtained before. He was simply jubilant." This galaxy was racing outward at more than a twentieth the speed of light. Caught up in the moment, Humason announced it was time to celebrate and promptly went down to his room, swung open his closet door, and took out a bottle of his mysterious "panther juice," an illicit alcoholic brew. After their toast at dawn and a brief nap, the two colleagues went for breakfast at the Monastery and called up Hubble on the phone with the news of the record-breaking redshift. "Milt," replied Hubble, "you are now using the 100-inch telescope the way it should be used." The link between a galaxy's velocity and its distance had been made even stronger. "You can't imagine how electric the atmosphere was," said Mayall. "So many things were happening in astronomy and physics—they all came to focus at that time and place."

Humason's velocities were so astounding that some astronomers were finding it difficult to believe that he could measure them at all. But Humason had the benefit of experience. By first pegging redshifts at relatively low velocities, he became as familiar with the spectral lines as with old friends, and found it easy to spot them as they shifted red-

PLATE III

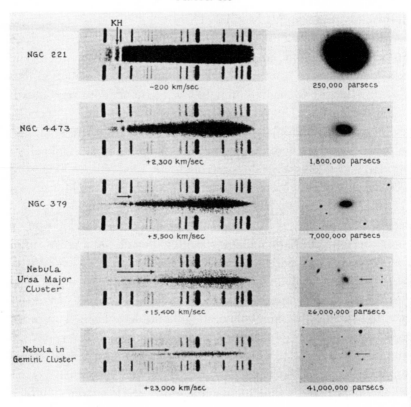

RED-SHIFTS IN THE SPECTRA OF EXTRA-GALACTIC NEBULAE

A selection from Milton Humason's measurements showing how the spectral lines for calcium (marked as KH) move farther to the right (red end of the spectrum) as both the distance for the galaxy and its velocity increase (1 parsec = 3.26 light-years) (*From* Astrophysical Journal 83 [1936]: 10–22, Plate III, *courtesy of the American Astronomical Society*) ward and got fainter once he pursued galaxies farther and farther into space.

So important did the Mount Wilson administration consider this endeavor that Hubble and Humason got almost all the "dark time" on the 100-inch throughout the 1930s, to the dismay of other galaxy researchers. Only during those precious few nights each month when the Moon, with its disruptive light, remained hidden below the horizon could the two investigators carry out the measurements on their extremely faint targets. "The intense publicity that swirled around Mount Wilson's nebular department, with Hubble the bright star at its core," noted Allan Sandage, "was anathema to the spectroscopists [at Mount Wilson]." Most astronomers at this time, in the United States and elsewhere, were focused on determining the life histories of stars. "Here they were," continued Sandage, "toiling away on stellar astrophysics—the most exciting and exotic facet of contemporary astronomy, to their eyes—yet the public seemed to find them, well, *boring.*" Even though less than 5 percent of Mount Wilson's major publications in this era involved cosmology, the topic dominated the news stories coming out of the observatory. "Some spectroscopists began to feel resentful," said Sandage. Even to this day, the legend persists that the Mount Wilson Observatory's sole focus at that time was galaxies, so great was the attention focused on Hubble and his accomplishments.

But what did it all mean? What was causing the galaxies to flee from the Milky Way in such a methodical way? Were these swift velocities even genuine? It was easy to equate the redshifts with velocity, as that was the simplest interpretation and the most straightforward way to talk about the phenomenon in scientific papers. Everyone used the terms interchangeably. But perhaps some new law of physics was at work and the galaxies weren't truly racing away after all. Maybe the retreat was entirely a chimera.

Hubble, the consummate observer, did not consider that question his main concern. He was reluctant to speculate. He wanted only to weigh the data that the universe provided him. Given that leaning, Hubble devoted most of his 1929 paper to establishing the link between a galaxy's distance and its redshift, its six pages filled with tables of numbers, a few equations, and a single graph. Only in the very last paragraph did he bring up a potential explanation. "The outstanding feature," he wrote, "is the possibility that the velocity-distance relation may represent the de Sitter effect," the most active model then in play. Maybe the light waves were lengthening as they traveled, setting up the illusion of movement; or maybe matter was truly scattering outward due to the weird nature of de Sitter space. More significant to Hubble was that bona fide data could now be offered in discussions of cosmological models. For centuries cosmology was a realm of speculation and imagination alone. Anybody's vision of the cosmos could be entertained—its origin, its behavior, its structure—simply because there was no way to refute it. But now actual cosmic measurements could be brought to the debate. Theory now had to meet the test of observation. It was one of the triumphs of twentieth-century astronomy, and Hubble initiated the endeavor. The galaxies became his "great beacons scattered through space," luminous markers for mapping the topography of the universe.

For the most part, Hubble remained focused solely on his observations, leaving theory to others. "The interpretation," he told de Sitter at one point, "should be left to you and the very few others who are competent to discuss the matter with authority." Hubble obviously had some serious doubts about what it all meant. "It is difficult to believe that the velocities are real—that all matter is actually scattering away from our region of space," he told a *Los Angeles Times* reporter in 1929. In the very first paragraph of his discovery paper on this subject he referred to the velocities of the galaxies as "apparent." He maintained this conceit for the rest of his career. As long as various theoretical explanations were under scrutiny, he didn't want to be caught on the wrong side. He was never comfortable in the robe of a theorist and so published his data in such a way that the measurements could remain unsullied, no matter what the interpretation. "Not until the empirical sources are exhausted, need we pass on to the dreamy realms of speculation," he mused. This attitude rubbed off on his loyal helpmate as well. "I have always been rather happy that . . . my part in the work was, you might say, fundamental; it can never be changed—no matter what the decision is as to what it means," said Humason.

Hubble was quite possessive of their legacy and kept close watch on it. When de Sitter in a 1930 review article casually referred to the link between velocity and distance ("It has been remarked by several astronomers that there appears to be a linear correlation"), Hubble immediately picked up his pen and reminded de Sitter who should be getting the lion's share of the credit. "The possibility of a velocity-distance relation among nebulae has been in the air for years—you, I believe, were the first to mention it," he wrote. "But . . . I consider the velocity-distance relation, its formulation, testing and confirmation, as a Mount Wilson contribution and I am deeply concerned in its recognition as such."

Hubble conveniently forgot to tell de Sitter that most of the galaxy velocities he first drew upon in his 1929 paper were actually Slipher's data, which Hubble used without direct citation or acknowledgment, a serious breach of scientific protocol. Hubble partially made up for this nefarious deed by briefly referring in his next big paper on the redshift law, published in 1931, to the "great pioneer work of V. M. Slipher at the Lowell Observatory." More gracious amends were made in 1953. That year, as Hubble was preparing a talk on the "Law of Red-Shifts" to be given in England (the prestigious George Darwin Lecture of the Royal Astronomical Society), he wrote Slipher asking for some slides of his first 1912 spectrum of the radial velocity of the Andromeda nebula and in this letter at last gave the Lowell Observatory astronomer due credit for his initial breakthrough (albeit more than two decades late). "I regard such first steps as by far the most important of all," wrote Hubble. "Once the field is opened, others can follow." In the lecture itself, Hubble professed that his discovery "emerged from a combination of radial velocities measured by Slipher at Flagstaff with distances derived at Mount Wilson. . . . Slipher worked almost alone, and ten years later . . . had contributed 42 out of the 46 nebular velocities then available."

Privately, Slipher was bitter that he didn't get more immediate public credit but was too humble and reserved to demand his share of the glory in 1929. He was at least honored by his peers for his contributions. The Royal Astronomical Society presented its highest award, the Gold Medal, to him in 1933, with its president, Frederick Stratton, amusingly announcing that "if cosmogonists to-day have to deal with a Universe that is expanding in fact as well as in fancy, at a rate which offers them special difficulties, a great part of the initial blame must be borne by our medallist." In many ways, Slipher's accomplishment resembled that of Arno Penzias and Robert Wilson several decades later. In 1964 the two Bell Laboratory researchers were calibrating a massive horn-shaped antenna in New Jersey in preparation for some radio astronomy observations and registered an unexpected cosmic radio noise wherever they looked on the sky, spending months trying to discover its source. Just as

Slipher revealed a remarkable phenomenon that took others time to fully interpret, so too did Penzias and Wilson need fellow astronomers to tell them what they had found, that they had been listening to the faint reverberation of the Big Bang all along. But there any resemblance between the two cases ends. While Penzias and Wilson received the Nobel Prize for their serendipitous discovery, firming their rank among the scientific elite, Slipher as the years passed was reduced in the public's eye to a secondary role in the momentous saga of the fleeing galaxies.

[15]

Your Calculations Are Correct, but Your Physical Insight Is Abominable

Ifty-four years after its founding in 1820, the Royal Astronomical Society began to hold its monthly meetings at new headquarters in the west wing of Burlington House, a former private Palladian mansion that houses a number of British learned societies off Piccadilly in the heart of London. At the society's gathering on January 10, 1930, after a report on the current performance of two clocks at the Royal Observatory in Greenwich, the chairman called upon Willem de Sitter, then visiting England, to give an account of his latest research. De Sitter rose and spoke that evening about his own attempts to link the velocity of a galaxy to its distance. Just as Hubble had demonstrated the previous year, de Sitter too graphed a straight line through his points, making use of data obtained by Hubble, Lundmark, and Shapley. But could he explain this orderly recession of the galaxies? "I am not sure that I can," de Sitter told his audience. The Dutch astronomer was coming to appreciate that his cosmological model was inadequate, not a good approximation of the observed universe at all. His solution depended on the cosmos being empty, but the universe was undoubtedly chock-full of matter.

In the ensuing discussion, Arthur Eddington casually wondered aloud why only two cosmological models—Einstein's and de Sitter's—had so far come out of general relativity to describe the universe. Were other solutions possible, ready for plucking within Einstein's equations? A number of respected mathematicians had been sporadically tinkering with the models, offering up modifications, but none generated wide interest. Was that the end of the road?

Einstein and de Sitter had each started with different simplifying assumptions and so arrived at different solutions. But they did have one thing in common: Both took for granted that the overall structure of space-time was static—fixed and rigid. "I suppose the trouble is that people look [only] for static solutions," noted Eddington at the meeting. From one perspective, de Sitter's solution could be viewed as nonstatic, if you considered any matter in it as immediately flying off, "but as there isn't any matter in it that does not matter," argued Eddington.

Much more was at stake in Eddington's question. It was easy to imagine a massive object like a star indenting space-time in a very local and specific location, but could the entire fabric of the cosmos, across the span of the universe, be changing as the eons passed? Could the universe itself be dynamic? It seemed more realistic and plausible to imagine the galaxies traveling *through* space rather than space-time itself varying, so everyone insisted on a cosmic space that did not move. "From the point of view of cosmologists in the 1920s," writes science historian Helge Kragh, a dynamic universe "was a concept outside their mental framework, something not to be considered, or, if it was considered, to be resisted." But, just in case, Eddington already had a research assistant looking into such a formulation.

What Eddington forgot was that this additional cosmological model had already been conceived and presented to him. This solution had been around for years and meshed nicely with Hubble's observations. It wasn't Einstein's universe, and it wasn't de Sitter's. Like Goldilocks and her chairs, this new cosmic model was something in between—and just right.

The novel solution was the brainchild of Eddington's former pupil, Abbé Georges Lemaître, both a physicist and Jesuit priest. A member of the faculty at the Catholic University of Louvain in Belgium, Lemaître soon read the remarks Eddington made at the London meeting, published in the latest issue of the *Observatory*, and quickly sent off a letter reminding Eddington of a paper he had written three years earlier, which provided the answer Eddington craved. Few had seen the article, titled "A Homogeneous Universe of Constant Mass and Increasing Radius Accounting for the Radial Velocities of Extra-Galactic Nebulae," because for some unknown reason Lemaître had published it in an obscure Belgian journal, *Annales de la Société Scientifique de Bruxelles* (*Annals of the Brussels Scientific Society*) rather than a publication on every astronomer's must-read list. Eddington had either put Lemaître's

paper aside, never getting around to reading it, or simply didn't comprehend its importance at the time. In any case, all memory of it had vanished from his mind. After receiving Lemaître's message, he was a bit
shamefaced at the lapse. Looking back over the 1927 paper, he at last
recognized its significance and with great enthusiasm made up for his
blunder. He speedily sent de Sitter a copy of Lemaître's article, writing at the top, "This seems a complete answer to the problem we were
discussing." De Sitter as well grasped the brilliance of Lemaître's
approach, calling it "ingenious" and immediately abandoning his own
solution. Eddington soon arranged for Lemaître's paper to be translated
and reprinted in the March 1931 issue of the *Monthly Notices of the
Royal Astronomical Society*, where it could at last be given a proper
showcase.

Originally trained in engineering, Lemaître had switched to mathematics for graduate work and upon receiving his doctorate enrolled in a
seminary and was ordained a priest in 1923. Becoming fascinated with
the mathematical beauty of general relativity, he went to Cambridge
University for postdoctoral studies to broaden his understanding of Einstein's equations under the guidance of the eminent Eddington, who
soon noticed Lemaître's talents. With his dark hair combed straight
back and a cherubic face framed by round glasses, Lemaître could easily be spotted on campus because of his attire, either a black suit or an
ankle-length cassock, set off by a stiff white clerical collar. Others could
find him just by pursuing the sound of his full, loud laugh, which was
readily aroused. Eddington told Shapley that the young Belgian, then
turning thirty, was "exceptionally brilliant . . . quite remarkable both for
his insight into physical significance of problems, and for his manipula-
tion of intractable formulae."

After a year in England, Lemaître traveled to the United States for
further study and soon became aware of—and very interested in—the
application of general relativity to cosmological questions. He made
sure to attend the 1925 Washington meeting of the American Astronomical Society and was in the audience when Russell read Hubble's
paper on the existence of other galaxies. While others in the room were
focused on Hubble having ended the "Great Debate," Lemaître was
two jumps ahead. Though new to astronomy, he quickly realized that
Hubble's discovery could also be applied to fashioning models of the
universe. The newfound galaxies could be used as markers to test the
condition of the universe as predicted by general relativity. Later that

year, while at MIT to complete an additional PhD, he began modifying de Sitter's cosmological model. Before returning to Belgium, he visited Slipher at the Lowell Observatory, in Arizona, and also journeyed to sunny California, in order to meet Hubble and learn of the latest distance measurements of the spiral nebulae.

What Lemaître did not know during this interlude was that another researcher had already completed a similar modification. The Russian mathematician Aleksandr Friedmann had done this while Lemaître was still preparing for the priesthood. Trained in pure and applied mathematics, Friedmann specialized in the physics of the atmosphere, working at an aerological observatory and applying his expertise at the Russian front during World War I. After the war he returned to St. Petersburg to work at a geophysics observatory. There, among his diverse interests, he began investigating new solutions to Einstein's general theory of relativity, which had not been known to Russian scientists until after the war and the ensuing Russian civil strife.

The rival theories of Einstein and de Sitter were, in a way, complementary rather than competitive. In de Sitter's universe there was no matter to provide a gravitational attraction, but the cosmological repulsion allowed for movement. Einstein's universe, on the other hand, included matter, which provided enough of a gravitational force to oppose the repulsion. With enough matter, all was in perfect balance. Einstein's universe remained motionless. Friedmann blended the best aspects of these universes. He brought the two extremes under one mathematical roof, providing a model that better described the universe as we observe it: containing matter and yet also moving.

What Friedmann did most of all was introduce *time* into the deliberations. In papers written in 1922 and 1924 Friedmann began to play, in a sense, with Einstein's cosmological model. He wanted to see how curvatures in space-time might change over time—to "demonstrate the possibility," as he put it. To Friedmann, this was purely a mathematical enterprise, not astronomy at all. His sole goal was to try out possible solutions to Einstein's equations when applied to the entire cosmos. Like Einstein, he too filled his model universe with matter, but this time had it rapidly moving as the eons passed. Moreover, depending on the amount of matter, this movement of space-time could be an expansion, a contraction, or even an oscillation between the two states. "We shall call this universe the *periodic world*," he wrote in his report to the *Zeitschrift für Physik*. Friedmann even computed an age for the uni-

verse, a first in the annals of astronomy. He arrived at a figure of ten billion years, not far from today's consensus of nearly fourteen billion
years, although Friedmann considered his estimate more a curiosity. He
made sure to note the age could also be infinite. But, all in all, his paper
was predominantly an exercise in relativistic mathematics rather than
cosmology, which is why it received so little attention at the time. Friedmann made no mention of nebulae, radiation, or redshifts, nor did he
promote a cosmic expansion over a contraction. The journal in fact had
indexed his article under relativity theory, making no reference that it
dealt with cosmology, which is why it was easily overlooked.

Einstein was certainly aware of the Russian's paper, though. He
promptly dismissed the solution, thinking it had no physical significance whatsoever. In a letter to the *Zeitschrift*, sent off right before he
went on tour in Japan, he wrote that Friedmann's results "appear to me
suspicious." Friedmann, unfortunately, had little chance to either
defend or champion his intriguing idea. In 1925, he became ill with
typhoid, just a month after conducting a record-breaking balloon ascent
(an altitude of 4.6 miles) to make meteorological and medical observations. He soon died at the age of thirty-seven. In a way, Friedmann had
offered his solution too early. At this stage, most general relativists
weren't terribly interested in astronomy, and astronomers who had
more at stake in this quest didn't yet make the connection, believing
that such models of the universe were more like mathematical toys, fun
to fiddle with but hardly attached to the real world. They didn't take
them seriously.

Lemaître was the exception. From the very start of his independent
calculations in the mid-1920s, he kept astronomy foremost in his mind,
unlike Friedmann. De Sitter's universe could explain the redshifted
nebulae but required the universe be nearly empty (which it was not).
Einstein's universe could be filled with matter but couldn't account for
the fleeing nebulae. Lemaître declared that his aim was to "combine
the advantages of both." Returning to Belgium and a professorship at
Louvain, Lemaître continued working on the problem, at last publishing his final result in 1927. Two full years before Hubble provided the
definitive observational proof, Lemaître unveiled a cosmological model
in which the radius of the universe increases and galaxies surf outward
on the wave. The receding galaxies, as Lemaître described it in his
paper, "are a cosmical effect of the expansion of the universe."

From our perspective, it appears that all the galaxies in the universe are rushing away from *us*—that we are somehow situated at the very center of the cosmic action—but in reality you would observe the same dash outward from any other galaxy in the universe. Lemaître was the first to say directly that the galaxies are fleeing from us because space-time at each and every point throughout the cosmos is continually stretching. The galaxies are not rushing *through* space but instead are being carried along as space-time inflates without end. The embedded galaxies are simply going along for the ride. That's why the recession occurs in a specific way: A galaxy twice as far from us recedes twice as fast; a galaxy three times farther travels three times faster, and so on. Lemaître even estimated a rate of cosmic expansion (625 kilometers per second per megaparsec, based on the galactic velocity and distance data then available) that was close to the figure of 500 that Hubble would later calculate.

This was a tremendous accomplishment and offered an astounding vision of how the universe operates. But no one noticed—no one at all. Lemaître's paper, like Friedmann's earlier, was completely ignored. It was as if the article had never been published. Lemaître traveled in Europe and the United States afterward but inexplicably did not widely discuss this latest idea with his colleagues, either in person or in letters. Throughout his deliberations, he had been in contact with astronomers who would have been tremendously interested in his new take on the universe, such as Shapley, Slipher, and Hubble. Yet he apparently kept silent. Either he still had doubts about his new cosmic model or his ardor was dampened by encounters he had with the architects of the leading cosmological models. Though outwardly an extrovert, Lemaître was still quite sensitive to the smallest slight. In October 1927, just six months after his paper came out in the Belgian journal, he met with Einstein during the Fifth Solvay Congress in Brussels, a triennial meeting of the world's top physicists, and the two had a brief chat about Lemaître's breakthrough in the city's Leopold Park. It was at this time that Lemaître first heard from Einstein about Friedmann's similar solution. By then Einstein no longer had any objection to the mathematics in either man's model (his initial rejection of Friedmann's work had been based on an error in his own calculations), but he was still repelled by the image of the cosmos that the models of both Friedmann and Lemaître conveyed. "Your calculations are correct, but your physical insight is abominable," asserted Einstein, who could not (and would not) imagine a universe in motion. Later, while accompanying Einstein

Georges Lemaître with Albert Einstein in 1933 at the
California Institute of Technology (*Courtesy of the Archives,
California Institute of Technology*)

on a university lab tour, the Belgian cleric continued to press his case,
talking about the latest evidence on the galaxies' speeding away from
Earth. But in the end he came away from the meeting with the impression that Einstein was "not current with the astronomical facts." Nine
months later at the 1928 General Assembly of the International Astronomical Union, de Sitter was equally dismissive of the little-known
priest. As one commentator noted, de Sitter seemingly had "no time for
an unassuming theorist without proper international credentials."

The impasse held until Hubble and Humason verified that the
galaxies were truly moving outward in a uniform way and Lemaître's
model, circulated more prominently in the *Monthly Notices* in 1931,
could at last explain it as the fabric of space-time stretching outward,
carrying the galaxies ever farther apart. Now it was no longer Einstein's

universe or de Sitter's universe, but the *expanding* universe, and Lemaître became the toast of the cosmological town for being one of its primary creators. Hubble did not really discover the expanding universe in 1929, as written up in textbooks and commonly presumed these days. That realization did not actually occur until Hubble's data could be viewed with Lemaître's model firmly in mind. Lemaître, far more than Friedmann, had linked his model with ongoing astronomical observations. His solution was described as a "brilliant discovery." Top mathematical theorists began to flock to the new field of relativistic cosmology, both to extend the model and to produce variations on Lemaître's original theme of a universe in bloom. In preparing a review paper for a physics journal in the early 1930s, Princeton theorist Howard P. Robertson, himself a leading expert in this new endeavor, noted, "Imagine my surprise on being able to rustle together more than 150 references on relativistic cosmology! It seems to me that some of our highlights . . . are going off the deep end."

Astronomers and theorists alike were thunderstruck by this radically new picture of the universe, which was reported as breathtaking in its grandeur and terrifying in its implications. "The theory of the expanding universe is in some respects so preposterous," said Eddington, "that we naturally hesitate before committing ourselves to it. It contains elements apparently so incredible that I feel almost an indignation that anyone should believe in it—except myself." That's because by then he knew that it was rooted in the most powerful idea to be introduced in the world of physics since Isaac Newton—Einstein's general theory of relativity—and test after test was proving it true.

James Jeans, a prolific writer as well as theorist, employed the iconic description of the cosmic expansion used to this day. "On the face of it," he said, "this looks as though the whole universe were uniformly expanding, like the surface of a balloon while it is being inflated, with a speed that doubles its size every 1,400 million years. . . . If Einstein's relativity cosmology is sound, the nebulae have no alternative—the properties of the space in which they exist compel them to scatter." Eddington first devised this picture when he introduced his colleagues to Lemaître's solution in a 1930 paper to the *Monthly Notices of the Royal Astronomical Society*. Paint dots on that balloon and, as it expands, every dot will move farther from every other dot in a regular fashion. Similarly, wrote Eddington, in the expanding universe the galaxies appear to be "embedded in the surface of a balloon which is steadily inflating." Every galaxy in the cosmos thus sees its neighbors receding into distant space.

Though Hubble left such interpretations of the velocity-distance relationship to others, he did participate in the discussions, hoping to glean what data needed to be gathered to select between competing theories. Astronomers and theorists previously resided in separate domains, but now he got them talking. Grace Hubble recalled the commotion it created in her household, shortly after Lemaître's model got wide circulation: "About every two weeks some of the men from Mount Wilson and Cal Tech came to the house in the evening . . . astronomers, physicists, mathematicians. They brought a blackboard from Cal Tech and put it up on the living-room wall. In the dining-room were sandwiches, beer, whiskey and sodawater; they strolled in and helped themselves. Sitting around the fire, smoking pipes, they talked over various approaches to problems, questioned, compared and contrasted their points of view. Someone would write equations on the blackboard and talk for a bit, and a discussion would follow."

There was much to argue about. Those still skeptical of general relativity were offering other explanations for the outward march of the galaxies. British cosmologist E. Arthur Milne, for example, posited that the expansion of space-time was merely an illusion. Space was steady as a rock, but the spiral nebulae upon forming started moving in random directions and with different velocities. Over the eons, the nebulae with the fastest speeds naturally moved farther out, setting up the appearance of a cosmic expansion. It was a model that philosophically pleased Milne, who didn't believe space could possibly curve, bend, or move.

Caltech astronomer Fritz Zwicky proposed that light waves, as they traveled through space, could be interacting with matter, setting up a sort of gravitational drag. The more a light wave traveled, the more it lost energy, shifting its wavelength toward the red end of the spectrum. It resembled the de Sitter effect, only this time matter was doing the work. This could explain why the nebulae farthest out displayed the largest redshifts. Space wasn't expanding at all; the photons of light were simply getting weaker and weaker in their journey through a matter-filled cosmos. Hence, this model came to be known as the "tired photon" theory. There was no natural way to explain how this would happen; it required a new law of physics, but that didn't deter Zwicky at all. He was a legend among astronomers for his chutzpah. He felt his explanation might be pointing to a new physical phenomenon.

Hubble worked for a number of years with Caltech theorist Richard Tolman on how to test these competing models of the universe. They wanted to see which one was most compatible with the data arriving at the telescope. Their effort eventually came to naught. Given the state of astronomy at the time and the instruments available, there were simply too many uncertainties—too much guesswork—to reliably choose one cosmological model over another. Their initial data, though, seemed to better support some alternative theories, like Zwicky's "tired photon" scheme. But Hubble made the call that his data were too uncertain, which kept the expanding universe in play. "We cannot assume that our knowledge of physical principles is yet complete," he wrote, "nevertheless, we should not replace a known, familiar principle, by an ad hoc explanation unless we are forced to that step by actual observations." To back away from Einstein, the proof for Hubble had to be overwhelming. On the other hand, the uncertainty of it all likely reinforced his qualms at advocating any particular interpretation.

Lick astronomer C. Donald Shane, in talks with Hubble in the 1930s, actually got the impression that Hubble had "a desire to show that the red shift was not an expansion . . . because he seemed always to be seeking some other explanation for it." Perusing Hubble's writings on the idea of an expanding universe, you immediately detect that he was uncomfortable with it. He acceded that theorists were "fully justified" in interpreting the galaxy redshifts as a movement outward; it was the most reasonable explanation that required no new laws of physics. But then he would invariably sneak an "on the other hand" into his script. He deemed a static and infinite universe more "plausible" and "familiar," like a pair of old shoes he found difficult to throw out. In his Rhodes Memorial Lectures, delivered at Oxford in the autumn of 1936, Hubble reaffirmed his vacillation over the interpretation of the redshifts. Their "significance is still uncertain," he stated. With the recent introduction of both quantum mechanics and relativity, which demonstrated quite explicitly that scientists' understanding of nature can change abruptly and in surprising ways, perhaps Hubble's caution was understandable. In his lecture Hubble went on to describe the expanding universe as a "dubious world," though still conceding it was the more likely interpretation of the redshifts. But with alternate explanations still in play, he concluded that astronomers were in "a dilemma [whose] resolution must await improved observations or improved theory or both."

What seemed to disturb Hubble most were the enormous velocities.

The farther he and Humason extended their searches into space, the faster and faster the galaxies were retreating. Near the absolute limit of Humason's spectrograph, he recorded velocities of about 25,000 miles per second, "around the earth in a second, out to the moon in 10 seconds, out to the sun in just over an hour . . . the notion is rather startling," noted Hubble.

As late as 1950, responding to a Kansas professor's written inquiry about redshifts, Hubble asserted that they "represent either actual recession (expanding universe) or some hitherto unknown principle of nature. I believe that the choice of these alternatives will be determined with the 200-inch [telescope on California's Palomar Mountain] within a few years." Maintaining his lawyerly ways, Hubble covered all the bases when making a public statement.

Others, such as Eddington, were confounded by such equivocation. "I just don't understand this eagerness to find some other theory than the expanding universe," he wrote in a letter to a colleague. "It arose out of difficulties . . . in Einstein's theory. If you do away with it, you throw back relativity theory into the infantile diseases of 25 years ago. And why the fact that the solution then found has received remarkable confirmation by observation should lead people to seek desperately for ways to avoid it, I cannot imagine."

While Hubble remained overly cautious, Shapley came to embrace the idea of an expansion lock, stock, and cosmic barrel. It's as if the two astronomers were magnets with the same polarity, always repulsing each other to opposite sides of a question. The ultimate imprimatur, though, was provided when Einstein arrived in Pasadena in 1931 in order to consult with the high priests of cosmology at both Caltech and Mount Wilson.

[16]

Started Off with a Bang

On November 30, 1930, Einstein, his wife, Elsa, his secretary, and a scientific assistant left Berlin for Antwerp, where they embarked on the steamer *Belgenland*. It was Einstein's second visit to the United States but his first journey to America's West Coast. Before leaving, Frau Einstein made a last-minute shopping trip to purchase a raincoat for the father of relativity. "Would it not be more practical to have the herr professor come here so we can give him an exact fit?" said the clothing store salesman. "If you knew how hard it was even to persuade my husband he needed a new coat, you wouldn't expect me to fetch him here. I wish you had my worries," she replied. It was teasingly said that Einstein was going to Pasadena to hunt for the sole twelve men in the world who could understand him.

The revered physicist arrived in New York on December 11, where he and Elsa were greeted by a barrage of journalists, photographers, and newsreel men, a chaotic scene that greatly discomfited Einstein. "This reminds me of a Punch and Judy show, all of you standing there watching us so intently," he remarked in German. The press described him that day as small, bright-eyed, his almost white hair trained back in a bushy pompadour, and "his face . . . as smooth as a girl's except for the tiny wrinkles about his eyes." Out on the deck, a cold damp wind soon blew through his locks, swiftly turning the carefully groomed pompadour into his well-known disheveled hairstyle. After a four-day stay in New York, he and his party continued their voyage on the *Belgenland* for California, by way of the Panama Canal.

Arthur Fleming, a member of the California Institute of Technology's executive council, first extended the invitation to visit, extolling his town's summery climate and rich scientific atmosphere. Einstein, then looking for a good rest among men who spoke the language of mathematics, eagerly accepted. For one, it was an opportunity for him to meet Albert A. Michelson, the physicist whose inexplicable failure to measure a predicted change in the speed of light due to Earth's motion through an "ether" permeating space was at last explained by Einstein's special theory of relativity, which did away with the ether altogether.

Aware of Einstein's dislike for publicity, his California hosts tried to dispense with an official welcome, as in New York, but to no avail. Upon docking in San Diego on New Year's Eve, the German visitors had to endure four hours of speeches, presentations, tours, and a radio talk. Only after all the hoopla had ceased were Einstein and Elsa finally taken northward by car, eventually settling into a small Pasadena bungalow specially renovated and furnished for their stay. While shunning many public events over their two-month visit, the Einsteins enjoyed a steady round of private engagements. Over the ensuing weeks, they hosted a dinner for the director of the Los Angeles Philharmonic (with Einstein briefly playing the violin for his guest), visited a Hollywood studio, had dinner at the home of film comedian Charlie Chaplin, and motored out to Palm Springs for a four-day holiday. They did put up with the glare of the celebrity spotlight on one special occasion. The couple, he decked out in tuxedo and she in full-length evening gown, attended the premiere of Chaplin's latest movie, *City Lights*, where Einstein laughed like a little boy. There was a simple reason for this exceptional night on the town: Chaplin, instantly recognizable throughout the world, was Elsa's matinee idol. "They cheer me because they all understand me, and they cheer you because no one understands you," Chaplin told Einstein as they walked into the theater to shouts and clapping that night.

Einstein's days, though, were solely devoted to research, with visits to either Caltech or Mount Wilson's Pasadena headquarters for talks and consultations with fellow scientists. For his convenience, he had a small army of chauffeurs at his beck and call, including Grace Hubble. When driving Einstein to an engagement one day, he turned to her and said, "Your husband's work is beautiful—and he has a beautiful spirit." Einstein had been given a room at Mount Wilson's main offices right across from Hubble's. The observatory made every attempt to shelter him from the press and allow him maximum time to interact with his colleagues,

Einstein and his wife, Elsa, with Charlie Chaplin at the premiere
of Chaplin's film *City Lights*, January 1931 (*Copyright Jewish
Chronicle Ltd/HIP/The Image Works*)

even keeping the doors locked at the headquarters and issuing keys.
Hale, though, stayed away from all the partying. "I have kept completely
out of the Einstein excitement," he told a friend, "and have not seen
him at all until he dropped into my lab the other day, fortunately with
no reporter. He is very simple and agreeable and greatly dislikes all the
newspaper notoriety. But as the town is swarming with reporters, several
of them sent out here for the occasion by eastern papers, he cannot
escape entirely."

That was certainly the case on January 29, 1931, when a carefully
orchestrated expedition was arranged for Einstein. That morning the
world's premier physicist and Hubble, its foremost astronomer, settled
into the plush leather seats of a sleek Pierce-Arrow touring car and traveled, along with a number of other observatory staffers, up to the site of
Hubble's astronomical triumphs—the sprawling telescope complex
atop Mount Wilson. Despite warnings from his doctor to avoid high elevations, Einstein was eager to make the trek, so he could view up close
the machinery that had had such a direct bearing on his theoretical
investigations.

This event was considered so noteworthy that a young filmmaker named Frank Capra, still three years away from his first Academy Award for the screwball comedy *It Happened One Night*, came along to document Einstein's every move on the mountain that day. Clambering with a few others into an open steel box, operated by cables, Einstein was first carried to the top of the 150-foot-high tower telescope, used exclusively for the study of the Sun. After admiring the view of southern California and duly photographed in the cold, stiff breeze, he again went aboard the miniature elevator back to the ground. "And here he comes," said the announcer in the newsreel's opening, "down from the sun tower, after a hard morning, looking a few million miles into his favorite space."

After lunch came the opportunity to visit the 100-inch telescope, where Einstein again dutifully posed for Capra, peering through the eyepiece while Walter Adams stiffly spoke, directly to the camera. "This hundred-inch reflector was completed about thirteen years ago and has

For the cameras Einstein pretends to peer through the 100-inch telescope during his visit to Mount Wilson. Edwin Hubble (center) smokes his pipe and observatory director Walter Adams (right) looks on. (*Courtesy of the Archives, California Institute of Technology*)

contributed in three or four notable ways to progress in astronomy," he droned. All the while Hubble was also in the frame, wearing his sporting plus-fours (golf trousers cut four inches below the knee) and silently puffing away on his ever-present pipe. Away from the camera, Einstein delighted in the telescope's instruments. This was his first view of a large reflecting telescope, and he was quick to grasp the intricacies in its construction and operation. Like a child at play, the fifty-one-year-old physicist scrambled about the framework, to the consternation of his hosts. Nearby was Einstein's wife. Told that the giant reflector was used to determine the universe's shape, Elsa reportedly replied with wifely pride, "Well, my husband does that on the back of an old envelope."

After an early dinner the party returned to the 100-inch telescope, when Einstein was at last able to do some real observing, peering at Jupiter, Mars, the asteroid Eros, several spiral nebulae, and the faint companion of the star Sirius. He remained in the dome until after one o'clock, finally retiring under protest and with the stipulation that he be called in time to see the sunrise. Everyone returned to Pasadena at about ten o'clock that same morning.

Five days later, astronomers and theorists gathered in the spacious library of the observatory's Pasadena offices, books lining the walls from floor to ceiling, to hear Einstein's assessment of what he had learned and absorbed from his visit to the mountain. Up to this point, he had been very wary of considering a universe in restless motion, curtly dismissing the models fashioned by both Friedmann and Lemaître. Einstein, by far, preferred a universe that stayed put. But on that day he at last conceded that the secret of the cosmos had undoubtedly been revealed by Hubble's observations. Einstein at last let go of his spherical universe. "A gasp of astonishment swept through the library," according to an Associated Press reporter in attendance. At a follow-up session a week later, Einstein went further and announced that "the red shift of distant nebulae has smashed my old construction like a hammer blow," swiftly swinging down his hand to illustrate the point to his audience. Einstein at this stage recognized that he no longer needed his cosmological constant to describe this dynamic universe. His original equations could handle the cosmic expansion just fine, which pleased him immensely. From the start, he had had qualms about the ad hoc addition, believing the constant tarnished the formal beauty of his theory. Tacking on the extra term, he reportedly said, was the "biggest blunder"

Einstein with Hubble (second from the left) and others from Caltech
and the observatory outside the dome of the 100-inch telescope during
his visit to Mount Wilson on January 29, 1931 (*Courtesy of the
Archives, California Institute of Technology*)

he ever made in his life. The cocky kid was getting older. If he had
trusted his equations from the start, he could have predicted that space-
time was in motion years before Hubble and Humason confirmed it,
which would have rocketed Einstein's reputation, towering as it was,
into the stratosphere.

Given his role in this turnabout, Hubble was soon revered as the man who "made Einstein change his mind." Aside from perhaps receiving a Nobel Prize, there was no higher accolade in science at the time.

A few weeks before Einstein roamed over the summit of Mount Wilson, Eddington delivered an address to the British Mathematical Association, where he called attention to the notorious elephant in the room, present ever since Lemaître first introduced the concept of an expanding universe. In his masterly 1927 journal article, Lemaître had coyly asked the question that likely arose in the mind of anyone reading the paper: How did this expansion get started? "It remains to find the cause," he answered at the time.

Eddington in his January 5 talk to the British mathematicians faced this conundrum head-on. In his mind's eye, he mentally put the expansion of space-time into reverse and pondered the condition of the universe at earlier and earlier epochs, back to the very launch of space, time, and all of creation. Could you reach a "beginning of time," he asked, when all matter and energy had the highest degree of organization possible? Eddington was horrified by this thought. The Cambridge theorist concluded that "philosophically, the notion of a beginning of the present order of Nature is repugnant to me. . . . By sweeping it far enough away from the sphere of our current physical problems, we fancy we have got rid of it. It is only when some of us are so misguided as to try to get back billions of years into the past that we find the sweepings all piled up like a high wall and forming a boundary—a beginning of time—which we cannot climb over." A few years earlier, before the reason for the retreating galaxies was even known and he was simply contemplating an early universe with more energy and order, Eddington had already declared that he did "not believe that the present order of things started off with a bang" (a precursor to British astronomer Fred Hoyle using a similar description on a 1949 BBC radio program, this time with an added adjective, which secured the scientific name— Big Bang—for the moment of creation). Eddington, though, preferred a commencement less abrupt and more restrained. "I picture . . . an even distribution of protons and electrons, extremely diffuse and filling all (spherical) space, remaining nearly balanced for an exceedingly long time until its inherent instability prevails. . . . There is no hurry for anything to begin to happen. But at last small irregular tendencies accumulate, and evolution gets under way. . . . As the matter drew closer together in the condensations, the various evolutionary processes followed—evolution of stars, evolution of the more complex elements, evolution of planets and life." The universe, in effect, eased into its expansion, like a massive train starting up slowly and then gaining speed.

Lemaître, however, was far bolder and had no hesitation at all in contemplating a more dramatic genesis. In response to Eddington's repulsion at an abrupt cosmic beginning, Lemaître submitted a short note to the journal *Nature* with the splendiferous title: "The Beginning of the World from the Point of View of Quantum Theory." "If we go back in the course of time," replied Lemaître, ". . . we find all the energy of the universe packed in a few or even in a unique quantum. . . . If this suggestion is correct, the beginning of the world happened a little before the beginning of space and time. I think that such a beginning of the world is far enough from the present order of Nature to be not at all repugnant. . . . We could conceive the beginning of the universe in the form of a unique atom, the atomic weight of which is the total mass of the universe. This highly unstable atom would divide in smaller and smaller atoms by a kind of super-radioactive process." He called his initial compact cauldron the "primeval atom." Today's stars and galaxies, he surmised, were constructed from the fragments blasted outward from this original superatom.

Lemaître was spurred by the revelations of atomic physics in the early decades of the twentieth century, where radioactive elements were seen to endure over times similar to the age then calculated for the universe, a few billion years. "The evolution of the world can be compared to a display of fireworks that has just ended: some few red wisps, ashes, and smoke," the Belgian cleric would later write. "Standing on a well-chilled cinder, we see the slow fading of the suns, and try to recall the vanished brilliance of the origin of the worlds." This idea would later be revised by others to show how our universe evolved, not from a super-atom, but from a cosmic seed of pure energy. From Lemaître's poetic scenario arose today's vision of the Big Bang, the cosmological model that shapes and directs the thoughts of cosmologists today as strongly as Ptolemy's crystalline spheres influenced natural philosophers in the Middle Ages.

Though ordained as an abbé, later rising to the rank of monsignor, Lemaître did not endure the fate of Galileo in contemplating a scientific explanation for heaven's workings, in this case the universe's creation. As Helge Kragh has noted, "Lemaître believed that God would hide nothing from the human mind, not even the physical nature of the very early universe." Times had assuredly changed—while Galileo was condemned by church officials to house arrest for his defense of a Sun-centered universe, Lemaître was lauded by the Church for his cosmological breakthrough. However, nothing could upset Lemaître more than assuming his cosmological model had been inspired by the biblical story of Genesis. His contemplation of the origin of space and time, he persistently asserted, arrived exclusively from the equations before him. As a scientist/priest, Lemaître religiously kept his physics and theology in separate, unattached compartments.

But the Big Bang model faced a number of challenges before it could be fully accepted. The biggest hurdle was the estimate of the universe's age, based on early (and incorrect) measurements of the rate of cosmic expansion. Hubble's initial rate, calculated from a relatively small sample of galaxies, suggested that the universe originated just two billion years ago, but astronomers already knew of stars around ten billion years old. Looking closer to home, it was also less than the estimated age of Earth. Geologic evidence at the time indicated that Earth's crust was at least three billion years old, likely more. This paradox posed a dilemma for the model for quite a while. How could Earth possibly be older than the universe?

There were other loose ends. For one, the Milky Way still appeared to be far larger than the other galaxies. The Andromeda galaxy, the closest spiral to us, shared so many features with the Milky Way—the same disk of stars, the same system of globular clusters arranged in a halo around it, the same variable stars blinking on and off—and yet all these objects appeared fainter than those in the Milky Way, based on Hubble's initial distance measurement. More than that, Andromeda was smaller. This greatly bothered astronomers, who were now readily applying the Copernican rule to the entire universe: It is unlikely that we occupy a privileged place in the cosmos.

This puzzle persisted until 1952, just when Hubble's long reign as the emperor of cosmology was coming to an end. Having gone into military work during World War II, Hubble had a lot of catching up to do at the war's end, but ill health prevented him from getting back on top. By then Walter Baade, a gifted observer, was beginning to overshadow Hubble with his revelatory work that at last put the universe (and the

Big Bang) in better shape. Using both the 100-inch telescope during the war and later the 200-inch launched in 1948 on Palomar Mountain in California, Baade was able to prove that there were two distinct kinds of Cepheid stars. The Cepheids that Hubble used to determine the distance to Andromeda and other galaxies were actually more luminous than the Cepheids that Shapley used to determine his distances to the globular clusters surrounding the Milky Way. Consequently, Hubble had been underestimating his distances to Andromeda and the other galaxies. Hubble's distances had to be completely reworked. Andromeda, for example, was actually twice as far out, which also meant it was bigger than anyone had perceived and so made it more like the Milky Way's twin. Andromeda wasn't smaller or fainter at all, just more distant than previously thought—and that meant the Milky Way was no longer the special kid on the block. Those who desired nature to be uniform breathed a huge sigh of relief. This adjustment had to be made to all the galaxies then measured, essentially doubling the size and age of the universe. This modification at last took the Big Bang model out from under its cloud. With Hubble's rate of expansion determined more accurately as well, as more and more galaxies were measured, the estimated age of the universe was further increased, which at last allowed enough time for all the stars and planets to form after an explosive birth of space and time.

"Never in all the history of science," said Willem de Sitter in 1931 at a Boston lecture, "has there been a period when new theories and hypotheses arose, flourished, and were abandoned in so quick succession as in the last fifteen or twenty years." And perhaps never again will astronomy face such a dramatic shift in its conception of the universe. It took only three short decades—from 1900 to 1930, virtual seconds into our past when weighed against humanity's life span—to make this mind-altering transition. The Milky Way, once the universe's lone inhabitant floating in an ocean of darkness, was suddenly joined by billions of other star-filled islands, arranged outward as far as telescopes could peer. Earth turned out to be less than a speck, the cosmic equivalent of a subatomic particle hovering within an immensity still difficult to grasp. It didn't stop there. Astronomers barely had time to adjust to this astounding celestial vastness when they were faced with the knowledge that space-time, the universe's very fabric, was expanding in all directions, carrying the galaxies with it. It was a rapid one-two punch from which astronomy is still reeling, as observers and theorists alike try to make sense of all its details: how the Big Bang was ignited, how the myriad galaxies were born and evolve, how (and if) the expansion will end.

James Keeler could not possibly have imagined where his pioneering explorations were going to lead when in 1898 he first walked over to the Crossley reflector, a pip-squeak of a telescope compared to Lick Observatory's grand refractor, and made the simple decision to focus his research on the spiral nebulae. But his observations from Ptolemy Ridge had broad repercussions. He at last made the professional astronomical community sit up and take notice of celestial objects other than planets and stars. Reigniting up the cause after Keeler's death, Heber Curtis generated even more momentum. The arsenal of data he gathered throughout the 1910s with the Crossley supported a very strong case that the spirals were no less than separate galaxies. Though it was all circumstantial evidence, Curtis's observations laid down a substantial foundation that made it far easier for Hubble to place the final capstone, his distance measurements to the closest spirals, that at last persuaded his fellow astronomers. Both Keeler and Curtis were vital pathfinders, carving out a route that led to Hubble's ultimate triumph.

In a similar fashion, Vesto Slipher spent many lonely hours at his Lowell Observatory telescope, year after year, building up the reservoir of galaxy velocities that Hubble then used to establish his historic link between a galaxy's redshift and its distance, a systematic pattern that served as powerful proof for the expanding universe predicted by Georges Lemaître. Yet Hubble remained remarkably silent about the meaning of what he and Humason had found. Neither in his personal conversations nor in his writings did Hubble discuss the implications of his finding on ideas concerning either the evolution of the universe from a primitive state or the necessity of a creation event. That would come from others. Hubble was not comfortable with imaginative speculation, beyond what his observations could plainly demonstrate. Hubble was always the skeptical scientist, forever the questioning lawyer.

Nevertheless this image of Hubble, someone aloof and hesitant to embrace a dynamic universe, slowly faded and was replaced by another portrait entirely. Over time, the story of the expanding universe's discovery evolved. Particularly after Hubble's death, more and more references were made to him as the sole discoverer of the universe's expansion. Poor Humason was shoved off to the shadowy sidelines of popular history, Slipher largely forgotten, and Lemaître's crucial theoretical interpretation diminished. Fine distinctions and the sharing of credit got lost. A rousing narrative is now usually drawn with Hubble as the main protagonist, even though in reality he was not the expanding universe's champion at all. But, as historians Kragh and Smith put it, "a growing community of American astronomers . . . by the 1960s were concentrating to an unprecedented degree on the study of galaxies [and] fashioned a hero, a founding father and a figure around whom they could drape a single version of the history of the discovery of the expanding universe." The public seems to yearn for heroes, and thus Hubble—so handsome, so manly, so erudite—easily joined the scientific pantheon, along with Newton and his apple, Galileo and his telescope, Darwin and his finches.

It is the victorious leader who is now best remembered in the public's mind, not his accomplished predecessors or productive partner. Humason became Sancho Panza to Hubble's Don Quixote. Only this time, the twirling windmills are replaced with spiraling nebulae, and the celestial man of la Mancha ends up conquering them all with dazzling success.

Whatever Happened to . . .

In 1900 **Charles Yerkes** moved to New York City, driven out of Chicago by anticorruption reformers. He went on to establish London's underground transit system. His fortune reduced, he died in 1905 at the age of sixty-eight, long estranged from his forty-seven-year-old wife, Mary Adelaide, who continued living in their Fifth Avenue mansion. Within a month, she married Wilson Mizner, a raconteur and scoundrel eighteen years younger, who was the basis for the character played by Clark Gable in the movie *San Francisco*. Mary divorced him a year and a half later.

To this day, the 40-inch telescope at the **Yerkes Observatory,** in southeast Wisconsin, maintains its status as the largest refractor in the world, although it is no longer used for professional research. Plans are under way to historically preserve the main building and convert it into a regional science center.

Percival Lowell, long a bachelor, at last succumbed to marriage in 1908 at the age of fifty-three. He married Constance Savage Keith, nine years his junior and for many years a neighbor in Boston. At the end of a long honeymoon in Europe, he and his bride took a balloon ride, ascending a mile above London. There he photographed the paths of Hyde Park to see if its linear paths, substitutes for the Martian canals, could be detected from a high altitude. When Lowell died at Mars Hill in 1916 at the age of sixty-one, the observatory spent a decade fighting in court with his widow for control of his estate, the bulk of which he had

Whirlpool galaxy (M51) taken by the Hubble Space Telescope
(*NASA, ESA, S. Beckwith [STScI], and the Hubble Heritage
Team [STScI/AURA]*)

intended to be used to carry on the observatory's work. Over that time,
she squandered half of the $2.3 million. Constance reportedly lived in
"opulent squalor" until her death in Massachusetts at the age of ninety
in 1954. The following decade, Lowell's exotic imaginings were finally
put to rest when a series of Mariner missions launched by the National
Aeronautics and Space Administration in 1965 and 1969 showed Mars
to be a completely barren world. When *Mariner 9* orbited the red planet
in 1971, though, it photographed ancient riverbeds with tributaries and
erosion patterns that appeared to have been carved by catastrophic
flooding episodes. There were Martian channels after all, but these
were forged by water flowing naturally in Mars's distant past rather than
constructed by present-day aliens.

Besides his beloved Mars, Percival Lowell had another passion: searching for "Planet X" beyond Neptune. Analyzing discrepancies in the
motions of Uranus and Neptune, he had come up with a predicted
location for the missing planet, in the farthest realm of the comets
some four billion miles from the Sun. The observatory's new director,
Vesto M. Slipher, continued to administer the search. In 1930 a newly

hired staff member, twenty-four-year-old Clyde Tombaugh, at last made the discovery. The new planet was named Pluto. The first two letters—*PL*—honored the man who initiated the planetary hunt. In 2006 Pluto, always considered an oddball because of its small size and eccentric orbit, was demoted to dwarf planet (a type of solar system body now called a plutoid), no longer one of the pantheon of classical planets.

Though detecting the swift speeds of the spiral nebulae was his most heralded accomplishment, Slipher made other notable discoveries during his long career. He played an important role in finding that interstellar space was not pristine but rather littered with faint wisps of gas and dust; connected certain features of auroras with solar activity; and accurately determined a number of planetary rotations. Slipher served as the Lowell Observatory's director for thirty-eight years. Esteemed by the townspeople, he prospered financially by shrewdly investing in ranch property, helping establish Flagstaff's community hotel, and running a retail furniture store at one point. His retirement in 1954 made the front page of the *Arizona Daily Sun*. He died in Flagstaff in 1969, three days before his ninety-fourth birthday.

After Lowell's death, the **Lowell Observatory** was often strapped for cash but survived, largely due to the astute administration (and added donations) of trustee Roger Lowell Putnam, Lowell's brother-in-law. Even then, the complex was surely headed for closure at the end of World War II, until the influx of federal funds into U.S. scientific research suddenly revived its resources. Today it continues its mission as a private, nonprofit education and research organization, carrying out studies on the solar system, comets, extrasolar planets, solar activity, and stars.

If **Heber Curtis** had stayed at the Lick Observatory, he might have had a chance of gathering the decisive proof that the spirals were island universes. But it's questionable that he would have extended his research to proving the cosmos was expanding. He was uncomfortable with Einstein's theory and participated in solar-eclipse tests hoping to prove general relativity wrong. In the 1930s he told Harlow Shapley that he wasn't keen on where the research on spiral nebulae was going: "I have so little confidence in the theories of Lemaître, Eddington, et al. in this field that I shall follow the safe if not sane course of just sitting tight." After spending ten years as director of the Allegheny Observatory, Curtis came full circle and finished up his career in the 1930s at the University of Michigan, where he had begun his undergraduate studies in the classics. He had hopes for erecting a big reflector for Michigan's use but the Depression intervened, dashing his plans. Curtis died in 1942. He always considered his work on the nebulae as his greatest contribution to astronomy.

The **Lick Observatory** continues to be owned and operated by the University of California. More than twenty families currently reside on the mountain, with the town maintaining its own police department and post office. While the Crossley reflector remains in operation for professional research, the Lick 36-inch refractor is primarily a popular attraction, used at scheduled times for public viewing. Since the 1920s, the observatory grounds have expanded to include nine research-grade telescopes, the largest being the 3-meter (120-inch) Shane Reflector.

George Ellery Hale died at the age of sixty-nine in 1938, a decade after he launched an endeavor to erect a 200-inch telescope atop California's Palomar Mountain, near San Diego. The telescope was at last dedicated in 1948. To venerate Hale's brilliant leadership in the telescope's design and construction and his achievements as the Mount Wilson Observatory director from 1904 to 1923, the 200-inch was named the Hale Telescope. One wonders what Hale's reaction might have been to this honor if he had lived to see the telescope in operation. "The truth is," he once noted, ". . . that I have been enjoying from boyhood the things I liked most to do, and why should one be praised for simply having a good time?" Six decades later, the Hale Telescope remains one of the larger optical telescopes in the world and continues to make major contributions to astronomical research.

Telescope designer **George Willis Ritchey,** who had supervised the optical work on the 100-inch Hooker Telescope, continued to bad-mouth the venture after Hale ordered its flawed disk polished and mounted. Once Hale nixed Ritchey's grandiose idea to replace the defective glass with a radically new type of mirror, the optician spread the word that the giant scope would fail. In an act of insubordination, Ritchey directly contacted Hooker, Hale's benefactor, to convince the businessman to take his side. Ritchey's gossip and unauthorized dealings—acts of disloyalty to Hale—ultimately led to his dismissal from Mount Wilson in 1919 at the age of fifty-four. Ritchey never got to use the Hooker telescope for his own astronomical investigations. He moved to his ranch east of Pasadena, growing lemons, oranges, and avocados and dreaming of designing ever-bigger telescopes, with mirrors up to 320 inches in width. In the 1920s he worked in France in an attempt to construct a telescope that would surpass the Hooker in size, until the French project was called off. In the early 1930s he had to settle for designing and constructing a 40-inch reflector for the U.S. Naval Observatory, then upgrading its equipment in Washington. He died in 1945, two months shy of eighty-one. He would never learn that his highly controversial design for the Naval Observatory scope, worked out earlier in collaboration with the French astronomer Henri Chrétien, would later be used in many giant telescopes built in the latter half of the twentieth century, including the Hubble Space Telescope.

Edwin Hubble's later work never quite equaled the amazing discoveries he made in the 1920s and early 1930s. His most productive days were behind him. His scientific life, in a way, came to a standstill as he awaited construction of a bigger telescope to advance his cosmic searches. During World War II, he was stationed at the U.S. Army's Ballistics Research Laboratory at Aberdeen, Maryland. There he applied his student training in orbital mechanics to calculating artillery-shell trajectories. Over the years, his noted arrogance tempered a bit. Astronomer George Abell, who briefly worked for Hubble while a graduate student in the early 1950s, remembered him as a "very gracious, kindly person, a real gentleman. . . . He always seemed to have time to talk to students and night assistants. . . . He may have mellowed in his old age." Hubble lived long enough to see the opening of the next great telescope after the 100-inch—the 200-inch telescope on Palomar Mountain. He was the first scheduled observer on the giant instrument in 1949 and got started by photographing the variable nebula NGC 2261, his good luck charm. The many snubs toward his colleagues over the years, though, ultimately kept him from his more cherished goal: becoming director of the newly combined Mount Wilson and Palomar observatories. Ira Bowen was appointed instead, a decision that simply stunned Hubble, who was certain the post was his for the taking. During the following summer, on a fishing trip near Grand Junction, Colorado, Hubble experienced a major heart attack and was hospitalized.

On September 28, 1953, the Hubbles were returning to their San Marino home by car, with Grace driving. Hubble was in the midst of preparations for going to Palomar for four nights of observing. When Grace was about to make a turn into their driveway, though, she noticed Edwin breathing shallowly. "Don't stop," he said. "Drive on in." By the time she parked in their front courtyard, he had died of a cerebral thrombosis. He was sixty-three. Grace lived for another twenty-seven years, vigilantly editing her husband's legacy.

Milton Humason, who had barely finished the eighth grade before dropping out of school to work at Mount Wilson, received an honorary doctorate from Sweden's Lund University in 1950 for his historic contributions to the discovery of the expanding universe, becoming that rare individual who went from elementary school directly to a PhD. By the end of his career, Humason had taken the spectra of more than six hundred galaxies. At his retirement, his son offered to buy him a small telescope to continue viewing the sky. "My God, Bill," he replied, "I've looked in an eyepiece all my life, I don't want to look in any more eyepieces." He went salmon fishing instead.

The Carnegie Institution of Washington continues to own and operate the **Mount Wilson Observatory,** although now in partnership with the Mount Wilson Institute, a nonprofit corporation established in 1985. The 100-inch Hooker Telescope was temporarily shut down in 1986 as a cost-cutting measure but brought back into operation in 1992. With the use of advanced technology instruments to analyze the light gathered by its mirror, the Hooker continues to carry out valuable research, such as searching for extrasolar planets and monitoring sunspot cycles on other stars.

The years that **Harlow Shapley** spent at Mount Wilson, proving our true place within the Milky Way, turned out to be the "high noon of his scientific life." After World War II, he sharply curtailed his astronomical research efforts and devoted more of his time to national and international affairs. An unabashed liberal, he played a leading role in the formation of UNESCO, the United Nations Educational, Scientific and Cultural Organization. His activities on behalf of world peace and his continuing contacts with Russian scientists brought him under investigation in 1946 by the infamous House Committee on Un-American

Activities. Senator Joseph McCarthy later accused him—wrongly—of being a Communist. After Shapley retired as director in 1952, the Harvard Observatory continued to be his academic home for yet another twenty years, until his death in 1972 at the age of eighty-six. He was buried in Sharon, New Hampshire, where he had lived for many years after his retirement. His grave is marked by a solid granite rock upon which is inscribed, "And We by His Triumph Are Lifted Level with the Skies," a quotation from the ancient Roman philosopher Lucretius.

Shapley's former boss and harshest critic, **Walter Adams,** succeeded Hale as director of the Mount Wilson Observatory in 1923 and remained at that post until his retirement in 1946. He continued to work at the Hale Solar Laboratory in Pasadena until his death ten years later. Staff astronomers on Mount Wilson noticed that Adams was more at ease once Shapley left the observatory, and the two actually reconciled a few years later. For Adams, Shapley was easier to take once he was firmly ensconced at Harvard. It is interesting to note, however, that when Adams wrote a thirty-nine-page memoir of his early days at Mount Wilson, published by the Astronomical Society of the Pacific in 1947, he made no mention of Shapley whatsoever.

Adriaan van Maanen was on staff at the Mount Wilson Observatory for thirty-four years. For a while, he hoped that his flawed spiral measures would still have value by at least demonstrating a spiral's direction of rotation. But in the early 1940s Hubble proved once and for all that van Maanen had been wrong about that as well; as others had seen earlier, a spiral's arms are trailing as they rotate, not leading. Van Maanen died of a heart attack in 1946. Just weeks before his death he finished the measurement of his five hundredth parallax field at the observatory's Pasadena headquarters. Though he was wrong on spiral rotations, van Maanen remained a world-class surveyor of stellar parallaxes.

Georges Lemaître made few notable contributions to cosmology after 1934 but continued to publish reviews and discussions. Although Einstein abandoned the cosmological constant λ in 1931, Lemaître continued to champion it. They had friendly arguments about this issue whenever they met, which led to the joke that "everywhere the two men went, the lambda was sure to go." Lemaître went on to do important work in celestial mechanics and pioneered the use of electronic computers for numerical calculations. He always hoped the explosive origin of the universe would be validated by astronomical observations and at last received news of the discovery of the cosmic microwave background, the remnant echo of the Big Bang, shortly before he died in 1966. His successor at Louvain, Odon Godart, brought the July 1, 1965, issue of the *Astrophysical Journal* that contained the Nobel Prize–winning report to Lemaître's hospital bed.

After his great surge of creativity between 1905 and 1917—the period when he generated both special and general relativity, introduced us to the particle of light called a photon, and fashioned the first relativistic model of the universe—**Albert Einstein** stepped away from further major developments in either quantum or cosmological theory and primarily tried, unsuccessfully, linking the forces of nature in one grand unified theory. He died in 1955, still thinking the cosmological constant was his biggest blunder. Ironically, astronomers have recently brought back the constant to help explain a universe that is not only expanding but accelerating, a behavior that Lemaître anticipated in the 1930s.

Notes

Abbreviations

AIP   Niels Bohr Library and Archives, American Institute of Physics, College Park, Maryland

CA   The Caltech Institute Archives, California Institute of Technology, Pasadena, California

HL   Henry Huntington Library, San Marino, California

HP   George Ellery Hale Papers, Caltech Institute Archives, California Institute of Technology, Pasadena, California. (There is also a microfilm edition of these papers at other libraries.)

HUA   Harvard University Archives, Harvard University, Cambridge, Massachusetts

HUB   Hubble Papers, Henry Huntington Library, San Marino, California

LOA   Mary Lea Shane Archives of the Lick Observatory, University of California, Santa Cruz, California

LPV   Plate Vault, Lick Observatory, Mount Hamilton, California

LWA   Lowell Observatory Archives, Flagstaff, Arizona

MWDF   Mount Wilson Observatory Director's Files, Henry Huntington Library, San Marino, California

NAS   The Archives of the National Academies, Washington, D.C.

Preface: January 1, 1925

ix   **"redolent of orchids and pleasant, cheerful snobbery"**: Fitzgerald (1925), p. 133.

ix   **some four thousand scientists descended upon Washington, D.C.**: "Thirty-Third Meeting" (1925), p. 245.

x   **uncharacteristically chatty**: According to Grace Coolidge, the president's wife, a young woman once sat next to her husband at a dinner party and bet the normally taciturn president that she could wring at least three words of conversation from him. Coolidge promptly responded, "You lose."

x   **"It has taken endless ages to create in men"**: "Welfare of World Depends on Science, Coolidge Declares" (1925), pp. 1, 9.

x "occurred an event which was marked on the program": "Thirty-Third Meeting of the American Astronomical Society" (1925), p. 159.

x give holiday sleds a good tryout: "Blanket of Snow Covers the City" (1925), p. 1.

x walked the short distance to the newly constructed Corcoran Hall: During World War II, with scientists working under a government contract designed to develop new technologies for the conflict, the basement of Corcoran Hall was the birthplace of the bazooka.

xi a paper modestly titled "Cepheids in Spiral Nebulae" was presented: "Thirty-third Meeting of the American Astronomical Society" (1925), p. 159.

xi the only spiral nebulae in the nighttime sky that can be seen with the naked eye: The center of the Triangulum galaxy can be seen with the naked eye only under exceptionally good conditions. Viewing the Andromeda nucleus without the aid of a telescope is easier.

xii Henry Norris Russell stood in for Hubble that morning: "Thirty-third Meeting of the American Astronomical Society" (1925), p. 159.

xii Could I possibly be wrong?: Sandage (2004), p. 528; Berendzen and Hoskin (1971), p. 11.

xiv "This was an era of extraordinary change": Frost (1933), p. 124.

xv "so far as astronomy is concerned . . . we do appear": Newcomb (1888), pp. 69–70.

xvi "Hubble's drive, scientific ability, and communication skills": Osterbrock, Brashear, and Gwinn (1990), p. 1.

xvii cosmos firma: An ancient Roman would more correctly have said cosmos firmus (for proper matching of masculine adjective to masculine noun), but I wanted to maintain the mellifluous sound and metaphorical connection to terra firma.

xviii book was labeled a "classic": Mayall (1937), p. 42.

xviii "[His] picture differs from today's only in details": From the "Forward" to the 1982 edition of Hubble's Realm of the Nebulae (1936), pp. xv–xvi.

1. The Little Republic of Science

3 An immense continent of rock . . . southward along California's coastline: J. McPhee (1998), pp. 125, 542.

4 "First on top" . . . "noble and true": Wright (2003), pp. 25–27.

4 "the public mind in this country": Ibid., p. 14.

4 the most innovative work at Lick: C. Donald Shane, Lick director in the 1950s, said that "the work [Keeler] did . . . with the Crossley was the most important work done on the mountain at that time." AIP, interview of C. Donald Shane by Helen Wright on July 11, 1967.

4 Ptolemy Ridge: Keeler (1900b), p. 326.

6 Keeler's celestial curiosity . . . lunar craters and the planets: "The New Director of Lick" (1898), p. 7. Also from Osterbrock (1984); Donald Osterbrock wrote the definitive biography of Keeler, and many of the details of Keeler's personal life were drawn from this outstanding work on nineteenth-century American astronomy.

6 a "constant succession of fire balls": Olmsted (1834), p. 365.

6 "the most remarkable in its appearance": Olmsted (1866), p. 223.

7 "extraordinary that a people": Trollope (1949), p. 158.

7 "lighthouse in the sky": White (1995), p. 124.

7 "Some Americans, haunted by a nagging sense": Miller (1970), p. 27.
7 "lankey green country boy" . . . "cracker drawl": Osterbrock (1984), pp. 8–10.
8 "Starting from essentially zero": Brush (1979), p. 48.
8 Lick earned his riches . . . 1906 earthquake: Ibid., pp. 36–37; Wright (2003), pp. 2, 5; Osterbrock, Gustafson, and Unruh (1988), pp. 3–4.
10 Without a legitimate heir: As an adult, Lick's illegitimate son, John Lick, came out to California to meet his father and stayed around for a number of years. They never got along, and Lick refused to acknowledge him as a son, leaving him only $3,000 in his will. After Lick's death, though, John filed suit for his father's fortune, claiming he was the rightful heir. After years of legal strife, Lick's board of trustees finally agreed to settle, giving John the sum of $533,000 for both himself and other contesting relatives. See Osterbrock (1984), pp. 40–43.
10 $4 million: Wright (2003), p. 6.
10 "If I had your wealth": Ibid., p. 7.
10 Louis Agassiz gave a widely reported lecture: Miller (1970), p. 100.
10 All these lessons . . . Fourth and Market: Osterbrock (1984), p. 38; Wright (2003), p. 28; Osterbrock, Gustafson, and Unruh (1988), p. 12.
10 "For the Air": Newton (1717), p. 98.
11 Over time Lick came to accept: Osterbrock (1984), p. 39.
12 "little Republic of Science": LOA, Keeler Papers, Box 31; Shinn (c. 1890).
12 "I intend to rot like a gentleman": Osterbrock (1984), p. 53; Wright (2003), p. 61.
12 The choice for director: Osterbrock (1984), p. 42.

2. A Rather Remarkable Number of Nebulae

13 There are some 360 switchbacks in all, and some were even given special names: AIP, interview of Douglas Aitken by David DeVorkin on July 23, 1977.
13 "The view from the observatory peak": LOA, Keeler Papers, Box 6, Folder 4.
14 "If he has the right ring": Osterbrock (1986), p. 53.
14 Often in the wintertime, storms would sweep over the mountain: Holden (1891), p. 73.
14 "a terrible old blow and grumbler" . . . "worthless": LOA, Keeler to Holden, January 6, 1888.
14 "no inconvenience was felt" . . . "spider's thread." Keeler (1888a, 1888b); LOA, Keeler to Holden, January 14, 1888. When the Voyager probe in the 1980s discovered a new separation in Saturn's rings, it was named the Keeler Gap in honor of the Lick astronomer.
15 displayed at the 1893 Chicago World's Fair: Osterbrock and Cruikshank (1983), p. 168.
15 "He was tolerant, amused and unwilling to take sides": Osterbrock (1984), p. 235.
15 "Beautiful and accurate": Barnard (1891), p. 546.
19 "as though it were a fort in hostile territory": AIP, interview of Lawrence Aller by David DeVorkin on August 18, 1979.
19 "I am a human being first": Osterbrock (1984), p. 108.
20 Saturn's rings were not solid: Maxwell (1983).
20 dispatched a report to the newly established Astrophysical Journal: Keeler (1895).

20 Crossley reflector: Keeler (1900b), p. 325.
20 Early telescopic mirrors: Osterbrock, Gustafson, and Unruh (1988), p. 22.
22 small zinc box: Babcock (1896).
22 "a pile of junk": Osterbrock (1984), p. 246.
22 "the czar," "the dictator": Ibid., pp. 233, 240.
22 went out to say good-bye: AIP, interview of C. Donald Shane by Elizabeth Calciano in 1969.
22 Keeler, by this time, was getting restless . . . raise his salary: Osterbrock (1984), pp. 239–44.
23 Keeler won the vote by 12 to 9: Ibid., p. 268.
23 "Stay with us, Keeler" LOA, Keeler Papers, Box 31, newspaper clipping.
23 telegraphed his acceptance: Osterbrock (1984), p. 270.
24 Keeler went back to Mount Hamilton . . . oiled dirt: Campbell (1971), pp. 9, 53–54, 66; Osterbrock (1984), pp. 278–79.
24 "It [was] like being shipwrecked on an island": Campbell (1971), p. 9.
24 If a hostess sent out an invitation for an evening gathering: Hussey (1903), p. 32.
24 Occasionally a ground squirrel would carry off a ball: Ibid., p. 30.
24 A biologist visiting Mount Hamilton: Shinn (c. 1890).
24 "There are no astronomical phenomena": Osterbrock (1984), p. 291.
25 "No member of the staff was asked": Campbell (1900a), p. 144.
25 acquired a stigma: Osterbrock (1984), p. 245.
26 Roberts had pioneered: Ibid., p. 169.
26 "hand down to our successors": Pang (1997), p. 177.
26 "No Work of Importance": Osterbrock (1984), p. 297.
26 innumerable engineering problems: LPV, Crossley Reflector Logbook, James F. Keeler, June 1, 1898, to April 10, 1899.
27 "The fainter stars" . . . "fairly good": Ibid.
27 upper wall was painted black: Keeler (1899d), p. 667.
27 "On the negative of November 10": Keeler (1898a), p. 289.
27 "Nebulous wisps . . .": Keeler (1898b), p. 246.
28 "The photographic power": Keeler (1899a), pp. 39–40.
28 "We know them so well today": Osterbrock (1984), p. 306.
28 "The [Crossley's] workmanship is poor": HP, Keeler to Hale, February 5, 1899.
28 first spiraling nebula on April 4 . . . "valueless": LPV, Crossley Reflector Logbook, James Keeler, June 1, 1898, to April 10, 1899.
29 "Everyone in the Observatory": LOA, Hale to Keeler, June 12, 1899.
30 "Several other faint nebulae": Keeler (1899b), p. 538.
30 just stood in front of Keeler's photographs: Osterbrock (1984), p. 309.
30 "on the successes rather than on the failures": LOA, Keeler to Campbell, June 14, 1900.
30 "The finest I have ever seen": Osterbrock (1984), p. 310.
31 "a rather remarkable number": Keeler (1899c), p. 128.
31 "there are nearly as many": Ibid.
31 "There are hundreds, if not thousands": Ibid.
31 only seventy-nine were identified as spirals: Dewhirst and Hoskin (1991), p. 263.
32 "a mirey climate for a great telescope": Osterbrock (1984), pp. 320–21.
32 "The spiral nebula has been regarded": Keeler (1900a), p. 1.
32 "from the great nebula in *Andromeda*": Keeler (1900b), p. 347.
33 "If . . . the spiral is the form": Ibid., p. 348.

33 "The heavens are full": LOA, "Abstract of Lecture at Stanford University," Keeler Papers, Box 31.
33 "Keeler . . . was a far better trained, more experienced spectroscopist": Osterbrock (1984), p. 357.
33 "follow up his remarkable beginnings": LOA, Hale to Campbell, September 14, 1900.
34 Keeler died unexpectedly: Osterbrock (1984), pp. 327–29; Tucker (1900), p. 399; Campbell (1900a), pp. 139–46.
34 "a hard cold": LPV, Crossley Reflector Logbook, Keeler, December 1, 1899, to July 24, 1900.
34 "nothing very serious": Osterbrock (1984), p. 327.
34 "incalculable": Campbell (1900b), p. 239.
34 "loss cannot be overestimated": Jones and Boyd (1971), pp. 428–29.
34 The journal *Science* ran a tribute: Hale (1900).
34 "The day of the refractor was over": Osterbrock (1984), p. 347.
35 he decided to build another 36-inch reflector: Ibid., pp. 345–46.
35 "The results obtained with the two-foot reflector": Ritchey (1901), pp. 232–33.

3. Grander Than the Truth

36 "Let us assume for the moment": Webb (1999), p. 9.
36 "the center of the universe is everywhere": Impey (2001), p. 38.
37 "If the Matter was evenly disposed": Kerszberg (1986), p. 79.
37 "There is a size at which dignity begins": T. Hardy (1883), p. 38.
38 "no other than a certain Effect": Wright (1750), p. 48.
40 "I don't mean to affirm": Ibid., p. 62.
40 "too remote for even our telescopes to reach": Ibid., p. 84.
40 "there may be innumerable other spheres": Swedenborg (1845), pp. 271–72.
40 what they *thought* he meant: See Hoskin (1970).
40 "just universes and, so to speak, Milky Ways": Kant (1900), p. 63.
41 Kant's manuscript was destroyed: Hetherington (1990b), p. 15.
41 "I easily persuaded myself": Kant (1900), p. 33.
41 "island universes": The phrase was never used by Kant. Humboldt first applied the term to describe Kant's theory in his book *Kosmos*, published in 1845. He wrote it in his native language as *Weltinsel*, "world island," which was later transformed into the more familiar expression.
41 Edmond Halley (of comet fame) counted six in all: Not all of the objects on Halley's list were true nebulae. The six are: (1) the Orion nebula, (2) the Andromeda nebula (now galaxy), (3) the globular cluster M22 in Sagittarius, (4) the globular cluster Omega Centauri, (5) the open star cluster M11 in Scutum, and (6) the globular cluster M13 in Hercules. In Halley's day, all appeared as unresolved clouds through a telescope.
42 "appear to the naked Eye": Halley (1714–16), p. 390.
42 Charles Messier published in France his famous list of more than one hundred nebulae: Messier (1781).
42 "I . . . saw, with the greatest pleasure": Herschel (1784b), pp. 439–40.
42 "These curious objects": Herschel (1789), p. 212.
43 "may well outvie our milky-way in grandeur": Herschel (1785), p. 260.
43 "When I read of the many charming discoveries": Bennett (1976), p. 75.

43   **Caroline, who had earlier joined him in England, fed him morsels of food by hand:** Caroline Herschel was more than her brother's handmaiden; she was an accomplished astronomer in her own right. A proficient comet hunter (she was the first woman to find one), she was awarded the Royal Astronomical Society's Gold Medal in 1828.

43   **"confirmed and established by a series of observations."** Herschel (1785), p. 220.

44   **capable of seeing out to cosmological distances:** Hoskin (1989), pp. 428–29.

45   **"I have seen double and treble nebulae":** Herschel (1784b), pp. 442–43, 448. Sixty years before Alexander von Humboldt originated the term *island universe,* William Herschel actually referred to the possibility that the Milky Way might be an "island" in his classic 1785 paper "On the Construction of the Heavens." "It is true," wrote Herschel, "that it would not be consistent confidently to affirm that we were on an island unless we had actually found ourselves every where bounded by the ocean, and therefore I shall go no further than the [gauges] will authorise; but considering the little depth of the stratum in all those places which have been actually [gauged] . . . there is but little room to expect a connection between our nebula and any of the neighbouring ones." See Herschel (1785), pp. 248–49.

45   **"The inhabitants of the planets that attend the stars":** Herschel (1785), p. 258.

45   **"A most singular phaenomenon!":** Belkora (2003), p. 109.

45   **"Cast your eye on this cloudy star":** Herschel (1791), pp. 73, 84.

45   **stars or a "shining fluid"—not both:** Ibid., p. 71.

46   **We were alone in the universe once again . . . at least for a while:** There are some qualifiers to this blunt statement. While others interpreted Herschel as having abandoned the thought of other universes, the great British astronomer did seem to maintain that certain nebulae, ones he had already resolved, were distant star systems. So his sense of the visible universe did extend beyond the Milky Way. (From Robert Smith, personal communication, May 5, 2008.)

46   **So big was the telescope tube:** Hetherington (1990b), p. 16.

46   **"to make a telescope of the largest dimensions possible":** "Report of the Council to the Forty-Ninth General Meeting of the Society," *Monthly Notices of the Royal Astronomical Society* 29 (February 1869): 124.

46   **to devote himself to a newfound career as a gentleman scientist:** The Parsons family had a rich engineering legacy. In 1884 Rosse's son Charles invented the first steam turbine that could convert the power of steam directly into electricity, a method adopted by power stations worldwide.

46   **a British reporter once caught him working at a vise:** Singh (2005), p. 181.

46   **"It is scarcely possible to preserve":** *Proceedings of the Royal Irish Academy* 2 (1844): 8.

47   **to resemble one of the ancient round towers of Ireland:** Clerke (1886), p. 151.

47   **"Sweeping down from the moat towards the lake":** Ball (1895), p. 193.

47   **"strange stellar cloudlets":** Proctor (1872), p. 64.

47   **"a structure and arrangement more wonderful and inexplicable":** "Report of the Council," p. 129.

47   **"With each successive increase of optical power":** Rosse (1850), p. 504.

48   **"existed only in the imagination of the astronomer":** MacPherson (1916), p. 132.

48   **"numerous firmaments":** Nichol (1840), p. 10.

48 "It is indeed wholly unlikely that our group. . . . THIRTY MILLIONS OF YEARS": Nichol (1846), pp. 17, 36–37.
49 This was a brave estimate. In 1831 British geologist Charles Lyell arrived at an age for Earth of 240 million years based on the fossils of marine mollusks, but it was still highly controversial. In 1836 Charles Darwin took a copy of Lyell's *Principles of Geology* along with him on his famous voyage on the *Beagle*, which greatly influenced his developing ideas on evolution.
49 "poised so skilfully . . ."; "perfectly wretched": Proctor (1872), pp. 64–67.
51 *what* they are instead of *where* they are: Keeler (1897), pp. 746, 749.
51 "coming upon a spring of water": Huggins (1897), p. 911.
51 "The chemistry of the solar system prevailed": Whiting (1915), p. 1.
51 "excited suspense . . . but a luminous gas": Huggins (1897), pp. 916–17.
52 "The nebular hypothesis made visible!": Turner (1911), p. 351.
52 "a planetary system at a somewhat advanced": Huggins and Huggins (1889), p. 60.
52 "a 'universe of stars,' like our own 'galactic cluster' ": Young (1891), p. 509.
52 "What is beyond the stellar system": Ibid., p. 512.
52 "This strange and beautiful object": Maunder (1885), p. 321.
52 "a scale of magnitude such as the imagination recoils": Clerke (1902), p. 403.
53 "[The nova] was in the heart of the Great Nebula": Frost (1933), p. 45.
53 "[I would deem] it a very great favor to be able to make use of your great harvest of new forms": LOA, Chamberlin to Keeler, January 30, 1900.
53 "The question whether nebulae are external galaxies" . . . "misleading": Clerke (1890), pp. 368, 373.
54 "That the spiral nebulae are star clusters is now raised to a certainty": Scheiner (1899), p. 150.
54 A further investigation was not undertaken until 1908: Fath (1908).
54 "The hypothesis that the central portion of a nebula": Ibid., p. 76.
54 perhaps because he was still a lowly graduate student: Osterbrock, Gustafson, and Unruh (1988), p. 188.
55 "stands or falls": Fath (1908), p. 77.

4. Such Is the Progress of Astronomy in the Wild and Wooly West

56 stockings on the gear of the giant telescope; Mitchell automobile: AIP, interview of Mary Lea Shane by Charles Weiner on July 15, 1967; interview of Charles Donald Shane by Bert Shapiro on February 11, 1977.
56 "a spectacular performance is kept up": LOA, Curtis Papers, unsigned letter to Curtis, August 9, 1905.
56 "wonderfully kind, jolly person, always smiling, always happy": AIP, interview of Mary Lea Shane by Charles Weiner on July 15, 1967.
56 feat once described as "remarkable": Trimble (1995), p. 1138.
57 These astronomers were specifying that the spirals' sizes and the brightness of their novae only made sense if they were milky ways at great distance: See Very (1911) and Wolf (1912).
57 "If the spiral nebulae are within the stellar system": Douglas (1957), pp. 26–27.
57 "in best harmony with known facts": Campbell (1917), p. 534.
57 a program that had not been a top priority since Keeler's death: Charles Perrine

took over the Crossley after Keeler's death and made some substantial improvements to its mount, drive, gears, and mirror system. While he did carry out some work on the nebulae, his most acclaimed accomplishment with the Crossley was discovering the sixth and seventh moons of Jupiter. See Osterbrock, Gustafson, and Unruh (1988), pp. 142–44.

57 a gifted mechanic: McMath (1944), pp. 246–47; Curtis (1914).
57 The mirror had already been remounted in 1904: See Perrine (1904).
58 "magnum opus": Stebbins (1950), p. 3.
59 student of the ancient languages: Aitken (1943), p. 276.
59 He hoped to continue at the Lick Observatory: LOA, Curtis to Keeler, March 24, 1900.
59 "ready and glad to be put at anything from a shovel up": LOA, Curtis to Campbell, April 11, 1900.
59 hired him on as an assistant: Osterbrock (1984), p. 342.
60 simply good training for a life on Mount Hamilton: LOA, Curtis to Campbell, June 9, 1902; AIP interview of Douglas Aitken by David DeVorkin on July 23, 1977.
60 covered in thick yellow dust: Stebbins (1950), p. 2.
60 saw three miles of fire-front, burning fiercely: Campbell (1971), pp. 62–64.
60 "And, naturally, the lens inverted everything": AIP, interview of Douglas Aitken by David DeVorkin on July 23, 1977.
60 "Queer how completely we seem to have taken root here": LOA, Curtis to Richard Tucker, March 23, 1909.
60 Halley's Comet: LOA, Curtis Papers, Folder 1, Halley report.
61 amassed a photographic library of around one hundred nebulae and clusters: Curtis (1912).
61 boosted that number to more than two hundred: LOA, Curtis Papers, "Report of Work from July 1, 1912, to July 1, 1913."
61 "Many of these nebulae show forms of unusual interest." Ibid.
61 rich diversity in their appearance: Curtis (1912).
61 "Crossley still has its old reputation": MWDF, Box 153, Curtis to Walter Adams, May 27, 1913.
61 "If you got a little bit sleepy at night": AIP, interview of Mary Lea Shane by Charles Weiner on July 15, 1967.
61 observe from a boat: This popular tale, often heard at the Lick Observatory, was told to me by Lick astronomer Tony Misch.
61 "of smooth nebulous material and also of soft star-like condensations or nebulous stars": Ritchey (1910b), p. 624.
62 "rotatory or otherwise. . . . As the spirals are undoubtedly in revolution": Curtis (1915), pp. 11–12.
63 "the Greek letter Φ . . . for lack of a better term": Curtis (1913), p. 43.
63 "shows dark lane down center" . . . "beautifully clear": LOA, Curtis Papers, Folder 1, "Edgewise or Greatly Elongated Spirals."
63 "due to the same general cause": Curtis (1918b), p. 49.
63 Not one spiral had ever been spotted in the thick of the Milky Way: For his doctoral research at the Lick Observatory, Roscoe Sanford searched the length and breadth of the Milky Way for signs of a spiral, using long exposures in hope of bringing to light faint nebulae previously hidden within the Milky Way. He didn't find any. See Sanford (1916–18).

63 "[The] great band of occulting matter in the plane of our galaxy": Curtis (1918b), p. 51.

64 "Were the Great Nebula in *Andromeda* situated five hundred times as far away": Curtis (1918a), p. 12.

65 nova in NGC 6946: Ritchey (1917).

65 was sure that the outbursts were not simply variable stars: Curtis (1917c), p. 108.

65 "That both these novae should have appeared in the *same* spiral": Ibid.

65 "must be regarded as having a very definite bearing": Curtis (1917b), p. 182.

65 "Such is the progress of Astronomy": HUA, Harlow Shapley to Henry Norris Russell, September 3, 1917, HUG 4773.10, Box 23C.

65 show off the plate: AIP, interview of C. Donald Shane by Helen Wright on July 11, 1967.

66 He said as much to the Associated Press: LOA, Newspaper Cuttings, Volume 9, 1905-1928, "Three New Stars Are Seen at Lick."

66 20 million light-years distant: Curtis was not far off the mark. NGC 4527, the location of the first nova he spotted, is currently estimated to be around 30 million light-years from Earth.

67 On one plate alone he counted 304 additional spirals: Curtis (1918a), p. 13.

67 "The great numbers of small spirals found on nearly all my plates": Ibid., pp. 12-14.

67 "Get up a collection of about 40 classy slides": LOA, Curtis Papers, Folder 3, 1919-20, Curtis to Campbell, February 6, 1919.

68 "The history of scientific discovery affords many instances": Curtis (1919), pp. 217-18.

68 Over the course of that March evening, Curtis laid out his arguments point by point: LOA, Curtis Papers, Folder 3, 1919-20, Lecture on "Modern Theories of the Spiral Nebulae."

69 "As to my staying here permanently, I have no idea whatever of doing that": LOA, Curtis Papers, Folder 2, Curtis to Campbell, December 8, 1918.

69 "The hypothesis of external galaxies is certainly a sublime and magnificent one": Crommelin (1917), p. 376.

5. My Regards to the Squashes

70 "is not without a considerable atmosphere": Herschel (1784a), p. 273.

71 "Considerable variations observed in the network of waterways": Pannekoek (1989), p. 378.

71 news story of the year: "Mars" (1907), p. 1.

71 who had made their fortunes creating the American cotton industry: Strauss (2001), p. 3.

72 "After lying dormant for many years": Lowell (1935), p. 5.

72 *occasionem cognosce*, "seize your opportunity": Hoyt (1996), p. 15.

72 he once listed his address as "cosmos": Strauss (2001), p. 5.

72 eventually fired one charter member of his observing staff: Hoyt (1996), pp. 123-24.

72 "The Strife of the Telescopes": Hoyt (1996), p. 112.

73 "as efficient as could be constructed": Hall (1970b), p. 162.

73 "I . . . take him only because I promised to do so": LWA, Lowell to W. A. Cogshall, July 7, 1901.

73 for many of America's greatest astronomers . . . red and blue ends of the spectrum: Smith (1994), pp. 45–48.

74 "When you shall have learnt all about the spectroscope": LWA, Lowell to Slipher, December 18, 1901.

74 "kept himself well insulated from public view": Hall (1970b), p. 161.

75 always wore a suit and tie to work when not observing: AIP, interview of Henry Giclas by Robert Smith on August 12, 1987.

75 "Don't observe sun much. It hurts lenses": LWA, Lowell to Slipher, January 11, 1902.

75 "Permit nobody whatever in observatory office": LWA, Lowell to Slipher, January 24, 1902.

75 "Will you kindly see if shredded wheat biscuit are to be got at Haychaff": LWA, Lowell to Slipher, January 4, 1903.

75 "How fare the squashes?"; "My regards to the squashes"; "You may when the squashes ripen send me one by express": LWA, Lowell to Slipher, October 7, 12, and 21, 1901.

75 "Why haven't I received squashes?": LWA, Lowell to Slipher, December 27, 1901.

75 "Thank you for taking so much pains with the garden!": LWA, Lowell to Slipher, May 26, 1902.

75 "Your vegetables came all right and delighted me hugely": LWA, Lowell to Slipher, July 7, 1902.

75 eventually becoming a virtuoso . . . watery Mars: Hoyt (1996), pp. 129–45.

76 no sign at all: Not until the 1960s did astronomers confirm that water vapor in the Martian atmosphere was more than a thousand times less than the amount found in Earth's atmosphere, far lower than what Slipher could possibly have measured in the early 1900s with his equipment.

77 gas existed in the seemingly empty space between the stars: Smith (1994), p. 52.

77 "Dear Mr. Slipher, I would like to have you take with your red sensitive plates the spectrum of a *white* nebula": LWA, Lowell to Slipher, February 8, 1909.

77 "I do not see much hope of our getting the spectrum": LWA, Slipher to Lowell, February 26, 1909.

77 Campbell at the Lick Observatory had recently written yet another article critical of the Lowell Observatory: The article was Campbell (1908), 560–62. According to John C. Duncan, then a graduate student at Lick working on his thesis, two astronomers at Lick had "charted several stars not seen there by Lowell . . . from what I can gather Campbell is preparing a bunch of fireworks to shoot off in the various periodicals. In all probability there will be much entertainment for those who enjoy scientific argument." (LWA, Duncan to Slipher, September 13, 1908.) Despite these occasional interobservatory tussles, Campbell and Slipher generally maintained a cordial correspondence, most often discussing equipment.

77 to see 173 stars in a given field of the sky, where Lick's 36-incher could see only 161: P. Lowell (1905), 391–92.

77 "I have come to the conclusion": LWA, Slipher to Miller, October 18, 1908.

78 "This plate of mine": LWA, Slipher to Lowell, December 3, 1910.

78 "there is no more pressing need at present": Smith (1994), p. 54.

79 "It is not really very good": LWA, Slipher to Lowell, September 26, 1912.

79 November 15 observation details: LWA, Spectrogram Record Book II, September 24, 1912, to July 28, 1913, pp. 34–37.

79 December 3 and 4 observation details: Ibid., pp. 61–62.

79 high-voltage induction coils: Hall (1970a), p. 85.

80 "encouraging results or (I should say) indications": LWA, Slipher to Lowell, December 19, 1912.

80 "I congratulate you on this fine bit of work": LWA, Lowell to Slipher, December 24, 1912.

80 "would doubtless impress all these observers": LWA, Slipher to Lowell, December 19, 1912.

81 On a scale from 1 to 10: LWA, Douglass to Lowell, January 14, 1895.

81 December 29–31 observation details: LWA, Spectrogram Record Book II, September 24, 1912, to July 28, 1913, pp. 69–70.

81 "I feel safe to say here that the velocity bids fair to come out unusually large": LWA, Slipher to Lowell, January 2, 1913.

81 spectrocomparator operation: Slipher (1917b), p. 405.

81 calculations to convert the measured shift: LWA, V. M. Slipher Working Papers, Box 4, Folder 4-9.

82 He also sent a print of the spectrum to Edward Fath: LWA, Slipher to Fath, January 18, 1913.

82 "the shift has no direct bearing": Fath (1908), p. 75.

82 today, with far better equipment, astronomers measure Andromeda approaching us at 301 kilometers per second: See I. D. Karachentsev and O. G. Kashibadze (2006), 7.

82 "agree as closely as could be expected": LWA, Slipher to Lowell, February 3, 1913.

82 publish his brief account: Slipher (1913).

82 "It looks to me as though you have found a gold mine": LWA, Miller to Slipher, June 9, 1913.

82 "beauty": LWA, Wolf to Slipher, February 21, 1913.

82 "It is hard to attribute it to anything but Doppler shift": LWA, Frost to Slipher, October 23, 1913.

82 "Your high velocity for [the] Andromeda Nebula is surprising in the extreme": LWA, Campbell to Slipher, April 9, 1913.

83 "I had planned to get at this work years ago": LWA, Wright to Slipher, August 19, 1914.

83 "It looks as if you had made a great discovery": LWA, Lowell to Slipher, February 8, 1913.

83 "Spectrograms of spiral nebulae are becoming more laborious": LWA, Slipher Papers, Hoyt-V. M. Box, Report F4, titled "Spectrographic Observations of Nebulae and Star Clusters."

83 "heavy and the accumulation of results slow": Slipher (1913), p. 57.

83 "telescopic object of great beauty": LWA, Slipher Working Papers, Box 4, Folder 4-4.

83 "no less than three times that of the great Andromeda Nebula": Ibid.

84 "When I got the velocity of the Andr. N. I went slow": LWA, Slipher to Miller, May 16, 1913.

84 dust clouds illuminated by reflected starlight: LWA, Slipher to J. C. Duncan, December 29, 1912.

84 "undergoing a strange disintegration": LWA, Slipher to E. Hertzsprung, May 8, 1914.

84 "more numerous in, rather than outside, the Galaxy": LWA, Slipher to Miller, May 16, 1913.

84 "I leaned against it": Hall (1970a), p. 85.

84 his exposures often ran twenty to forty hours: Slipher (1917b), p. 404.

84 "With such prolonged exposures the accumulation of plates": LWA, Slipher to Lowell, May 4, 1913.

84 "It is our problem now and I hope we can keep it": LWA, Slipher to Lowell, May 16, 1913.

85 "My harty [sic] congratulations": LWA, Hertzsprung to Slipher, March 14, 1914.

85 "It is a question in my mind": LWA, Slipher to Hertzsprung, May 8, 1914.

85 Slipher inwardly feared . . . Let the work speak for itself: Strauss (2001), p. 244.

85 confident of what he was seeing: AIP, interview of Henry Giclas by Robert Smith on August 12, 1987.

85 "Spectrographic Observations of Nebulae": *Popular Astronomy* 23 (1915): 21–24.

86 "about 25 times the average stellar velocity": Ibid., p. 23.

86 his fellow astronomers rose to their feet and gave him a resounding ovation: Smith (1982), p. 19.

86 "Let me congratulate you upon the success of your hard work": LWA, Campbell to Slipher, November 2, 1914.

87 "I am . . . glad to have your kind offer": LWA, Slipher to Edwin Frost, October 22, 1914.

87 enlisted the help of a mathematician: LWA, Slipher Working Papers, Box 4, Folder 4-16.

87 "It has for a long time been suggested that the spiral nebulae are stellar systems": Slipher (1917b), p. 409.

87 "scattering" in some way: Ibid., p. 407.

87 By 1925, forty-five spiral nebulae velocities were pegged with assurance, and it was Slipher who had measured nearly all of them: Sandage (2004), p. 499.

88 he noticed a particular progression to the stampede outward: Wirtz (1922).

88 a term they labeled *K*: Use of the *K* term in spiral nebulae redshift studies was introduced in 1916 by Lick Observatory astronomer George Paddock, who thought the correction would no longer be needed once a sufficient number of observations were made. Others, like Wirtz, swiftly adopted the convention. See Paddock (1916). The *K* term was actually first used by stellar astronomers. Astronomers were finding that the value for the motion of the Sun, its speed and direction through the galaxy, could change depending on the celestial object—a particular star or nebula—that was used to gauge it. To bring them into agreement, astronomers introduced the *K* correction term. By the 1960s, with improved measurements, this "*K*-effect" for stars silently disappeared from the astronomical literature.

6. It Is Worthy of Notice

90 **Ancient Persians called the biggest one Al Bakr:** The Large Magellanic Cloud was named Al Bakr by the noted Persian astronomer Al-Sûfi in his *Book of Fixed*

Stars, written in 964. While not visible from northern Persia, it was visible to Middle Eastern peoples farther south, near the strait of Bab el Mandeb.

90　"two clouds of mist": Nowell (1962), p. 127.

91　"capable of doing as much and as good routine work": Pickering (1898), p. 4.

92　These women "computers" . . . photographic magnitude: Jones and Boyd (1971), pp. 388–90.

92　"He treated [the computers] as equals in the astronomical world": Ibid., p. 390.

93　Leavitt grew up in Massachusetts, within a big and supportive family: Johnson (2005), pp. 25–26. Many of the personal details of Leavitt's life are drawn from George Johnson's excellent biography of Henrietta Leavitt, the most comprehensive review of her life to date.

93　"For light amusements, she appeared to care little": Bailey (1922), p. 197.

94　"For this I should be willing to pay thirty cents an hour": Johnson (2005), pp. 31–32.

95　"variable-star 'fiend' ": Ibid., p. 37.

95　one of the first and brightest discovered: The English astronomer John Goodricke first noticed the variable brightness of δ Cephei in 1784. An astronomy prodigy (and also deaf like Leavitt), he won the Royal Society's prestigious Copley medal at the age of nineteen for his work on eclipsing binary stars. He died three years later of pneumonia.

95　"As a rule, they are faint during the greater part of the time": Leavitt (1908), p. 107.

95　"It is worthy of notice": Ibid.

97　"A remarkable relation between the brightness of these variables and the length of their periods will be noticed": Leavitt and Pickering (1912), p. 1.

98　"masterpiece": Rubin (2005), p. 1817.

98　"to work, not to think": Payne-Gaposchkin (1984), p. 149.

98　"It is to be hoped, also, that the parallaxes [essentially, distances] of some variables of this type may be measured": Leavitt and Pickering (1912), p. 3.

98　the cold aggravated her hearing condition: Johnson (2005), p. 31.

98　observatory's prime function was to collect and classify data: Jones and Boyd (1971), p. 369.

98　quickly assigned Leavitt another task: Johnson (2005), pp. 56–57.

98　"a harsh decision, which condemned a brilliant scientist to uncongenial work": Payne-Gaposchkin (1984), p. 146.

99　"[It's] of enormous importance in the present discussions": HUA, Shapley to Leavitt, May 22, 1920.

99　"Miss Leavitt had no understudy competent to take up her work": HUA, Shapley to Frederick Seares, December 13, 1921.

99　nominate her for a Nobel Prize in physics: Johnson (2005), p. 118.

7. Empire Builder

103　at most around 20,000 to 30,000 light-years wide: Smith (1982), pp. 58–60.

104　"owe *all* to Hale and his dreams": Wright (1966), p. 14.

104　Pickering discovered women and Hale discovered money: Rubin (2005), p. 1817.

104　"astronomical research with a feeling of awe": Hale (1898), p. 651.

105　confirmed that the element carbon resided in the Sun: Wright (1966), p. 59.

105 "reaching up toward the heavens in the great dome": Ibid., p. 71.

105 "I would not consider [joining the faculty] *for a moment*": Ibid., p. 92.

107 "The donor could have no more enduring monument" . . . "send the bill to me": Ibid., pp. 96–98.

107 "the embodiment and representative of corruption in municipal affairs": Jones and Boyd (1971), p. 429.

107 "giants were plotting, fighting, dreaming on every hand": Dreiser and Booth (1916), p. 172.

107 "Mr. Yerkes, when he took the matter in hand . . ."; ". . . shortly be licked": Osterbrock (1984), p. 185.

108 "there may be some who view with disfavor": Keeler (1897), p. 749.

108 grinding the mirror and designing its support system: Ritchey (1897).

108 The descendant of Irish immigrant craftsmen . . . reflectors were the instruments of astronomy's future: Osterbrock (1993), pp. 33–37.

108 reputation as a cantankerous cuss: Sandage (2004), pp. 96–97.

109 "The possibility of having you for a neighbor": HP, Keeler to Hale, February 5, 1899.

109 "Wilson's Peak": Wilson's Peak was named after Benjamin Davis Wilson, who in the 1850s was the first nonnative to explore the mountain, which was situated near his orchards and winery. Wilson was the grandfather of General George S. Patton Jr.

109 "*the* place": Osterbrock (1984), p. 350.

109 Harvard briefly considered setting up a permanent telescope there: Wright (1966), p. 165.

110 "to encourage investigation, research and discovery": Ibid., p. 159.

110 "seemed almost too good to be true": Ibid.

110 surpassed the funds then endowed for research at all American universities combined: Hetherington (1996), p. 104.

110 start work on the mountain: Wright (1966), pp. 187–88.

110 His mistress, the Los Angeles socialite Alicia Mosgrove: Osterbrock (1993), p. 74.

110 "an inner excitement—a higher degree of interest—a higher degree of suffering": Wright (1966), p. 198.

110 only a few farmhouses and barns nearby: Adams (1947), p. 223.

110 Jasper, Pinto, Duck, and Maude: Ibid., p. 218.

111 hundreds of tons of material were hauled up: Sandage (2004), pp. 165–67.

111 no imperfection extended farther than two millionths of an inch: Wright (1966), p. 228.

112 Hale decided he would not follow the Lick Observatory model: Sheehan and Osterbrock (2000), p. 101.

112 "Hale was never so happy": Adams (1947), p. 223.

112 to see "the woods" instead of the trees: Wright, Warnow, and Weiner (1972), p. 273.

112 with all astronomers required to wear coat and tie: AIP, interview of Allan Sandage by Spencer Weart on May 22 and 23, 1978.

113 "partly because of the strong influence of Dr. Hale's remarkable personality": HL, Walter Adams Papers, Box 1, Folder 1.15, "Autobiographical Notes."

113 Harlow Shapley showed up fully prepped . . . "Please come to Mount Wilson": Shapley (1969), pp. 44–45.

8. The Solar System Is Off Center and Consequently Man Is Too

114 "I have looked at some cluster plates a little": HUA, Shapley to Russell, May 20, 1914.

114 "He is much more venturesome": HUA, Hale to A. Lawrence Lowell, March 29, 1920.

115 "keep the rhythm going": Shapley (1969), p. 11.

115 "The St. Louis *Globe-Democrat* was our chief contact": Ibid., p. 5.

115 refused admission . . . had always desired: Ibid., p. 12. On May 3, 1963, the town of Carthage, Missouri, celebrated "Harlow Shapley Day," in honor of its most famous citizen. Along with a parade of thirty floats and fourteen marching bands, the high school, which had rejected Shapley's admission fifty-seven years earlier, gave him an honorary diploma. See Hoagland (1965), pp. 424–25.

115 "So there I was": Shapley (1969), p. 17. Martha Shapley, in some remembrances after her husband's death, said that "the story about 'Archaeology/Astronomy' in the catalogue was a H.S. joke." HUA, Martha Shapley's Notes on His [Shapley's] Life.

115 He actually was in need of a job . . . with honors: Shapley (1969), pp. 17–21.

115 "*thinks* about what he is doing": DeVorkin (2000), p. 104.

116 accept this rising star: Ibid.

116 specialized in eclipsing binaries: Shapley (1969), p. 25.

116 "wild Missourian": Ibid., p. 31.

116 "worse than log tables": HL, Seares Papers, Shapley to Seares, December 26, 1912.

117 "his cane to sweep the undergraduates out of their path": DeVorkin (2000), p. 105.

117 helped open doors for Shapley to become a staff astronomer: HL, Seares Papers, Seares to Shapley, April 27, 1912.

117 salary of $90 a month, plus free board on the mountain: HUA, Hale to Shapley, November 7, 1912.

117 happily computed eclipsing binary orbits: Shapley (1969), p. 49.

117 "Just killed a 3 ft. rattlesnake": Hoge (2005), p. 4.

117 "We had to be rugged in those days": Shapley (1969), p. 51.

117 Adriaan van Maanen, the latter of whom first arrived at Mount Wilson in 1911 as a volunteer assistant and remained on as a staff member for thirty-five years: Adams (1947), p. 294.

117 "A discussion with him was like a rousing game of ping-pong": Payne-Gaposchkin (1984), p. 155.

118 "A generous supporter, a stimulating companion": Ibid., p. 156.

118 so regular in his habits . . . cigarette stand: Sutton (1933b).

118 "I feel very sure that if I should go away": HUA, Shapley to George Monk, January 28, 1918.

118 "to make measures of stars in globular clusters": Shapley (1969), p. 41.

118 he had begun discovering large numbers of variables: Historian Horace Smith suggests that Bailey found enough Cepheids in Omega Centauri, the largest globular cluster in the Milky Way, to have discerned a crude period-luminosity relation six years before Henrietta Leavitt's first suggestion of such a rule. But Bailey was more focused on gathering data than interpreting it and so never made the connection. See Smith (2000), pp. 190–91.

118 "became synonyms": Shapley (1969), p. 90.
118 "I have not intended to intrude upon your field": HUA, Shapley to S. I. Bailey, January 30, 1917.
119 "I hope you will appreciate the fact": HUA, Bailey to Shapley, February 15, 1917.
119 Bailey was primarily a data gatherer: Smith (2000), pp. 194–95.
119 a trait enhanced during his apprenticeship with Russell, who advocated problem-driven research: See DeVorkin (2000).
119 was not known until the 1600s: The German astronomer Johann Abraham Ihle in 1665 discovered the first globular cluster, later labeled M22 by Charles Messier, while observing Saturn. The cluster is situated within the constellation Sagittarius.
119 "It is quite obvious that a globular cluster": Shapley (1915a), p. 213.
120 These included Omega Centauri (the biggest of them all): Recent evidence suggests that Omega Centauri, which was always atypical, is not a true globular cluster but rather a dwarf galaxy stripped of its outermost stars.
121 "Her discovery . . . is destined to be one of the most significant results of stellar astronomy": HUA, Shapley to Pickering, September 24, 1917.
121 too small to be discernible by ground-based telescopes: Space telescopes, specially designed to do parallax work, have extended distance measurements out farther.
122 30,000 light-years distant: Hertzsprung (1914), p. 204. Hertzsprung's estimate, when published in the German journal *Astronomische Nachrichten* in 1914, was first printed as a much reduced 3,000 light-years, which diminished the impact of his finding. It was a clumsy arithmetical error on Hertzsprung's part. That it was meant to be around 30,000 light-years (10,000 parsecs) is seen in a typed, unsigned note in which either Walter Adams or George Hale remarks that by an "ingenious argument" Hertzsprung has found the distance of the Magellanic Cloud "to be 10,000 parsecs—the greatest distance we have yet had occasion to mention." But the published error may have contributed to the delay in recognizing that other galaxies reside outside the boundaries of the Milky Way. CA, Hale Papers, Box 2, Hale/Adams correspondence. See also Sandage (2004), p. 361.
122 demonstrated for the first time: Fernie (1969), p. 708.
122 "I had not thought of making the very pretty use": Smith (1982), p. 72.
122 concluded that they were giant stars: Russell (1913).
122 arriving at 80,000 light-years: Smith (1982), p. 72.
123 "improved and extended": Shapley (1918a), p. 108.
123 Shapley tried mightily to check with Leavitt on this question: Shapley was still concerned late in his project. "I notice that a great many of the hundreds of [variables in the Small Magellanic Cloud] are fainter. Does Miss Leavitt know if they have shorter periods[?] . . . The matter is of much importance, as you know, because of the relation between periods and brightness," he wrote Pickering in 1917. (HUA, Shapley to Pickering, August 27, 1917.) Leavitt was then away on an extended vacation and could not provide an immediate answer.
123 "Routine stuff": HUA, Russell to Shapley, November 26, 1920.
123 "This proposition scarcely needs proof": Shapley (1914), p. 449.
123 "The whole line of reasoning . . . was brilliant": Sandage (2004), p. 303.
124 "definite conclusions from these data cannot be safely made": Bailey (1919), p. 250.
124 With the assistance of Edison Hoge, he took some three hundred photographs: Shapley (1918b), p. 156.

124 "the work on clusters goes on monotonously": Gingerich (1975), p. 346.
124 Hale had convinced him to stay at his job: HUA, Shapley to Russell, July 22, 1918.
125 With the first hint of dawn in the east . . . settle any squabbles: Sandage (2004), pp. 181, 195.
125 "The most unwarranted fun of all comes from *bugs*": HUA, Shapley to Oliver D. Kellogg, December 31, 1918.
125 "Another method is to read your thermometer": Shapley (1969), p. 66.
125 His findings were published in scientific journals: For example, see H. Shapley (1924), pp. 436–39.
125 further rest and relaxation: HUA, Shapley to Russell, September 3, 1917.
125 some well-known star clusters *within* the Milky Way were at least 50,000 light-years distant: Smith (2006), p. 319.
126 "This is a peculiar universe": HUA, Shapley to Russell, October 31, 1917.
126 "the minimum distance of the Andromeda Nebula": Shapley (1917b), p. 216.
127 "like a winding spring": Slipher (1917a), p. 62.
127 "V. M. does a little, Hale a little more, and I much": HUA, Shapley to Russell, September 3, 1917.
127 "inclined to believe in the reality of the [spirals'] internal proper motions": HUA, Russell to Shapley, November 8, 1917.
127 "word was law": Payne-Gaposchkin (1984), p. 177.
127 "the general plan of the sidereal system . . . bearing on the structure of the universe": Shapley (1918a), p. 92.
128 "striking": Shapley (1918b), p. 168.
128 "impossible to count every star shown": Melotte (1915): 168.
128 around 20,000 parsecs . . . away: Shapley (1919d), p. 313.
128 In 1909 the Swedish astronomer Karl Bohlin even dared to suggest that the center of the galaxy was in that direction: K. Bohlin, *Kungliga Svenska Vetenskapsakademiens handlingar* 43:10 (1909).
128 couldn't wait that long to spread the news: It should be noted that Shapley sketched out his results earlier in smaller publications, but the full details were presented in both the *Astrophysical Journal* and *Contributions from the Mount Wilson Observatory.*
128 "now, with startling suddenness and definiteness": HUA, Shapley to Eddington, January 8, 1918.
129 "You may have been completely prepared for the result": Ibid.
129 "While I cannot pretend to have anticipated the view of the stellar system": HUA, Eddington to Shapley, February 25, 1918.
129 "May I impose upon your time for a little while": HP, Shapley to Hale, January 19, 1918.
129 "Start a messenger on a light-wave down the main highway from the center": Ibid.
130 "the nearby spirals to either side much as the prow of a moving boat cuts through the waves": Shapley (1920), p. 100.
130 "I believe the evidence is quite against the island universe theory of spirals": HUA, Shapley to MacPherson, May 6, 1919.
130 "The observational problems opened up are unlimited": HP, Shapley to Hale, January 19, 1918.
131 "The solar system is off center and consequently man is too": Shapley (1969), pp. 59–60.

131 "this marks an epoch in the history of astronomy": HUA, Eddington to Shapley, October 24, 1918.

131 "simply amazing": Russell (1918), p. 412.

131 "certainly changing our ideas of the universe at a great rate": HUA, Jeans to Shapley, April 6, 1919.

131 "always admired the way in which Shapley finished this whole problem": Baade (1963), p. 9.

131 "super–Milky Way": "Universe Multiplied a Thousand Times by Harvard Astronomer's Calculations," New York Times, May 31, 1921, p. 1.

132 "Personally I am glad to see man sink into such physical nothingness": Ibid.

132 says Shapley in the article: Shapley wrote Henry Norris Russell two weeks later that the Times' interview with him was actually a "fake . . . evidently a rehash of last year's news about the Hale lecture [Great Debate]. It was served up new because of my shift East, of which they had just heard." HUA, Shapley to Russell, June 16, 1921.

132 Earth, proclaimed the headline, was now a "Rube": Chicago Daily Tribune, May 31, 1921, p. 1.

132 "You have struck a trail of great promise. . . . I think you are right in making daring hypotheses": HP, Hale to Shapley, March 14, 1918.

132 Though not possessing a good telescope, he organized a massive effort to measure the positions of hundreds of thousands of stars on plates taken at other observatories, partially with the help of state prisoners: Hetherington (1990b), p. 28.

132 roughly 30,000 light-years wide and 4,000 light-years thick: The full dimensions in Kapteyn's 1920 model were officially 60,000 light-years wide and 7,800 light-years thick, but the stellar distributions out in those more distant regions were extremely low, making a precise border difficult to define. See Paul (1993), p. 155. Many references cite the 30,000-light-year width.

133 difficult for Kapteyn and his colleagues: Smith (1982), p. 69.

133 "building from above, while we are up from below": Gingerich (2000), p. 201.

133 "carnival barker's certainty of truth": Sandage (2004), p. 288.

133 quick to jump to conclusions based on meager observations: AIP, interview of Harry Plaskett by David DeVorkin on March 29, 1978.

133 "two different breeds of cats": Smith (1982), p. 124.

133 "has never given the credit where it belongs": MWDF, Adams to Hale, December 10, 1917.

133 "I have never seen a quicker mind": Whitney (1971), p. 218.

134 Once Lindblad worked out the theory, Oort rounded up the evidence: Smith (1982), p. 157.

134 "With the plan of the sidereal system here outlined": Shapley (1918d), p. 53.

134 "We may compare our galactic system to a continent": MacPherson (1919), p. 334.

9. He Surely Looks Like the Fourth Dimension!

135 "the discovery of a universal formal principle": Schilpp (1949), p. 53.

136 "It does not seem that something like that can exist!": Fölsing (1997), p. 46.

136 "Newton, forgive me": Schilpp (1949), p. 31.

137 "In all my life I have labored not nearly as hard": Pais (1982), p. 216.

137 "I was beside myself with ecstasy for days": Hoffmann (1972), p. 125.

138 "Spacetime tells mass how to move": Ciufolini and Wheeler (1995), p. 13.

138 "When a blind beetle crawls over the surface of a curved branch": Isaacson (2007), p. 196.

139 "Whether the theory ultimately proves to be correct or not": Douglas (1957), p. 39.

140 "Newton's plant, which had outgrown its pot, and transplanted it to a more open field": Ibid., p. 118.

140 "people seem to forget that I am an astronomer": Ibid., p. 115.

140 "he couldn't talk at all": AIP, interview of Hermann Bondi by David DeVorkin on March 20, 1978.

140 declared valuable to the "national interest": Douglas (1957), p. 92.

141 Einstein was the first to do this: Einstein was actually prompted to do this after a discussion of general relativity with de Sitter in the fall of 1916. Kragh (2007), p. 131.

141 "Cosmological Considerations Arising from the General Theory of Relativity": Einstein (1917).

141 "I compare space to a cloth": Kahn and Kahn (1975), p. 452.

141 "It exposes me to the danger of being confined to a madhouse": Isaacson (2007), p. 252.

142 "as required by the fact of the small velocities of the stars": Translated in Lorentz, Einstein, Minkowski, and Weyl (1923), p. 188.

142 discussions in fact that inspired Einstein to conceive his spherical universe: Kerszberg (1989), pp. 99, 172.

143 "the frequency of light-vibrations diminishes": De Sitter (1917), p. 26.

143 "amongst the most distant objects we know": Ibid., p. 27.

143 "Einstein's universe contains matter but no motion": Eddington (1933), p. 46.

143 "does not make sense to me": Kahn and Kahn (1975), p. 453.

144 "systematically": De Sitter (1917), p. 28.

145 "it will always remain beyond my grasp": Smith (1982), p. 173.

145 he had early on suggested a specific test: Einstein (1911).

146 "This should serve for an ample verification": Dyson (1917), p. 447.

146 "What will it mean . . . if we get double the Einstein deflection?": Douglas (1957), p. 40.

146 "We are conscious only of the weird half-light of the landscape": Eddington (1920), p. 115.

147 "Cottingham, you won't have to go home alone": Douglas (1957), p. 40.

147 "One thing is certain, and the rest debate": Ibid., p. 44.

147 These were the results that Eddington and Dyson stressed in their reports: See Dyson, Eddington, and Davidson (1920).

147 "LIGHTS ALL ASKEW IN THE HEAVENS": New York Times, November 10, 1919, p. 17.

147 Eddington admitted he was unscientifically rooting for Einstein: Eddington (1920), p. 116.

148 "I hoped it would not be true": Douglas (1957), p. 44.

148 "We met in quick succession Their Eminences": LOA, Curtis Papers, Curtis to Campbell, May 11, 1921.

148 "He surely looks like the fourth dimension!" Ibid.

148 "bombshell . . . which quite blew up the meeting of the Academy": HUA, Shapley to Russell, May 4, 1925.

148 "I am really getting pretty tired of the fundamentalist's attitude of the opponents of relativity": HUA, Russell to Shapley, May 21, 1925.

10. Go at Each Other "Hammer and Tongs"

149 The year 1920 was one of achievements: My thanks to Virginia Trimble for pointing out some of these interesting facts in a review of the debate written for its seventy-fifth anniversary in 1995. See Trimble (1995) and also Streissguth (2001), p. 42.

150 "homeric fight": De Sitter (1932), p. 86.

150 "done to death": NAS, Abbot to Hale, January 3, 1920.

150 "I pray to God that the progress of science will send relativity to some region of space beyond the fourth dimension": HP, Abbot to Hale, January 20, 1920.

150 Abbot wondered . . . island-universe theory: Hoskin (1976a), p. 169; Smith (1983), p. 28; NAS, Abbot to Hale, January 3, 1920.

150 "daring innovator . . ."; ". . . and more often concluded 'not proven' than 'not so' ": Struve (1960), p. 398.

151 "Perhaps Harvard is amateurish, compared with Mount Wilson": HUA, Shapley to Russell, February 12, 1919.

151 worried how he would come across: Shapley was increasingly uncomfortable at Mount Wilson, where he didn't get along with deputy director Walter Adams. Adams strongly criticized Shapley's model of the galaxy when it first came out, questioning the way Shapley cut corners in reaching his conclusions. Shapley blamed Adams's disapproval on "professional jealousies." (See HUA, Director's Correspondence, Seth Nicholson to Shapley, November 6, 1921.)

151 Curtis was known to be a dynamic lecturer; Shapley feared he would look bad by comparison: The role of the Harvard appointment on Shapley's performance at the debate was first discussed by British historian Michael Hoskin. Historic accounts of the Great Debate previous to Hoskin were based solely on the printed publication of the debate. Hoskin was the first to unearth archival materials on both the session and its background. See Hoskin (1976a).

151 "I am sure that we could be just as good friends if we did go at each other 'hammer and tongs' ": HUA, Curtis to Shapley, February 26, 1920.

151 " 'take the lid off' and definitely attach each other's view-point": Ibid.

151 "I have neither time nor data nor very good arguments": HUA, Shapley to Russell, March 31, 1920.

152 "two talks on the same subject": HP, Shapley to Hale, February 19, 1920.

152 "My sympathies are with the audience": HUA, Shapley to Abbot, March 12, 1920.

152 "We could scarcely get warmed up in 35 minutes": HP, Curtis to Hale, March 9, 1920.

152 compromised at forty minutes: HUA, Abbot to Shapley, March 18, 1920.

152 "If you or he wish to answer points made by the other": HUA, Shapley Papers, Hale to Curtis, March 3, 1920.

152 For Curtis it was $2 for the stagecoach to San Jose, then another $100 for the round-trip railroad ticket: LOA, Curtis Papers, Curtis to Campbell, April 8, 1920.

152 When the train broke down . . . to collect a few native ants: AIP, interview of Harlow Shapley by Charles Weiner and Helen Wright on August 8, 1966.

152 "growth and development" . . . in weather forecasting: NAS, Program of Scientific Sessions, Annual Meeting, April 26, 27, 28, 1920.

153 "Dr. Harlow Shapley, of the Mount Wilson solar observatory": "Scientists Gather for 1920 Conclave" (1920), p. 38.

153 two friends of Harvard president A. Lawrence Lowell . . . were in the audience to size him up: Bok (1978), p. 250.

153 "just got a new theory of Eternity": Shapley (1969), p. 78.

153 conference dinner was the following night: NAS, Academy press release, "America's Academicians Meet in Washington," April 19, 1920.

153 Shapley did save the typescript of his talk: Subsequent Shapley lecture quotes are taken from HUA, Shapley Papers, "Debate MS."

154 "so much greater weight" . . . "be used as checks or as secondary standards": Shapley (1918d), p. 43.

154 wondering whether he should change his approach on the fly: HUA, Curtis to Shapley, June 13, 1920.

154 some of his slides, displaying his essential points, do survive: All the major points are discussed in Hoskin (1976a), pp. 178–81.

155 eleven "miserable" Cepheids: HUA, Shapley to Russell, March 31, 1920.

155 Everyone in essence went home maintaining the beliefs they held: Fernie (1995), p. 412.

156 "came out considerably in front": Hoskin (1976a), p. 174.

156 "gift of the gab": Ibid. There's some evidence that Shapley got wind of this gossip about his poor speaking skills. Once at Harvard, he wrote his old boss George Hale that he was planning a series of lectures. "It turns out that I have some of the knacks of entertaining a general audience (as I rather suspected would be the case if I got a little experience)—not too much dignity, you know, some enthusiasm, and an increasing confidence." HL, Walter Adams Papers, Shapley to Hale, October 3, 1921.

156 "He has . . . a some what peculiar and nervous personality" . . . "more balance more force and a broader mental range": HUA, G. R. Agassiz to Lowell, April 28, 1920.

156 "Yes, I guess mine was too technical": HUA, Curtis to Shapley, June 13, 1920.

156 At first Curtis wasn't keen on publishing his comments: HUA, Curtis to Shapley, June 13, 1920.

156 "generally observed in composing telegrams" . . . "shoot our arrows into the air": HUA, Curtis to Shapley, August 2, 1920.

157 "ten pages of buncombe": HUA, Shapley to Curtis, July 27, 1920.

157 "Should I go ahead, shoot my shot (or wad)": Ibid.

157 "at least a brief statement of how you explain them if not island universes": HUA, Curtis to Shapley, September 8, 1920.

157 "appear fatal to such an interpretation": Shapley and Curtis (1921), p. 192.

157 "I see no reason for thinking them stellar or universes": HUA, Shapley to Russell, September 30, 1920.

157 "the island universe theory must be definitely abandoned": Shapley and Curtis (1921), p. 214.

159 Van Maanen was the descendant of an aristocratic family . . . a rare find at the time: Berendzen and Shamieh (1973), p. 582, and Seares (1946).

159 "One always returns to one's first love," he scribbled on the title page of a 1944 paper on stellar parallaxes: Sandage (2004), p. 127, and van Maanen (1944).

159 "Do not use this stereocomparator without consulting A. van Maanen": Trimble (1995), p. 1138.

159 played a good game of tennis: AIP, interview of Nicholas U. Mayall, June 3, 1976.

160 "He could go to a dinner and soon have the whole table laughing": Shapley (1969), p. 56.

160 An accomplished chef: Sandage (2004), p. 129.

160 "Van Maanen and I are in ill-favor because we do or try to do too much": HUA, Shapley to G. Monk, January 28, 1918.

160 van Maanen always seemed to see this effect: Hetherington (1990b), p. 30.

160 Ritchey was then using Mount Wilson's 60-inch telescope ... details never before captured: Ibid., pp. 31–33.

160 at first measured no variation but got permission from Ritchey to keep the plates to study them further: HP, van Maanen to Hale, May 2, 1916; Hale to Chamberlin, December 28, 1915.

160 he chose thirty-two stars ... would be negligible: Hetherington (1990b), p. 35.

161 "If the results ... could be taken at their face value": Van Maanen (1916), pp. 219–20. John Duncan, just appointed director of the Wellesley College Observatory, in Massachusetts, took a long trip west to visit observatories in the summer of 1916. There he assisted in giving the new 100-inch mirror its first coat of silver and wrote Slipher that "van Maanen, who is a very enthusiastic Dutchman, has measured with the Blink some photographs of Messier 101 made some years apart and gets what seems to be certain evidence of a motion *along* the arms of the spiral." LWA, Duncan to Slipher, July 14, 1916.

161 meant ... the nebula's edge had to be traveling faster than the speed of light: Shapley (1919e), p. 266.

161 van Maanen followed all the precautions: Hetherington (1990b), p. 37.

161 "While the recent revival of the notion that spiral nebulae are mere distant constellations": HP, Chamberlain to Hale, January 31, 1916.

161 "might indicate that these bodies are not as distant as is usually supposed to be the case": Hetherington (1974b), pp. 52–53.

161 "So that we do not know yet if this is an island universe!": HP, van Maanen to Hale, December 17, 1917.

162 "His wide experience in astrometric work": HL, Walter Adams Papers, Adams to John C. Merriam, August 15, 1935.

162 "a much greater time interval will probably be necessary before nebular rotations can be definitely established": Hetherington (1990b), p. 26.

162 "The mean of five measures each of which is not worth a damn": LOA, Curtis Papers, Curtis to Campbell, July 11, 1922.

162 "entirely in agreement with some speculations in which I have recently been indulging": Jeans (1917a), p. 60.

162 both van Maanen and Jeans began to calculate higher masses for the spirals: Smith (1982), p. 40.

164 seemed to imply his methods were valid: Hetherington (1990b), p. 42.

164 "would be so bold as to question the authenticity of the internal motions": Smart (1924), p. 334.

164 "I finished ... my measures of M51": HUA, van Maanen to Shapley, May 23, 1921.

164 "Congratulations on the nebulous results!": HUA, Shapley to van Maanen, June 8, 1921.

164　"I think that your nebular motions are taken seriously now": HUA, Shapley to van Maanen, September 8, 1921.

164　"raise a strong objection to the 'island-universe' hypothesis": Van Maanen (1921), p. 1.

164　"which, obviously, are extremely improbable": Ibid., p. 5.

164　"a great number of very distant stars . . . crowded together [to] give the impression of nebulous objects": Lundmark (1921), p. 324.

165　"speak for a large distance": Ibid., p. 326.

165　Shapley began to feel sizable pressure: After Lundmark published a paper in 1922 criticizing some of Shapley's research, Shapley undiplomatically wrote Lundmark that "there will be little gain if either of us . . . strive to pick to pieces small and irrelevant points. . . . Think how many flaws or hasty conclusions you or I might find in your big paper on the distances of globular clusters." HUA, Shapley to Lundmark, July 15, 1922. Lundmark was deeply upset by Shapley's remarks and did stop his criticism of van Maanen's work for a while, lest others start putting his own findings under a microscope. HUA, van Maanen to Shapley, October 21, 1922. Robert Smith points out that Lundmark had the opportunity to remeasure van Maanen's plates during a stay at Mount Wilson in the early 1920s and was briefly convinced that van Maanen had detected some real motions in the spirals, which made him deem the island-universe theory "rather hopeless." But by 1924 additional study convinced Lundmark he had been wrong, returning him to the island-universe fold. See Smith (1982), p. 108.

165　"celestial speed champion" . . . "many millions of light years" away: Slipher (1921), p. 6.

165　"increases the probability": Öpik (1922), p. 410.

165　"Shapley couldn't swing the thing alone" . . . "and I might keep Shapley from too riotous an imagination,—in print": HP, Russell to Hale, June 13, 1920.

166　"I would rather do astronomy": DeVorkin (2000), p. 169.

166　"Chief Observer or something of the sort": HUA, Julian L. Coolidge to Shapley, November 24, 1920.

166　He, a bit miffed, curtly turned it down: HUA, Shapley to A. Lawrence Lowell, December 10, 1920.

166　try him out for a year as chief of staff: George Hale first made this suggestion in a letter to Harvard president Lawrence Lowell. "You might give Dr. Shapley for a year some position such as you recently offered him for a longer period," he wrote. "This would enable you to test his scientific and personal qualifications, with the purpose of appointing him Director in the case of a favorable outcome. . . . I am willing to give him a leave of absence for a year if you wish to try this plan." HP, Hale to Lowell, December 11, 1920. Complete behind-the-scenes details on Shapley's struggle to garner the Harvard appointment is found in Gingerich (1988).

166　"a kind of rotating galaxy for ideas": Hoagland (1965), p. 429.

166　bounding up the stairs two steps at a time: Payne-Gaposchkin (1984), p. 155.

166　"He cast spells over people": AIP, interview with Helen Sawyer Hogg by David DeVorkin on August 17, 1979.

166　band of enthusiastic workers: AIP, interview of Harry Plaskett by David DeVorkin on March 29, 1978.

166　"he inspired us all": AIP, interview of Leo Goldberg by Spencer Weart on May 16, 1978.

166　He also stubbornly ignored new scientific data at times: AIP, interview with Jesse Greenstein by Paul Wright on July 31, 1974.

166 "I thought I told you that I left Mount Wilson just to avoid this ordeal": HL, Walter Adams Papers, Shapley to Gianetti, July 29, 1921.

166 tendered his resignation ten days before the Washington debate took place: LOA, Curtis to Campbell, April 16, 1920.

167 "the biggest mistake he ever made": AIP, interview with C. Donald Shane by Elizabeth Calciano in 1969.

167 "the California combination of instruments PLUS climate": Osterbrock, Gustafson, and Unruh (1988), p. 146.

167 "You play golf don't you? Well, this is my golf": Stebbins (1950), June 24.

167 "memorable set-to" . . . "I have always thought that the clubs we wielded at each other. . . ."; "watching the strife with interest": HUA, Curtis to Shapley, July 10, 1922.

168 "photographing, photographing. . . ." . . . "hunt for novae and variables": LOA, Curtis to Aitken, January 2, 1925.

168 "I am copying that instrument in my design far more than any other": LOA, Curtis to Aitken, March 16, 1934.

11. Adonis

169 "Adonis": HUB, Box 7, Grace's memoirs.

169 "Had we been casting": HUB, Box 8, Anita Loos remembrance.

169 adding dubious credentials to his curriculum vitae: This may have been a family trait. Hubble's father was described by his family as working at certain positions, which it was later discovered he never held. See Christianson (1995), p. 12.

170 And the longer time went on, said astronomer Nicholas Mayall, who once worked with Hubble, the higher the pedestal got: AIP, interview of Nicholas U. Mayall by Bert Shapiro, February 13, 1977.

170 he ruled his domestic realm with a firm puritanical hand, a strictness that was balanced by the more forgiving and accessible mother: HUB, Box 8, Helen Hubble memoir.

170 permitted to stay up past his bedtime: Ibid.

170 In high school: Facts concerning Hubble's high school accomplishments come from HUB, Box 2.

170 "He always seemed to be looking for an audience to which he could expound some theory or other": Christianson (1995), p. 31.

171 "outlandish" career choice: Ibid., p. 40.

171 Hubble compromised by taking science classes . . . as well as . . . classics: HUB, Box 25, undergraduate course book.

172 "Motor cars, at last, were successfully competing with horses": HUB, Box 1, Folder 23, pp. 1–2.

172 "whiz" at calculus, who "often utterly dumfounded" the professor: HUB, Box 19, John Schommer to Grace Hubble, May 15, 1958.

172 best physics student: HUB, Box 25, "The Daily Maroon," January 26, 1910.

172 Chicago promoters were eager for him to turn professional: HUB, Box 7, "University of Chicago, 1906–1910, 1914–1917," p. 3.

172 Good in academics but not "mere bookworms" . . . "moral force of character": Encyclopaedia Britannica (1911).

173 "man of magnificent physique, admirable scholarship, and worthy and lovable character": HUB, Box 15, Millikan to Edmund James, January 8, 1910.

173 three years on an annual stipend of fifteen hundred dollars: HUB, Box 25, "The Daily Maroon," January 26, 1910.

173 "considerable ability. Manly": Osterbrock, Brashear, and Gwinn (1990), p. 4.

173 "had transformed [Hubble], seemingly, into a phony Englishman, as phony as his accent": Christianson (1995), p. 64.

173 "I sometimes feel that there is within me, to do what the average man would not do": Ibid., p. 67.

173 "Why not be first in Rome?": HUB, Box 8, Grace's memoirs.

174 translating what may have been legal correspondence: Christianson (1995), p. 86.

174 All this time he was actually teaching at the high school in New Albany, Indiana . . . dedicated the school's 1914 yearbook to him: HUB, Box 22A.

174 "So I chucked the law": HUB, Box 7, "Hubble: A Biographical Memoir." Hubble was eventually awarded an honorary doctor of laws degree from the University of California in 1949.

175 "splendid specimen," who showed "exceptional ability": Osterbrock, Brashear, and Gwinn (1990), p. 5.

176 "Send us three hundred words expressing your ideas on the habitability of Mars": Frost (1933), p. 217.

176 "Those who have visited a large observatory on such a night": Ibid., p. 205.

176 "So you say that each of those points of light is a sun": Ibid., p. 207.

177 Frost himself was slowly losing his eyesight due to cataracts: Christianson (1995), p. 95.

178 "Suppose them to be extra-sidereal [outside the Milky Way] and perhaps we see clusters of galaxies": Hubble (1920), p. 75.

178 "But it shows clearly the hand of a great scientist groping toward the solution of great problems": Osterbrock, Brashear, and Gwinn (1990), p. 7.

178 "questions await their answers for instruments more powerful than those we now possess": Hubble (1920), p. 69.

178 "I have offered Hubbell [sic] a position with us at $1200. per year": HP, Hale to Adams, November 1, 1916.

179 he didn't have the money to offer his graduating student a well-paid position: HP, Henry Gale to Adams, April 4, 1917.

179 Within days Hubble asked Frost for a letter of recommendation . . . a military reservation on Lake Michigan, north of Chicago: Osterbrock, Brashear, and Gwinn (1990), pp. 8–9.

179 "scimpy": Christianson (1995), p. 101.

179 Hubble had already sent a letter: MWDF, Hubble to Hale, April 10, 1917.

179 "to renew as soon as you are able to accept it": MWDF, Hale to Hubble, April 19, 1917.

180 "Stirring times": Osterbrock, Brashear, and Gwinn (1990), p. 9.

180 rendered unconscious at one point by a shell exploding nearby: HUB, Box 7, Grace's memoir.

180 no "wound chevrons" were authorized: HUB, Box 25, discharge certificate.

180 "I barely got under fire": Christianson (1995), p. 109.

180 posh dinner hosted by the best and the brightest of British astronomy: Ibid., p. 110.

181 "My interest has for the most part been with nebulae especially photographic study of the fainter ones": MWDF, Box 159, Hubble to Hale, May 12, 1919.

181 "I had been hoping" . . . "as we expect to get the 100-inch telescope into commission very soon": MWDF, Hale to Hubble, June 9, 1919.
181 arrived in New York on August 10: Osterbrock, Brashear, and Gwinn (1990), p. 11.
181 "Just demobilized. Will proceed Pasadena at once unless you advise to contrary": MWDF, Hubble to Hale, August 22, 1919.
181 September 11, 1919: Christianson (1995), p. 122.

12. On the Brink of a Big Discovery—or Maybe a Big Paradox

182 He was a man of endless enthusiasms: It's been suggested that Hale suffered a severe form of manic depression, a psychiatric syndrome marked by periods of elevated mood, physical restlessness, and sharpened creative thinking, interlaced with bouts of depression. See Sheehan and Osterbrock (2000).
182 "a driving power which was given no rest until it had brought his plans and schemes to fruition": Wright (1966), p. 17.
182 "He has reached a place where scientific work and honors are not enough": Osterbrock (1993), p. 157.
182 In the summer of 1906 he spent a weekend at the home of John Hooker . . . secret of their mysterious nature: Wright (1966), pp. 252–53; Osterbrock (1993), p. 92.
183 Hale's younger brother, Will, once called George the greatest gambler in the world: Wright (1966), p. 184.
183 "We don't pay for this!": Ibid., p. 254.
183 "that glass was in the bottom of the ocean": Wright (1966), p. 263. Evelina Hale through these times fiercely protected her husband and wished the 100-inch glass disk gone in a letter dated December 24, 1910, to astronomer Walter Adams, who served as the Mount Wilson Observatory's acting director in Hale's absence. In that message she beseeched Adams to send no bad news to Hale during his recovery.
184 made a good case: See Sheehan and Osterbrock (2000), p. 105.
184 he initiated the grinding: Osterbrock (1993), p. 142.
184 "there was more publicity . . . than was desirable": MWDF, Adams to Hale, July 5, 1917.
185 "To add to the gloom": Adams (1947), p. 301.
185 first Hale then Adams returned . . . at 2:30 in the morning: Wright (1966), pp. 318–20.
185 "High in heaven it shone": Noyes (1922), pp. 2–3.
186 "Very little has been done with it . . . because of the war contracts in the shop": HUA, Shapley to R. G. Aitken, October 14, 1918.
186 Ritchey, for example, had to turn his attention to making lenses and prisms: Osterbrock (1993), pp. 144–45.
186 "In such an embarrassment of riches": Hale (1922), p. 33.
187 took about an hour then to make the journey in a motorcar: HUB, Box 7, "Hubble: A Biographical Memoir."
187 Seven days later Hubble tried out the 60-inch telescope . . . "striking changes have happened [in it] since 1916": HUB, Box 29, Logbook; HUB, Box 7, "Hubble: A Biographical Memoir."

187 "He was photographing at the Newtonian focus of the 60-inch": Humason (1954), p. 291.
187 what he called his "magic mirror": HUB, Box 1, "The Exploration of Space" lecture.
187 Hubble's first night on 100-inch: HUB, Box 29, Logbook.
188 The variable nebula soon became his observational "mascot": This is according to Milton Humason. HUB, Box 7, "Hubble: A Biographical Memoir."
188 each was entered into his official Observing Book: HUB, Box 7, "Hubble: A Biographical Memoir."
188 got a paper published fairly quickly: Seares and Hubble (1920).
188 "to determine the relation of nebulae to the universe": LWA, Hubble to Slipher, April 4, 1923.
188 "We are on the brink of a big discovery—or maybe a big paradox": HUA, Russell to Shapley, September 17, 1920.
188 "I have just gone into the lecture room, pressed a button, and heard records": LOA, Curtis Papers, Curtis to Campbell, January 26, 1922.
189 Hubble cultivated an air of sophistication and restraint: AIP, interview with Nicholas U. Mayall, June 3, 1976.
189 occasionally blow smoke rings out into the room: HUB, Box 7, Grace's memoir.
189 "stuffed shirt": CA, interview with Jesse L. Greenstein by Rachel Prud'homme, February 25, March 16, and March 23, 1982.
189 "write an inter-office memo": AIP, interview of Halton Arp by Paul Wright, July 29, 1975.
189 wearing jodhpurs, leather puttees, and a beret while observing: AIP, interview of Horace Babcock by Spencer Weart on July 25, 1977.
189 "Bah Jove" . . . "Missourian tongue" . . . "Hubble disliked van Maanen from the time he himself arrived on Mount Wilson" . . . "Hubble just didn't like people": Shapley (1969), p. 57.
190 "conscientious slacker": AIP, interview of Dorritt Hoffleit by David DeVorkin on August 4, 1979.
191 "lend some color to the hypothesis that the spirals are stellar systems": Hubble (1920), p. 77.
191 the term *non-galactic* didn't mean the spirals were necessarily "outside our galaxy": Hubble (1922), p. 166.
191 "half a dozen of the largest spirals in addition to Andromeda should be followed carefully for novae": LWA, Hubble to Slipher, February 23, 1922.
192 "I must confess that I am rather dazed by [Hubble's] letter": LWA, Wright to Slipher, March 7, 1922.
192 particularly fired up about a nebula classification scheme: LWA, Hubble to Slipher, February 23, 1922.
192 "pathologically shy around colleagues with whom he had little . . . contact": Sandage (2004), p. 525.
193 "What a powerful instrument the 100-inch is in bringing out those desperately faint nebulae": HUA, Shapley to Hubble, August 3, 1923.
194 "It appears to be a great star cloud that is at least three or four times as far away as the most distant of known globular clusters": Shapley (1923b), p. 2.
194 "the most distant object seen by man, another universe of stars": "A Dist Universe of Stars" (1924), p. x.
194 "neither galactic in size nor stellar in composition": Shapley (19⁻

194 "Eleven . . . are clearly Cepheids": Hubble (1925b), p. 412.
194 "N.G.C. 6822 lies far outside the limits of the galactic system": Ibid., p. 410.
194 "This was the astronomical observing experience at its best": Sandage (2004), p. 178.
195 "You begin with deskwork": Mayall (1954), p. 80.
195 Hale considered coffee "unwholesome": HUB, Box 7, "Hubble: A Biographical Memoir."
195 were offered two pieces of bread, two eggs, butter and jam: This regimen continued until 1955. Sandage described it as their "starvation rations." Sandage (2004), pp. 191–92.
195 he washed his own dish afterward: Christianson (1995), p. 123.
196 "You knew where you stood with him": HUB, Box 7, Grace Hubble interview with Humason.
196 Messier objects were as familiar to him as the alphabet: Mayall (1954), p. 80.
196 wispy nebula that Shapley had reported seeing on two occasions: HL, Adams Papers, Shapley to Adams, July 12, 1923.
196 "Shapley object is probably an accident": HUB, 100-inch Logbook.

13. Countless Whole Worlds . . . Strewn All Over the Sky

199 a whine, a series of loud clicks, and then a final *clang* as the instrument was secured into place: HUB, Box 7, "Hubble: A Biographical Memoir."
199 The famous Andromeda nebula, the target of choice: All details of Hubble's observations of Andromeda are taken from HUB, 100-inch Logbook.
200 numbering each nova and variable: HUB, Box 1, Hubble Addenda.
202 "Dear Shapley:—You will be interested to hear that I have found a Cepheid variable in the Andromeda Nebula": HUA, Hubble to Shapley, February 19, 1924.
203 figuring out early on, soon after he arrived at Mount Wilson, that they were pulsating stars: Shapley (1914).
204 "Here is the letter that has destroyed my universe": Payne-Gaposchkin (1984), p. 209.
204 "the crop of novae and of the two variable stars in the direction of the Andromeda nebula": HUA, Shapley to Hubble, February 27, 1924.
205 "He was standing at the laboratory window, looking at a plate of Orion": HUB, Box 7, Grace's memoir.
205 Hubble and Grace, now widowed, renewed their acquaintance: Osterbrock, Brashear, and Gwinn (1990), p. 14.
205 "Do you think you can stay up later than an astronomer?": HUB, Box 7, "Hubble: A Biographical Memoir."
206 with none of Hubble's family members in attendance: Over the succeeding years, Hubble withdrew even further from his family, as if wishing his midwestern roots would just wither away and die. His younger brother, Bill, a dairy farmer, took responsibility for his mother's care, allowing Edwin to pursue his dreams unimpeded. See Christianson (1995), pp. 98–99, 166.
206 liked to mingle with the elite of Hollywood society rather than astronomers: Dunaway (1989), p. 69.
207 "A stranger could drop raspberry soufflé on the rug without hearing a murmur": Ibid.

207   "quite out of the common": A comment made by Susan Ertz, a friend of Grace's
      from elementary school. HUB, Box 1, Folder 3.
207   "was Watson to his Sherlock Holmes": HUB, Box 7, "Hubble: A Biographical
      Memoir."
207   found even more variables: Hubble (1925a).
207   hundreds of pages now filed away in an archive: The Huntington Library in San
      Marino, California.
207   he sometimes cut corners in the darkroom: AIP, interview of Nicholas U. Mayall
      by Bert Shapiro, February 13, 1977; interview of Martin Schwarzschild by
      Spencer Weart on June 3, 1977.
208   "one lump of beauty mixed with lots of incredible boredom and discomfort":
      AIP, interview of Jesse Greenstein by Paul Wright on July 31, 1974.
209   "You . . . may be interested to hear that variable stars are now being found":
      LWA, Hubble to Slipher, July 14, 1924.
209   Slipher had already heard: LWA, Slipher to Hubble, August 8, 1924.
209   The news was rapidly spreading on the astronomical grapevine: HUB, Box 1,
      "Edwin Hubble and the Existence of External Galaxies" by Michael Hoskin.
209   "What do you think of Hubble's Cepheids": HUA, van Maanen to Shapley,
      March 14, 1924.
209   "I feel it is still premature to base conclusions on these variables": HUA, Hub-
      ble to Shapley, August 25, 1924.
210   "exciting" . . . "What tremendous luck you are having": HUA, Shapley to Hub-
      ble, September 5, 1924.
210   most boisterous promoter: Shapley soon published a popular article titled
      "Beyond the Bounds of the Milky Way." HP, Shapley to Hale, April 2, 1925.
210   "I am wasting a good deal of time": LWA, Hubble to Slipher, December 20,
      1924.
211   "Finds Spiral Nebulae Are Stellar Systems": New York Times, November 23,
      1924, p. 6.
211   "the rapid progress of knowledge, and the changing state of speculative theo-
      ries": Doig (1924), p. 99.
211   "undoubtedly among the most notable scientific advances of the year":
      Berendzen, Hart, and Seeley (1984), p. 134.
211   "Heartiest congratulations on your Cepheids in spiral nebulae!": HUB, Russell
      to Hubble, December 12, 1924.
212   "considerable interest" in the outcome: "Welfare of World Depends on Science,
      Coolidge Declares" (1925), p. 9.
212   "The real reason for my reluctance in hurrying to press": Hubble to Russell,
      February 19, 1925, in Berendzen and Hoskin (1971), p. 11.
212   "I believe the measured rotations must be abandoned": Ibid.
212   "an ass!!": HUB, Stebbins to Hubble, February 16, 1925.
212   "We walked back to the group in the lobby": Ibid.
213   "I have always believed that the spirals are island universes": LOA, Curtis to
      Aitken, January 2, 1925.
213   "Dr. Hubble . . . has found that the outer parts of the two most conspicuous
      nebulae": HUB, Box 9.
213   The Cepheids were fast becoming the gold standard for measuring distances:
      Russell (1925), p. 103.
213   "The great distances recently derived have made rapid rotation impossible":
      Luyten (1926), p. 388.

214 "van Maanen's measurements have to go": Berendzen, Hart, and Seeley (1984), p. 123.

214 Parasitologist Lemuel Cleveland of the School of Hygiene and Public Health at Johns Hopkins was also honored: "Honor for Dr. Edwin P. Hubble" (1925), pp. 100–101.

214 "To scientists, . . . the infinite and the infinitesimal are merely relative terms, alike in importance": "Infinite and Infinitesimal" (1925), p. B4.

215 The Hubbles had just bought an acre lot in San Marino: HUB, Box 7, Grace's memoir. In the late 1970s the Hubble home was placed on the National Register of Historic Places. See Pasadena *Star-News*, April 5, 1977.

215 "If an old scrap of paper, published within the sacred period": Blades (1930), p. J10.

215 Duncan found three variable stars within the Triangulum nebula, M33: Duncan (1922).

216 colleague at Mount Wilson, George Ritchey, had photographed thousands of "soft star-like condensations" in Andromeda: Ritchey (1910a), p. 32.

216 suggested that strict divisions were in place in Mount Wilson: Shapley (1969), p. 58.

216 "I faithfully went along with my friend van Maanen": Ibid., p. 80.

217 Shapley was so certain of his position that he proceeded to take a handkerchief out of his pocket and rub out the marks: This story was first published in Smith (1982), p. 144. Smith noted that he found no documentary proof but judged there were "some pointers to its possible truth." Allan Sandage elaborates on the tale in his history of Mount Wilson. Sandage (2004), pp. 495–98.

217 "spiral nebulae" were on his agenda and that "cosmogony" would be his future field: HUA, Shapley to Kellogg, June 10, 1920, and December 1, 1920.

217 "The work that Hubble did on galaxies was very largely using my methods": Shapley (1969), pp. 57–58.

217 "in the fields of observation": Louis Pasteur, Inaugural Lecture, University of Lillé, December 7, 1854.

218 "There is just not one universe": HUB, Box 28, Scrapbook.

218 catchiest headline: Ibid.

218 "more systems of stars than there are hairs in the whiskers of Santa Claus": Blades (1930), p. J10.

218 "Professor Edwin Hubble announces that he has found another universe": "The Universe, Inc." (1926), 133.

218 "Astronomy, as a matter of popular interest": "Crowd Jams Library for Hubble Talk" (1927).

218 "It is like looking at those lights": Blakeslee (1930).

218 did by chance discover "Comet Hubble" in August 1937: HUB, 100-inch Logbook.

218 "I am commuting to a spiral nebula": HUB, Box 8, biographical memoir.

219 "astronomy is a science in which exact truth is ever stranger than fiction": Jeans (1929), p. 8.

219 "How terrifying! . . . nothing at all!": HUB, Box 10, Folder HUB 195.

220 "I want to get away from both the words universe and nebula": HUA, Shapley to Hubble, May 29, 1929.

220 didn't see any pressing need to abolish the "venerable precedent" of preserving the word *galaxy*: HUA, Hubble to Shapley, May 15, 1929.

220 "The term *nebulae* offers the values of tradition": Hubble (1936), p. 18.

220 quickly pinpointed whether they came from the East or West Coast of the United States: Smith (1982), p. 151.

220 "I want to compare them with the novae in spirals": HUA, van Maanen to Shapley, February 18, 1925.

220 "I am completely at a loss to know what to believe": HUA, Shapley to van Maanen, March 8, 1925.

220 "what to think of your confounded spirals": HUA, Shapley to van Maanen, April 6, 1931.

221 "van Maanen's contradiction disturbed her husband so greatly": Sandage (2004), p. 528.

221 "a decided internal motion in the same direction": Hale, Adams, and Seares (1931), p. 200.

221 "They asked me to give him time. Well, I gave him time, I gave him ten years": HUB, Box 16, remembrance by Grace Hubble to Michael Hoskin, March 7, 1968.

221 Van Maanen was sure that Hubble had been heading up a cabal to deny him a fair share of time on the 100-inch. That's when van Maanen slapped his sign on the front of the Blink, warning others not to use the machine without his permission: Christianson (1995), p. 231.

221 The skirmish even extended into the dining room atop Mount Wilson: AIP, interview of Olin Wilson by David DeVorkin on July 11, 1978.

222 "Hubble skillfully employed trial tactics": Hetherington (1990a), p. 23.

222 "no evidence of motion": HUB, Box 3, Folder 52.

222 "Its language was intemperate in many places": HL, Adams Papers, Adams to Merriam, August 15, 1935.

222 resolution involved delicate diplomacy: Hetherington (1990a), p. 10; Sandage (2004), p. 215.

223 "no compromise, no compromise": AIP, interview of Nicholas U. Mayall, June 3, 1976.

223 "I do not feel that Hubble's attitude in this matter was in any way justified": HL, Adams Papers, Adams to Merriam, August 15, 1935.

223 "Print what you like, but print it elsewhere": HP, Seares to Hale, January 24, 1935. Historian Robert Smith was the first to track down the correspondence on this matter and bring this skirmish to light. See Smith (1982), pp. 135–36.

223 "The attitude of van Maanen in the matter was much superior to that of Hubble": HL, Adams to Merriam, August 15, 1935.

223 "recognized this curious 'blind spot' in almost every important dealing": HL, Adams to Merriam, February 19, 1936.

224 "Great men have to go their own way": Christianson (1995), p. 225.

224 "With Edwin, it was out of sight, out of mind": Ibid., p. 61.

224 accompanied by a paper by van Maanen in which he acknowledged the existence of possible errors: Van Maanen's first draft essentially just restated his results. There's considerable evidence that Adams then intervened and dictated some concessionary phrases, which van Maanen agreed to. Brashear and Hetherington (1991), pp. 419–20.

224 brief note came out in the May 1935 issue of the *Astrophysical Journal*: Hubble (1935).

224 had van Maanen's paper immediately follow: Van Maanen (1935).

224 whenever the two passed each other in the observatory hallways, they ex-

changed not a word: AIP, interviews of Nicholas U. Mayall, June 3, 1976, and February 13, 1977.

14. Using the 100-Inch Telescope the Way It Should Be Used

225 "shun us like a plague": Eddington (1928), p. 166.

225 more than three hundred delegates attended the gathering, where they were entertained with boat excursions down the city's noted canals: LWA, Lampland to Slipher, July 8, 1928.

225 "Most of the Americans appear to be over here this summer": LWA, Lampland to Slipher, August 8, 1928.

225 appointed acting chairman of the IAU Nebulae Commission: Stratton (1929), p. 250.

226 his magisterial 1926 paper: Hubble (1926).

226 "consistent with the marked tendency already observed": Humason (1927), p. 318.

226 de Sitter encouraged Hubble at this time to extend the redshift measurements of the spiral nebulae: This is according to Milton Humason as stated in HUB, Box 7, Grace's memoir.

226 "The Flagstaff assault on these objects": HUA, Shapley to Russell, May 22, 1929.

227 "I didn't feel much enthusiasm" . . . "test of endurance": HUB, Box 7, Grace's memoir.

227 one summer as a teenager . . . taking any astronomer who wanted to go with him: Sandage (2004), p. 527.

227 Mount Wilson hotel: The hotel was built in 1905 by the Mount Wilson Toll Road and Hotel Company. The original structure burned down in 1913 but was soon rebuilt and remained open until 1963. Sandage (2004), p. 24.

227 tracked the animal down and shot him between the eyes: Sutton (1933b), p. 14.

228 Humason was even put in charge of scheduling telescope time: Sandage (2004), p. 185.

229 He accomplished this feat by establishing a ladder of measurements: Hubble (1929a).

229 to directly obtain the distances to six relatively nearby galaxies: The sixth galaxy was actually done indirectly; it was a companion to one of the five and hence assumed to be a similar distance.

229 "Mr. Strömberg has very kindly checked the general order of these values. . . . Solutions of this sort have been published by Lundmark": Hubble (1929a), p. 171.

230 He didn't even like Lundmark: Smith (1982), p. 183.

230 Hubble held up publication of his data to make sure he had nailed down every argument: HUA, Hubble to Shapley, May 15, 1929.

230 "There is more to the advance of science than new observations and new theories": Hetherington (1996), p. 126.

231 "For such scanty material, so poorly distributed, the results are fairly definite": Hubble (1929a), p. 170.

232 "I agreed to try one exposure": AIP, interview of Milton Humason by Bert Shapiro, around 1965.

232 "that the mountain itself is rolling eastward with the earth at ten times an express train's speed": Sutton (1933a), p. G12.

232 MacCormack calculated a final velocity of 3,779 kilometers per second: Humason (1929), p. 167.

232 The success spurred Mount Wilson officials: AIP, interview of Milton Humason by Bert Shapiro, around 1965.

232 Humason was ready to quit: Ibid.

232 "The high velocity for N. G. C. 7619 derived from these plates": Humason (1929), p. 167.

233 "in part regretting a lost opportunity to pursue such a relation himself": Smith, (1982), p. 184.

233 "the speed of spiral nebulae is dependent to some extent upon apparent brightness, indicating the relation of speed to distance": Shapley (1929), p. 565.

233 Within two years, Hubble and Humason measured forty more galaxies: Hubble and Humason (1931).

233 "Humason's adventures were spectacular": HUB, Box 2, "The Law of Red-Shifts," George Darwin Lecture, May 8, 1953.

233 "My God, Nick, this is a big shift!": AIP, interview of Nicholas U. Mayall by Bert Shapiro, February 13, 1977.

233 "panther juice": Ibid. In the interview transcript, "juice" was substituted for a four-letter word that Mayall used in the oral interview.

233 "you are now using the 100-inch telescope the way it should be used": AIP, interview of Nicholas U. Mayall by Bert Shapiro, February 13, 1977.

234 "You can't imagine how electric the atmosphere was": Ibid.

234 Humason had the benefit of experience: AIP, interview of Nicholas U. Mayall, June 3, 1976.

235 "The intense publicity that swirled around Mount Wilson's nebular department": Sandage (2004), p. 284.

235 Even though less than 5 percent of Mount Wilson's major publications in this era involved cosmology: Allan Sandage tallied up the papers and, after discounting the inconsequential ones, found that only 33 out of 760 papers in the Mount Wilson Contribution series, which ran from 1906 to 1949, concerned either galaxies or the universe. Sandage (2004), p. 481.

235 "Some spectroscopists began to feel resentful": Ibid., p. 284.

235 "The outstanding feature": Hubble (1929a), p. 173.

236 "great beacons scattered through space": Hubble (1937), p. 15.

236 "The interpretation . . . should be left to you": HUB, Hubble to de Sitter, September 23, 1931.

236 "It is difficult to believe that the velocities are real": "Stranger Than Fiction" (1929), p. F4.

236 referred to the velocities of the galaxies as "apparent": Hubble (1929a), p. 168.

236 "Not until the empirical sources are exhausted": Hubble (1936), p. 202.

236 "I have always been rather happy that . . . my part in the work was, you might say, fundamental": AIP, interview of Milton Humason by Bert Shapiro around 1965.

236 "It has been remarked by several astronomers that there appears to be a linear correlation": De Sitter (1930), p. 169.

236 "The possibility of a velocity-distance relation among nebulae has been in the air for years": HUB, Hubble to de Sitter, August 21, 1930.

237 "great pioneer work of V. M. Slipher": Hubble and Humason (1931), pp. 57–58.
237 "I regard such first steps as by far the most important of all": LWA, Hubble to Slipher, March 6, 1953.
237 "emerged from a combination of radial velocities measured by Slipher at Flagstaff": Hubble (1953), p. 658.
237 "if cosmogonists to-day have to deal with a Universe that is expanding": Stratton (1933), p. 477.

15. Your Calculations Are Correct, but Your Physical Insight Is Abominable

239 "I am not sure that I can": "Report of the RAS Meeting in January 1930" (1930), p. 38.
240 "I suppose the trouble is that people look [only] for static solutions . . . that does not matter": Ibid., p. 39.
240 "a concept outside their mental framework": Kragh (2007), p. 139.
240 Lemaître soon read the remarks Eddington made: Eisenstaedt (1993), p. 361; McVittie (1967), p. 295.
241 "This seems a complete answer to the problem we were discussing": Smith (1982), p. 198.
241 calling it "ingenious": De Sitter (1930), p. 171.
241 find him just by pursuing the sound of his full, loud laugh: McCrea (1990), p. 204.
241 "exceptionally brilliant . . . quite remarkable both for his insight": HUA, Eddington to Shapley, May 3, 1924.
241 Lemaître traveled to the United States for further study . . . in order to meet Hubble and learn of the latest distance measurements of the spiral nebulae: Kragh (1987), pp. 118–19; Kragh (1990), p. 542.
242 introduce *time* into the deliberations: Other theorists began to try this out as well, making de Sitter's model nonstatic. It was a lively and active pursuit among theorists, who included Kornelius Lanczos in 1922, Hermann Weyl in 1923, and H. P. Robertson in 1928. All these transformations, however, were treated as mathematical solutions for largely academic purposes.
242 "demonstrate the possibility": Friedmann (1922), p. 377.
242 "We shall call this universe the *periodic world*": Ibid., p. 385.
243 "appear to me suspicious": Einstein (1922), p. 326. Several months later, Einstein realized that he had based his negative opinion on an error in his calculations. He immediately wrote to the *Zeitschrift für Physik* that "Mr. Friedmann's results are correct and shed new light." See Einstein (1923), p. 228.
243 They didn't take them seriously: AIP, interview of William McCrea by Robert Smith on September 22, 1978.
243 "combine the advantages of both": Lemaître (1931a), p. 483.
243 "are a cosmical effect of the expansion of the universe": Ibid., p. 489. The gravitational field of a galaxy, far stronger than the field outside it, keeps the galaxy intact during the expansion.
244 Lemaître even estimated a rate of cosmic expansion: Kragh (2007), p. 144.
244 inexplicably did not widely discuss this latest idea with his colleagues: Kragh (1987), p. 125.

244 "Your calculations are correct, but your physical insight is abominable": Smith (1990), p. 57.

245 "not current with the astronomical facts": Kragh (1987), p. 125.

245 "no time for an unassuming theorist without proper international credentials": Deprit (1984), p. 371.

246 "brilliant discovery": "Discussion on the Evolution of the Universe" (1932), p. 584.

246 "Imagine my surprise on being able to rustle together more than 150 references": CA, Robertson to R. C. Tolman, July 7, 1932. In 1929 Robertson had also derived a cosmological model similar to Friedmann's and Lemaître's but did not recognize the dynamic nature of the universe hidden within his equations. Though aware of Hubble's newfound law concerning distance and redshifts, he didn't recognize it as observational proof for an expanding universe at the time. See Kragh (2007), pp. 142, 146.

246 reported as breathtaking in its grandeur and terrifying in its implications: "A Prize for Lemaître" (1934), p. 16.

246 "The theory of the expanding universe is in some respects so preposterous": "Discussion on the Evolution of the Universe" (1932), p. 587.

246 "On the face of it": Jeans (1932), p. 563.

246 Eddington first devised this picture: Eddington (1930), p. 669.

246 "embedded in the surface of a balloon": Ibid.

247 "About every two weeks some of the men from Mount Wilson and Cal Tech came to the house": HUB, Box 7, Grace's memoir.

247 British cosmologist E. Arthur Milne, for example, posited that the expansion of space-time was merely an illusion: Milne (1932); Hetherington (1982), p. 46.

247 the "tired photon" theory: Zwicky (1929a and 1929b).

248 Hubble worked for a number of years with Caltech theorist Richard Tolman: Hubble and Tolman (1935).

248 Hubble made the call that his data were too uncertain, which kept the expanding universe in play: Hetherington (1996), pp. 163–70. Historian Norriss Hetherington first pointed out Hubble's philosophical preference for an expanding, homogeneous universe, despite the noted astronomer's public statements that he was objectively testing all models. In the end, he preferred the simplicity and beauty of general relativity to dreaming up new laws of physics to fit his observations, as Zwicky was doing. Zwicky did not take this verdict sitting down. He famously accused Hubble and the "sycophants" among his young assistants with doctoring "their observational data, to hide their shortcomings and to make the majority of the astronomers accept and believe in some of their most prejudicial and erroneous presentations and interpretations of facts."

248 "We cannot assume that our knowledge of physical principles is yet complete": Hubble (1937), p. 26.

248 "a desire to show that the red shift was not an expansion": AIP, interview with C. Donald Shane by Helen Wright on July 11, 1967.

248 Perusing Hubble's writings on the idea of an expanding universe: All quoted phrases in this paragraph are from Hubble (1937), pp. v and 26.

249 "around the earth in a second, out to the moon in 10 seconds": Ibid., pp. 29–30.

249 "represent either actual recession (expanding universe) or some hitherto

unknown principle of nature": HUB, Box 15, Hubble to Harvey Zinszer, July 21, 1950.

249   "I just don't understand this eagerness": Douglas (1957), p. 113.

16. Started Off with a Bang

250   "Would it not be more practical to have the herr professor come here": "Einsteins Start Trip to America" (1930), p. 5.

250   to hunt for the sole twelve men in the world: "Relativity" (1930), p. A4.

250   "This reminds me of a Punch and Judy show": "Einstein Battles 'Wolves' " (1930), p. 1.

250   "his face . . . as smooth as a girl's": Ibid., p. 2.

250   Arthur Fleming . . . first extended the invitation: Sutton (1930), p. A1.

251   steady round of private engagements: "Einstein's Date Book Crammed" (1931), p. A1; "Notables of World to Opening" (1931), p. B14; Feigl (1931).

251   Einstein laughed like a little boy: Hall (1931), p. 28.

251   "They cheer me because they all understand me": Isaacson (2007), p. 374.

251   "Your husband's work is beautiful": HUB, Box 8, "Biographical Memoir."

251   Einstein had been given a room at Mount Wilson's main offices . . . issuing keys: AIP, interview of Nicholas U. Mayall by Norriss S. Hetherington on June 3, 1976.

252   "I have kept completely out of the Einstein excitement": HP, Hale to Harry Manley Godwin, January 15, 1931.

252   carefully orchestrated expedition was arranged for Einstein: HL, Walter Adams Papers, Supplement Box 4, Folder 4.87.

253   young filmmaker named Frank Capra: In 1918 Capra had graduated from Throop Institute, later renamed the California Institute of Technology, with a BS degree in chemical engineering.

253   "And here he comes . . . down from the sun tower": CA, Einstein Film Footage, 1931.

253   "This hundred-inch reflector was completed about thirteen years ago": Ibid.

254   "Well, my husband does that on the back of an old envelope": Clark (1971), p. 434.

254   After an early dinner the party returned to the 100-inch telescope: HL, Walter Adams Papers, Supplement Box 4, Folder 4.87.

254   on that day he at last conceded: "Einstein Drops Idea of 'Closed' Universe" (1931), p. 1.

254   "A gasp of astonishment swept through the library": Christianson (1995), p. 210.

254   "the red shift of distant nebulae has smashed my old construction like a hammer blow": "Red Shift of Nebulae a Puzzle, Says Einstein" (1931), p. 15.

254   "biggest blunder": This is not a direct quote from Einstein. The Russian-American physicist George Gamow relayed this story in his autobiography, saying Einstein used the now-famous phrase while they were having a chat one day. Gamow (1970), p. 44. Ironically, at the start of the twenty-first century, astrophysicists reinserted the constant into their cosmological calculations to help them explain why the universe's expansion seems to be accelerating as the eons pass.

256   "made Einstein change his mind": "Hubble to Visit Oxford" (1934).

256 "It remains to find the cause": Lemaître (1931a), p. 489.

256 "beginning of time" . . . "philosophically, the notion of a beginning of the present order of Nature is repugnant to me": Eddington (1931), pp. 449–50.

256 "not believe that the present order of things started off with a bang": Eddington made this remark in a series of lectures given at the University of Edinburgh, later published as Eddington (1928). See p. 85.

256 Hoyle using a similar description: Hoyle's radio lectures on the cosmos, in which he first used the term *Big Bang*, were later published. See Hoyle (1950), pp. 119, 124.

256 "I picture . . . an even distribution of protons and electrons": Eddington (1933), pp. 56–57.

257 "If we go back in the course of time": Lemaître (1931b).

257 Lemaître was spurred by the revelations of atomic physics: Kragh (2007), pp. 152–53.

257 "The evolution of the world can be compared to a display of fireworks": Lemaître (1950), p. 78.

258 "Lemaître believed that God would hide nothing from the human mind": Kragh (1990), p. 542.

258 Times had assuredly changed: Though Lemaître was both scientist and priest, he believed that science and theology should remain separate entities. He disagreed when Pope Pius XII in 1951 announced that the Big Bang cosmology confirmed the fundamental doctrines of Christian theology. "As far as I can see," he said, "such a theory [of the primeval atom] remains entirely outside any metaphysical or religious question. It leaves the materialist free to deny any transcendental Being." See Kragh (1987), pp. 133–34.

259 Baade was able to prove that there were two distinct kinds of Cepheid stars: Baade (1952).

259 Those who desired nature to be uniform breathed a huge sigh of relief: The astronomical community was aghast when Harlow Shapley went to the press and attempted to claim that he, not Baade, had first discovered the correction to Hubble's distance scale. What he actually did was go back to some of his old observations and simply confirm Baade's discovery after the fact. Sandage (2004), p. 310.

259 "Never in all the history of science": De Sitter (1932), p. 3.

261 "a growing community of American astronomers . . . by the 1960s were concentrating to an unprecedented degree on the study of galaxies": Kragh and Smith (2003), p. 157.

Whatever Happened to . . .

262 In 1900 Charles Yerkes moved to New York City: Miller (1970), p. 110.

262 Within a month, she married Wilson Mizner: Franch (2006), pp. 318–23.

262 maintains its status as the largest refractor: A 49-inch refractor was displayed at the 1900 Paris Exposition but was never used professionally and ultimately dismantled.

262 At the end of a long honeymoon in Europe, he and his bride took a balloon ride: Hoyt (1996), p. 233.

262 the observatory spent a decade fighting in court with his widow for control of his estate. . . . "opulent squalor" until her death at the age of ninety in 1954:

The phrase "opulent squalor" was used by the Reverend Fay Lincoln Gemmell, who did chores for Constance while a theology student in the 1940s. Putnam (1994), p. 104.

264 "I have so little confidence in the theories of Lemaître, Eddington, et al. in this field that I shall follow the safe if not sane course of just sitting tight": HUA, Curtis to Shapley, August 24, 1932.

265 He had hopes for erecting a big reflector for Michigan's use: J. Stebbins (1950). A 36-inch reflecting telescope, dedicated as the Heber Curtis Memorial Telescope, was erected in 1950 on Peach Mountain, northwest of Ann Arbor. It was devoted to the study of galactic and extragalactic structure. In 1967 the telescope was moved to the Cerro Tololo Inter-American Observatory in Chile.

265 He always considered his work on the nebulae as his greatest contribution to astronomy: McMath (1942), p. 69.

265 "The truth is . . . that I have been enjoying from boyhood the things I liked most to do": Wright, Warnow, and Weiner (1972), p. 99.

266 He moved to his ranch east of Pasadena, growing lemons, oranges, and avocados and dreaming of designing ever-bigger telescopes, with mirrors up to 320 inches in width: Osterbrock (1993), pp. 160–64.

266 controversial design for the Naval Observatory scope, worked out earlier in collaboration with the French astronomer Henri Chrétien, would later be used in many giant telescopes: Ibid., p. 282.

266 "very gracious, kindly person, a real gentleman": AIP, interview of George Abell by Spencer Weart, November 14, 1977.

266 Ira Bowen was appointed instead, a decision that simply stunned Hubble: Sandage (2004), p. 530.

267 When Grace was about to make a turn into their driveway, though, she noticed Edwin breathing shallowly: Dunaway (1989), p. 247.

267 that rare individual who went from elementary school directly to a PhD: Sandage (2004), p. 192.

267 "My God, Bill," he replied, "I've looked in an eyepiece all my life, I don't want to look in any more eyepieces": AIP, interview of Milton Humason by Bert Shapiro around 1965.

267 "high noon of his scientific life": Kopal (1972), p. 429.

268 His grave is marked by a solid granite rock upon which is inscribed, "And We by His Triumph Are Lifted Level with the Skies": Bok (1978), pp. 254–58.

268 wrote a thirty-nine-page memoir: See Adams (1947).

268 in the early 1940s Hubble proved once and for all that . . . a spiral's arms are trailing as they rotate, not leading: Berendzen and Hart (1973), p. 91.

268 Just weeks before his death he finished the measurement of his five hundredth parallax field at the observatory's Pasadena headquarters: Seares (1946), p. 89.

268 "everywhere the two men went, the lambda was sure to go": "Amiable Abbe" (1961), p. 42.

269 at last received news of the discovery: Deprit (1984), p. 391.

Acknowledgments

My journey into this special moment in astronomical history began at archives located on both coasts of the United States. For their invaluable help during my research, deep appreciation is extended to archivists Dorothy Schaumberg and Cheryl Dandridge at the Mary Lea Shane Archives of the Lick Observatory; Kristen Sanders and Christine Bunting with the special collections at the library of the University of California, Santa Cruz; Charlotte ("Shelley") Erwin and Bonnie Ludt at the Caltech Institute Archives in Pasadena; Janice Goldblum at the National Academies Archives in Washington, D.C.; Melanie Brown, Julie Gass, Mark Matienzo, Jennifer Sullivan, and Spencer Weart with the Niels Bohr Library and Archives at the American Institute of Physics in College Park, Maryland; Nora Murphy at the MIT Archives and Brian Marsden with the Harvard-Smithsonian Center for Astrophysics, both in Cambridge, Massachusetts; the Harvard University Archives; and Meredith Berbée, Juan Gomez, Kate Henningsen, Laura Stalker, and Catherine Wehrey at the Henry Huntington Library in San Marino, California. There Dan Lewis, Huntington's curator for the History of Science and Technology, was particularly helpful in tracking down last-minute bits of information as the book was nearing completion.

A special thank-you goes to Antoinette Beiser, Lowell Observatory's archivist in Flagstaff, Arizona. Antoinette went to extraordinary lengths to unearth every letter, logbook, journal, and artifact connected to Vesto Slipher, which allowed me to deepen the record on this oft-forgotten astronomer. More than that, she and her friends, the "Thursday Night Wing Dingers," offered much-needed respite after hours.

I was certainly not the first to peruse these archives in search of the story behind the discovery of the modern universe. I am hugely indebted to the historians who went before me and blazed the trail. Several graciously offered sage advice and beneficial suggestions while reviewing sections of my work in progress, in particular David DeVorkin, curator for the history of astronomy at the National Air and Space Museum of the Smithsonian Institution in Washington, D.C., and Robert Smith, professor of history, University of Alberta, Canada. I am especially beholden to Norriss Hetherington, visiting scholar with the Office of the History of Science and Technology at the University of California, Berkeley, who provided guidance and feedback from the very start of my project until its finish. And I thank his wife, Edith, for her cordial hospitality while visiting the San Francisco area. Donald Osterbrock, former Lick Observatory director, provided similar counsel, until, sadly, he passed away at the age of eighty-two in 2007. I am grateful to his wife, Irene, who helped in his historical research, for reviewing the sections that involved his areas of expertise.

I would like to thank my engaging guides to three of the facilities crucial to this story: Kevin Schindler at the Lowell Observatory, Don Nicholson at the Mount Wilson Observatory, and Tony Misch at Lick Observatory, who also provided copies of historic photographs taken by both James Keeler and Heber Curtis.

And throughout this long venture, I was fortunate to receive continual encouragement from my colleagues in the MIT Graduate Program in Science Writing—Rob Kanigel, Shannon Larkin, Tom Levenson, Alan Lightman, Boyce Rensberger—as well as close friends and family who kept my spirits high with their unceasing interest in my progress. For this I thank Elizabeth Eaton, Linda and Steve Wohler, the McCabes—Tara, Paul, Ian, and Hugh—Elizabeth Maggio, Ike Ghozeil, Sarah and Peter Saulson, Ellen and Marty Shell, Eunice and Cliff Lowe, and my mother, who will celebrate her eighty-eighth birthday soon after this book is published. I am grateful to my agent, Russ Galen, in never wavering to get this project off the ground, and to my editor, Edward Kastenmeier, for his ever enthusiastic advocacy, superb insights, and invaluable suggestions.

I am indebted to my husband, Steve Lowe, who provided both gentle criticism and a keen editorial eye throughout the course of my research and writing. His love, encouragement, and expertise on matters scientific helped bring this book to fruition. Thank you, Steve, for always being there.

Lastly, I should mention that the immense pleasure I experienced in working on this book inspired me to name my new bearded collie puppy, both a champ and a scamp, Hubble.

Bibliography

Adams, W. S. 1929. "New Stellar Discoveries Amaze Science." *Los Angeles Examiner*, June 23, pp. 1, 8.

———. 1947. "Early Days at Mount Wilson." *Publications of the Astronomical Society of the Pacific* 59 (October): 213–31; (December): 285–304.

Aitken, R. G. 1943. "Biographical Memoir of Heber Doust Curtis." *Biographical Memoirs*, vol. 22. Washington, D.C.: National Academy of Sciences.

"Amiable Abbe." 1961. *Newsweek* 58 (September 4): 42.

Baade, W. 1952. "A Revision of the Extra-Galactic Distance Scale." *Transactions of the International Astronomical Union* 8: 397–98.

———. 1963. *Evolution of Stars and Galaxies*. Cambridge, Mass.: Harvard University Press.

Babcock, A. H. 1896. "Completion of the Big Crossley Reflector Dome for the Lick Observatory." *San Francisco Chronicle*, September 27.

Baida, P. 1986. "Dreiser's Fabulous Tycoon." *Forbes 400* (October 27): 97–102.

Bailey, S. I. 1919. "Variable Stars in the Cluster Messier 15." *Annals of the Astronomical Observatory of Harvard College* 78: 248–50.

———. 1922. "Henrietta Swan Leavitt." *Popular Astronomy* 30 (April): 197–99.

Ball, R. S. 1895. *The Great Astronomers*. London: Isbister.

Barnard, E. E. 1891. "Observations of the Planet Jupiter and His Satellites During 1890 with the 12-inch Equatorial of the Lick Observatory." *Monthly Notices of the Royal Astronomical Society* 51: 543–56.

Belkora, L. 2003. *Minding the Heavens*. Bristol: Institute of Physics Publishing.

Bennett, J. A. 1976. "On the Power of Penetrating into Space: The Telescopes of William Herschel." *Journal for the History of Astronomy* 7: 75–108.

Berendzen, R., and R. Hart. 1973. "Adriaan van Maanen's Influence on the Island Universe Theory." *Journal for the History of Astronomy* 4: 46–56, 73–98.

Berendzen, R., R. Hart, and D. Seeley. 1984. *Man Discovers the Galaxies*. New York: Columbia University Press.

Berendzen, R., and M. Hoskin. 1971. "Hubble's Announcement of Cepheids in Spiral Nebulae." *Astronomical Society of the Pacific Leaflets* 504: 1–15.

Berendzen, R., and C. Shamieh. 1973. "Adriaan van Maanen." *Dictionary of Scientific Biography*, vol. 8. New York: Scribner's.

Bertotti, B., R. Balbinot, S. Bergia, and A. Messina, eds. 1990. *Modern Cosmology in Retrospect*. Cambridge: Cambridge University Press.

Blades, B. 1930. "On the Trail of Star-Gazers." *Los Angeles Times*, August 10, p. J10.

Blakeslee, H. W. 1930. "Distance to Stars 75 Million Light-Years Away." Associated Press Service, November 17.

"Blanket of Snow Covers the City." 1925. *Washington Post*, January 1, p. 1.

Bohlin, K. 1909. *Kungliga Svenska Vetenskapsakademiens handlingar* 43:10.

Bok, B. J. 1974. "Harlow Shapley." *Quarterly Journal of the Royal Astronomical Society* 15: 53–57.

———. 1978. "Harlow Shapley." *Biographical Memoirs*, vol. 49. Washington, D.C.: National Academy of Sciences.

Bowler, P. J., and I. R. Morus. 2005. *Making Modern Science*. Chicago: University of Chicago Press.

Brashear, R. W., and N. S. Hetherington. 1991. "The Hubble–van Maanen Conflict over Internal Motions in Spiral Nebulae: Yet More New Information on an Already Old Topic." *Vistas in Astronomy* 34: 415–23.

Brush, S. G. 1979. "Looking Up: The Rise of Astronomy in America." *American Studies* 20: 41–67.

Campbell, K. 1971. *Life on Mount Hamilton, 1899–1913*, ed. Elizabeth Spedding Calciano. Santa Cruz: University of California Library.

Campbell, W. W. 1900a. "James Edward Keeler." *Publications of the Astronomical Society of the Pacific* 12: 139–46.

———. 1900b. "James Edward Keeler." *Astrophysical Journal* 12: 239–53.

———. 1908. "Comparative Power of the 36-Inch Refractor of the Lick Observatory." *Popular Astronomy* 16: 560–62.

———. 1917. "The Nebulae." *Science* 45: 513–48.

Cannon, Annie J. 1915. "The Henry Draper Memorial." *Journal of the Royal Astronomical Society of Canada* 9 (May–June): 203–15.

"Charles T. Yerkes Dead." 1905. *New York Times*, December 30, p. 4.

Christianson, G. E. 1995. *Edwin Hubble: Mariner of the Nebulae*. Chicago: University of Chicago Press.

Ciufolini, I., and J. A. Wheeler. 1995. *Gravitation and Inertia*. Princeton, N.J.: Princeton University Press.

Clark, D. H., and M. D. H. Clark. 2004. *Measuring the Cosmos*. New Brunswick, N.J.: Rutgers University Press.

Clark, R. W. 1971. *Einstein*. New York: World Publishing Company.

Clerke, A. M. 1886. *A Popular History of Astronomy During the Nineteenth Century*. Edinburgh: A. & C. Black.

———. 1890. *The System of the Stars*. London: Longmans, Green, and Company.

———. 1902. *A Popular History of Astronomy During the Nineteenth Century*. London: Adam and Charles Black.

Crommelin, A. C. D. 1917. "Are the Spiral Nebulae External Galaxies?" *Scientia* 21: 365–76.

Cropper, W. H. 2001. *Great Physicists: The Life and Times of Leading Physicists from Galileo to Hawking*. Oxford: Oxford University Press.

"Crowd Jams Library for Hubble Talk." 1927. *Los Angeles Examiner*, October 21.

Curtis, H. D. 1912. "Descriptions of 132 Nebulae and Clusters Photographed with the Crossley Reflector." *Lick Observatory Bulletin*, no. 219: 81–84.

———. 1913. "Descriptions of 109 Nebulae and Clusters Photographed with the Crossley Reflector: Second List." *Lick Observatory Bulletin*, no. 248: 43–46.

———. 1914. "Improvements in the Crossley Mounting." *Publications of the Astronomical Society of the Pacific* 26: 46–51.

———. 1915. "Preliminary Note on Nebular Proper Motions." *Proceedings of the National Academy of Sciences* 1 (January 15): 10–12.

———. 1917a. "A Study of Absorption Effects in the Spiral Nebulae." *Publications of the Astronomical Society of the Pacific* 29: 145–46.

———. 1917b. "New Stars in Spiral Nebulae." *Publications of the Astronomical Society of the Pacific* 29: 180–82.

———. 1917c. "Three Novae in Spiral Nebulae." *Lick Observatory Bulletin* 9 (300): 108–10.

———. 1917d. "Novae in Spiral Nebulae and the Island Universe Theory." *Publications of the Astronomical Society of the Pacific* 29: 206–7.

———. 1917e. "The Nebulae." *Publications of the Astronomical Society of the Pacific* 29: 91–103.

———. 1918a. "Descriptions of 762 Nebulae and Clusters Photographed with the Crossley Reflector." *Publications of the Lick Observatory* 13: 11–42.

———. 1918b. "A Study of Occulting Matter in the Spiral Nebulae." *Publications of the Lick Observatory* 13: 45–54.

———. 1919. "Modern Theories of the Spiral Nebulae." *Journal of the Washington Academy of Sciences* 9: 217–27.

———. 1924. "The Spiral Nebulae and the Constitution of the Universe." *Scientia* 35: 1–9.

De Lapparent, V., M. J. Geller, and J. P. Huchra. 1986. "A Slice of the Universe." *Astrophysical Journal* 302 (March 1): L1–L5.

Deprit, A. 1984. "Monsignor Georges Lemaître." In *The Big Bang and Georges Lemaître*, ed. A. Berger. Dordrecht, Holland: D. Reidel.

De Sitter, W. 1917. "On Einstein's Theory of Gravitation, and Its Astronomical Consequences. Third Paper." *Monthly Notices of the Royal Astronomical Society* 78: 3–28.

———. 1930. "On the Magnitudes, Diameters and Distances of the Extragalactic Nebulae, and Their Apparent Radial Velocities." *Bulletin of the Astronomical Institutes of the Netherlands* 5 (May 26): 157–71.

———. 1932. *Kosmos: A Course of Six Lectures.* Cambridge, Mass.: Harvard University Press.

DeVorkin, D. H. 2000. *Henry Norris Russell.* Princeton, N.J.: Princeton University Press.

Dewhirst, D. W., and M. Hoskin. 1991. "The Rosse Spirals." *Journal for the History of Astronomy* 22: 257–66.

"Discussion on the Evolution of the Universe." 1932. *British Association for the Advancement of Science. Report of the Centenary Meeting. London — 1931.* London: Office of the British Association.

"A Distant Universe of Stars." 1924. *Science* 59 (January 18): x.

Doig, P. 1924. "The Spiral Nebulae." *Journal of the British Astronomical Association* 35 (December): 99–105.

Douglas, A. V. 1957. *The Life of Arthur Stanley Eddington.* London: Thomas Nelson and Sons Ltd.

Dreiser, T., and F. Booth. 1916. *A Hoosier Holiday.* New York: John Lane.

Dunaway, D. K. 1989. *Huxley in Hollywood.* New York: Anchor Books.

Duncan, J. C. 1922. "Three Variable Stars and a Suspected Nova in the Spiral Nebula M 33 Trianguli." *Publications of the Astronomical Society of the Pacific* 34: 290–91.

———. 1923. "Photographic Studies of Nebulae. Third Paper." *Contributions from the Mount Wilson Observatory,* no. 256: 9–20.

Dyson, F. W. 1917. "On the Opportunity Afforded by the Eclipse of 1919 May 29 of Verifying Einstein's Theory of Gravitation." *Monthly Notices of the Royal Astronomical Society* 77: 445–47.

Dyson, F. W., A. S. Eddington, and C. Davidson. 1920. "A Determination of the Deflection of Light by the Sun's Gravitational Field, from Observations Made at the Total Eclipse of May 29, 1919." *Philosophical Transactions of the Royal Society of London* 220: 291–333.

Eddington, A. S. 1916. "The Nature of Globular Clusters." *Observatory* 39: 513–14.

———. 1920. *Space, Time, and Gravitation.* Cambridge: Cambridge University Press.

———. 1928. *The Nature of the Physical World.* New York: Macmillan.

———. 1930. "On the Instability of Einstein's Spherical World." *Monthly Notices of the Royal Astronomical Society* 90: 668–78.

———. 1931. "The End of the World: From the Standpoint of Mathematical Physics." *Nature* 127 (March 21): 447–53.

———. 1933. *The Expanding Universe.* Cambridge: Cambridge University Press.

Einstein, A. 1911. "On the Influence of Gravity on the Propagation of Light." *Annalen der Physik* 35: 898–908.

———. 1917. "Kosmologische Betrachtungen zur allgemeinen Relativitätstheorie." *Sitzungsberichte der Königlich Preußischen Akademie der Wissenschaften zu Berlin* 6: 142–52.

———. 1922. "Bemerkung zu der Arbeit von A. Friedmann 'Über die Krümmung des Raumes.'" *Zeitschrift für Physik* 11: 326.

———. 1923. "Notiz zu der Arbeit von A. Friedmann." *Zeitschrift für Physik* 16: 228.

Einstein, A., and W. de Sitter. 1932. "On the Relation Between the Expansion and the Mean Density of the Universe." *Proceedings of the National Academy of Sciences* 18 (March 15): 213–14.

"Einstein Battles 'Wolves.'" 1930. *Los Angeles Times*, December 12, p. 1.

"Einstein Drops Idea of 'Closed' Universe." 1931. *New York Times*, February 5, p. 1.

"Einstein Guest at Mt. Wilson." 1931. *Los Angeles Times*, January 30, p. A1.

"Einstein's Date Book Crammed." 1931. *Los Angeles Times*, January 14, p. A1.

"Einsteins Start Trip to America." 1930. *Los Angeles Times*, December 1, p. 5.

Eisenstaedt, J. 1993. "Lemaître and the Schwarzschild Solution." In *The Attraction of Gravitation*, ed. J. Earman, M. Janssen, and J. D. Norton. Boston: Birkhäuser.

Encyclopaedia Britannica. 1911. "Rhodes, Cecil John."

Fath, E. A. 1908. "The Spectra of Some Spiral Nebulae and Globular Star Clusters." *Lick Observatory Bulletin* 149: 71–77.

Feigl, A. 1931. "Frau Professor Einstein." *Los Angeles Times*, February 1, p. A1.

Fernie, J. D. 1969. "The Period-Luminosity Relation: A Historical Review." *Publications of the Astronomical Society of the Pacific* 81 (December): 707–31.

———. 1970. "The Historical Quest for the Nature of the Spiral Nebulae." *Publications of the Astronomical Society of the Pacific* 82 (December): 1189–1230.

———. 1995. "The Great Debate." *American Scientist* 83 (September–October): 410–13.

"Finds Spiral Nebulae Are Stellar Systems." 1924. *New York Times*, November 23, p. 6.

Fitzgerald, F. S. 1925. *The Great Gatsby.* New York: Scribner's.

Fölsing, A. 1997. *Albert Einstein: A Biography.* New York: Viking Press.

Franch, J. 2006. *Robber Baron: The Life of Charles Tyson Yerkes.* Urbana: University of Illinois Press.

Friedmann, A. 1922. "Über die Krümmung des Raumes." *Zeitschrift für Physik* 10: 377–86.

Frost, E. B. 1933. *An Astronomer's Life*. Boston: Houghton Mifflin.

Gamow, G. 1970. *My World Line*. New York: Viking Press.

Gingerich, O. 1975. "Harlow Shapley." *Dictionary of Scientific Biography*, vol. 12. New York: Scribner's.

———. 1978. "James Lick's Observatory." *Pacific Discovery* 31: 1–10.

———. 1987. "The Mysterious Nebulae, 1610–1924." *Journal of the Royal Astronomical Society of Canada* 81: 113–27.

———. 1988. "How Shapley Came to Harvard; or, Snatching the Prize from the Jaws of Debate." *Journal for the History of Astronomy* 19: 201–7.

———. 1990a. "Through Rugged Ways to the Galaxies." *Journal for the History of Astronomy* 21: 77–88.

———. 1990b. "Shapley, Hubble, and Cosmology." In *Evolution of the Universe of Galaxies: Edwin Hubble Centennial Symposium*, ed. Richard G. Kron. San Francisco: Astronomical Society of the Pacific.

———. 2000. "Kapteyn, Shapley and Their Universes." In *The Legacy of J. C. Kapteyn: Studies on Kapteyn and the Development of Modern Astronomy*, ed. P. C. Van Der Kruit and K. Berkel. Dordrecht: Kluwer Academic Publishers.

Gordon, K. J. 1969. "History of Our Understanding of a Spiral Galaxy: Messier 33." *Quarterly Journal of the Royal Astronomical Society* 10: 293–307.

Grigorian, A. T. 1972. "Aleksandr Friedmann." *Dictionary of Scientific Biography*, vol. 5. New York: Scribner's.

Hale, G. E. 1898. "The Function of Large Telescopes." *Science* 7 (May 13): 650–62.

———. 1900. "James Edward Keeler." *Science* 12 (September 7): 353–57.

———. 1915. *Ten Years' Work of a Mountain Observatory*. Washington, D.C.: Carnegie Institution of Washington.

———. 1922. *The New Heavens*. New York: Charles Scribner's Sons.

Hale, G. E., W. S. Adams, and F. H. Seares. 1931. "Mount Wilson Observatory." *Carnegie Institution of Washington Year Book* 30: 171–221.

Hall, J. S. 1970a. "V. M. Slipher's Trailblazing Career." *Sky & Telescope* 39 (February): 84–86.

———. 1970b. "Vesto Melvin Slipher." *Year Book of the American Philosophical Society*: 161–66.

Hall, M. 1931. "Chaplin Here to See Silent Film Open." *New York Times*, February 5, p. 28.

Halley, E. 1714–16. "An Account of Several Nebulae or Lucid Spots Like Clouds, Lately Discovered Among the Fixt Stars by Help of the Telescope." *Philosophical Transactions* 29: 390–92.

Hardy, T. 1883. *Two on a Tower*, 3rd ed. London: Simpson Low.

Hart, R., and R. Berendzen. 1971. "Hubble, Lundmark and the Classification of Non-Galactic Nebulae." *Journal for the History of Astronomy* 2: 200.

Herschel, W. 1784a. "On the Remarkable Appearances at the Polar Regions of the Planet Mars; the Inclination of Its Axis, the Position of Its Poles, and Its Spheroidical Figure; with a Few Hints Relating to Its Real Diameter and Atmosphere." *Philosophical Transactions of the Royal Society of London* 74: 233–73.

———. 1784b. "Account of Some Observations Tending to Investigate the Construction of the Heavens." *Philosophical Transactions of the Royal Society of London* 74: 437–51.

———. 1785. "On the Construction of the Heavens." *Philosophical Transactions of the Royal Society of London* 75: 213–66.

———. 1789. "Catalogue of a Second Thousand of New Nebulae and Clusters of Stars; with a Few Introductory Remarks on the Construction of the Heavens." *Philosophical Transactions of the Royal Society of London* 79: 212–55.

———. 1791. "On Nebulous Stars, Properly So Called." *Philosophical Transactions of the Royal Society of London* 81: 71–88.

———. 1811. "Astronomical Observations Relating to the Construction of the Heavens, Arranged for the Purpose of a Critical Examination, the Result of Which Appears to Throw Some New Light upon the Organization of the Celestial Bodies." *Philosophical Transactions of the Royal Society of London* 101: 269–336.

Hertzsprung, E. 1914. "Über die räumliche Verteilung der Veränderlichen vom δ Cephei-Typus [On the Spatial Distribution of Variables of the δ Cephei Type]." *Astronomische Nachrichten* 196: 201–8.

Hetherington, N. S. 1971. "The Measurement of Radial Velocities of Spiral Nebulae." *Isis* 62 (September): 309–13.

———. 1973. "The Delayed Response to Suggestions of an Expanding Universe." *Journal of the British Astronomical Association* 84: 22–28.

———. 1974a. "Edwin Hubble's Examination of Internal Motions of Spiral Nebulae." *Quarterly Journal of the Royal Astronomical Society* 15: 392–418.

———. 1974b. "Adriaan van Maanen on the Significance of Internal Motions in Spiral Nebulae." *Journal for the History of Astronomy* 5: 52–53.

———. 1975. "The Simultaneous 'Discovery' of Internal Motions in Spiral Nebulae." *Journal for the History of Astronomy* 6: 115–25.

———. 1982. "Philosophical Values and Observation in Edwin Hubble's Choice of a Model of the Universe." *Historical Studies in the Physical Sciences* 13: 41–67.

———. 1983. "Mid-Nineteenth-Century American Astronomy: Science in a Developing Nation." *Annals of Science* 40: 61–80.

———. 1990a. "Edwin Hubble's Cosmology." In *Evolution of the Universe of Galaxies: Edwin Hubble Centennial Symposium*, ed. R. G. Kron. San Francisco: Astronomical Society of the Pacific.

———, ed. 1990b. *The Edwin Hubble Papers*. Tucson: Pachart Publishing House.

———. 1996. *Hubble's Cosmology*. Tucson: Pachart Publishing House.

Hetherington, N. S., and R. S. Brashear. 1992. "Walter S. Adams and the Imposed Settlement between Edwin Hubble and Adriaan van Maanen." *Journal for the History of Astronomy* 23: 53–56.

Hoagland, H. 1965. "Harlow Shapley—Some Recollections." *Publications of the Astronomical Society of the Pacific* 77: 422–30.

Hoffmann, B. 1972. *Albert Einstein: Creator and Rebel*. New York: Viking Press.

Hoge, V. 2005. "Wendell and Edison Hoge on Mount Wilson." *Reflections* [Mount Wilson Observatory Association Newsletter] (June): 3–6.

Holden, E. S. 1891. "Life at the Lick Observatory." *Scientific American* 64 (January): 73.

"Honor for Dr. Edwin P. Hubble." 1925. *Publications of the Astronomical Society of the Pacific* 37: 100–101.

Hoskin, M. A. 1967. "Apparatus and Ideas in Mid-Nineteenth-Century Cosmology." *Vistas in Astronomy* 9: 79–85.

———. 1970. "The Cosmology of Thomas Wright of Durham." *Journal for the History of Astronomy* 1: 44–52.

——. 1976a. "The 'Great Debate': What Really Happened." *Journal for the History of Astronomy* 7: 169–82.

——. 1976b. "Ritchey, Curtis and the Discovery of Novae in Spiral Nebulae." *Journal for the History of Astronomy* 7: 47–53.

——. 1989. "William Herschel and the Construction of the Heavens." *Proceedings of the American Philosophical Society* 133: 427–32.

——. 2002. "The Leviathan of Parsontown: Ambitions and Achievements." *Journal for the History of Astronomy* 33: 57–70.

Hoyle, F. 1950. *The Nature of the Universe*. New York: Harper.

Hoyt, W. G. 1980. "Vesto Melvin Slipher." In *Biographical Memoirs*, vol. 52. Washington, D.C.: National Academy Press.

——. 1996. *Lowell and Mars*. Tucson: University of Arizona Press.

Hubble, E. P. 1920. "Photographic Investigations of Faint Nebulae." *Publications of the Yerkes Observatory* 4: 69–85.

——. 1922. "A General Study of Diffuse Galactic Nebulae." *Astrophysical Journal* 56: 162–99.

——. 1925a. "Cepheids in Spiral Nebulae." *Publications of the American Astronomical Society* 5: 261–64.

——. 1925b. "N.G.C. 6822, a Remote Stellar System." *Astrophysical Journal* 62: 409–33.

——. 1926. "Extra-Galactic Nebulae." *Astrophysical Journal* 64: 321–69.

——. 1928. "Ten Million Worlds in Sky Census." *Los Angeles Examiner*, October 28, pp. 1–2.

——. 1929a. "A Relation Between Distance and Radial Velocity Among Extra-Galactic Nebulae." *Proceedings of the National Academy of Sciences* 15 (March 15): 168–73.

——. 1929b. "On the Curvature of Space." *Carnegie Institution of Washington News Service Bulletin*, no. 13: 77–78.

——. 1935. "Angular Rotations of Spiral Nebulae." *Astrophysical Journal* 81: 334–35.

——. 1936. *The Realm of the Nebulae*. New Haven, Conn.: Yale University Press.

——. 1937. *The Observational Approach to Cosmology*. Oxford: Clarendon Press.

——. 1953. "The Law of Red-Shifts." *Monthly Notices of the Royal Astronomical Society* 113: 658–66.

Hubble, E., and M. L. Humason. 1931. "The Velocity-Distance Relation Among Extra-Galactic Nebulae." *Astrophysical Journal* 74: 43–80.

Hubble, E., and R. C. Tolman. 1935. "Two Methods of Investigating the Nature of the Nebular Red-Shift." *Astrophysical Journal* 82: 302–37.

"Hubble to Visit Oxford." 1934. *San Francisco Chronicle*, May 6.

Huggins, W. 1897. "The New Astronomy." *The Nineteenth Century* 41 (June): 907–29.

Huggins, W., and Mrs. Huggins. 1889. "On the Spectrum, Visible and Photographic, of the Great Nebula in Orion." *Proceedings of the Royal Society of London* 46: 40–60.

Humason, M. 1927. "Radial Velocities in Two Nebulae." *Publications of the Astronomical Society of the Pacific* 39: 317–18.

——. 1929. "The Large Radial Velocity of N. G. C. 7619." *Proceedings of the National Academy of Sciences* 15 (March): 167–68.

——. 1954. "Obituary Notices." *Monthly Notices of the Royal Astronomical Society* 114: 291–95.

Hussey, E. F. 1903. "Life at a Mountain Observatory." *Atlantic Monthly* 92 (July): 29–32.

Impey, C. 2001. "Reacting to the Size and the Shape of the Universe." *Mercury* (January–February): 36–40.

"Infinite and Infinitesimal." 1925. *Los Angeles Times*, March 22, p. B4.

International Astronomical Union. 1928. "Report of the Commission on Nebulae and Star Clusters." Commission no. 28.

Isaacson, W. 2007. *Einstein: His Life and Universe*. New York: Simon & Schuster. .

Jeans, J. H. 1917a. "Internal Motion in Spiral Nebulae." *Observatory* 40: 60–61.

———. 1917b. "On the Structure of Our Local Universe." *Observatory* 40: 406–7.

———. 1919. *Problems of Cosmogony and Stellar Dynamics*. Cambridge: Cambridge University Press.

———. 1923. "Internal Motions in Spiral Nebulae." *Monthly Notices of the Royal Astronomical Society* 84: 60–76.

———. 1929. *Eos or the Wider Aspects of Cosmogony*. New York: E. P. Dutton.

———. 1930. *The Mysterious Universe*. New York: Macmillan.

———. 1932. "Beyond the Milky Way." *British Association for the Advancement of Science. Report of the Centenary Meeting. London, 1931.* London: Office of the British Association.

Johnson, G. 2005. *Miss Leavitt's Stars*. New York: W. W. Norton.

Jones, B. Z., and L. G. Boyd. 1971. *The Harvard College Observatory: The First Four Directorships, 1839–1919*. Cambridge, Mass.: Harvard University Press.

Jones, K. G. 1976. "S Andromedae, 1885: An Analysis of Contemporary Reports and a Reconstruction." *Journal for the History of Astronomy* 7: 27–40.

Kahn, C., and F. Kahn. 1975. "Letters from Einstein to de Sitter on the Nature of the Universe." *Nature* 257 (October 9): 451–54.

Kant, I. 1900. *Kant's Cosmogony as in His Essay on the Retardation of the Rotation of the Earth and His Natural History and Theory of the Heavens*, ed. and trans. W. Hastie. Glasgow: James Maclehose and Sons.

Karachentsev, I. D., and O. G. Kashibadze. 2006. "Masses of the Local Group and of the M81 Group Estimated from Distortions in the Local Velocity Field." *Astrophysics* 49 (January): 7.

Keeler, J. E. 1888a. "The First Observations of Saturn with the Great Telescope." *San Francisco Examiner*, January 10.

———. 1888b. "First Observations of Saturn with the 36-Inch Equatorial of Lick Observatory." *The Sidereal Messenger*, no. 62.

———. 1895. "A Spectroscopic Proof of the Meteoric Constitution of Saturn's Rings." *Astrophysical Journal* 1: 416–27.

———. 1897. "The Importance of Astrophysical Research and the Relation of Astrophysics to Other Physical Sciences." *Science* 6 (November 19): 745–55.

———. 1898a. "Photographs of Comet I, 1898 (Brooks), Made with the Crossley Reflector of the Lick Observatory." *Astrophysical Journal* 8: 287–90.

———. 1898b. "The Small Bright Nebula Near *Merope*." *Publications of the Astronomical Society of the Pacific* 10: 245–46.

———. 1899a. "Photograph of the Great Nebula in *Orion*, Taken with the Crossley Reflector of the Lick Observatory." *Publications of the Astronomical Society of the Pacific* 11: 39–40.

———. 1899b. "Small Nebulae Discovered with the Crossley Reflector of the Lick Observatory." *Monthly Notices of the Royal Astronomical Society* 59: 537–38.

———. 1899c. "New Nebulae Discovered Photographically with the Crossley Reflector of the Lick Observatory." *Monthly Notices of the Royal Astronomical Society* 60: 128.

——. 1899d. "Scientific Work of the Lick Observatory." *Science* 10: 665–70.

——. 1900a. "On the Predominance of Spiral Forms Among the Nebulae." *Astronomische Nachrichten* 151: 1.

——. 1900b. "The Crossley Reflector of the Lick Observatory." *Astrophysical Journal* 11: 325–49.

Kerszberg, P. 1986. "The Cosmological Question in Newton's Science." *Osiris*, 2nd series, 2: 69–106.

——. 1989. *The Invented Universe: The Einstein–de Sitter Controversy (1916–17) and the Rise of Relativistic Cosmology*. Oxford: Clarendon Press.

Kopal, Z. 1972. "Dr. Harlow Shapley." *Nature* 240: 429–30.

Kostinsky, S. 1916. "Probable Motions in the Spiral Nebula Messier 51 (Canes Venatici) Found with the Stereo-Comparator." *Monthly Notices of the Royal Astronomical Society* 77: 233–34.

Kragh, H. 1987. "The Beginning of the World: Georges Lemaître and the Expanding Universe." *Centaurus* 32: 114–39.

——. 1990. "Georges Lemaître." *Dictionary of Scientific Biography*, vol. 18, suppl. 2. New York: Scribner's.

——. 1996. *Cosmology and Controversy*. Princeton: Princeton University Press, 1996.

——. 2007. *Conceptions of Cosmos*. Oxford: Oxford University Press.

Kragh, H., and R. W. Smith. 2003. "Who Discovered the Expanding Universe?" *History of Science* 41: 141–62.

Kreiken, E. A. 1920. "On the Differential Measurement of Proper Motion." *Observatory* 43: 255–60.

Kron, R. G., ed. 1990. *Evolution of the Universe of Galaxies: Edwin Hubble Centennial Symposium*. Astronomical Society of the Pacific Conference Series, vol. 10. San Francisco: Astronomical Society of the Pacific.

Lankford, J. 1997. *American Astronomy: Community, Careers, and Power, 1859–1940*. Chicago: University of Chicago Press.

Leavitt, H. S. 1908. "1777 Variables in the Magellanic Clouds." *Annals of the Astronomical Observatory of Harvard College* 60: 87–108.

Leavitt, H., and E. C. Pickering. 1912. "Periods of 25 Variable Stars in the Small Magellanic Cloud." *Harvard College Observatory Circular* no. 173: 1–3.

Lemaître, G. 1931a. "A Homogeneous Universe of Constant Mass and Increasing Radius Accounting for the Radial Velocity of Extra-Galactic Nebulae." *Monthly Notices of the Royal Astronomical Society* 91: 483–89.

——. 1931b. "The Beginning of the World from the Point of View of Quantum Theory." *Nature* 127: 706.

——. 1950. *The Primeval Atom*. New York: Van Nostrand.

Lorentz, H. A., A. Einstein, H. Minkowski, and H. Weyl. 1923. *The Principle of Relativity*. Trans. W. Perrett and G. B. Jeffery. London: Methuen and Company.

Lowell, A. L. 1935. *Biography of Percival Lowell*. New York: Macmillan.

Lowell, P. 1905. "Chart of Faint Stars Visible at the Lowell Observatory." *Popular Astronomy* 13: 391–92.

Lundmark, K. 1919. "Die Stellung der kugelförmigen Sternhaufen und Spiralnebel zu unserem Sternsystem." *Astronomische Nachrichten* 209: 369.

——. 1921. "The Spiral Nebula Messier 33." *Publications of the Astronomical Society of the Pacific* 33: 324–27.

——. 1922. "On the Motions of Spirals." *Publications of the Astronomical Society of the Pacific* 34: 108–15.

Luyten, W. J. 1926. "Island Universes." *Natural History* 26: 386–91.

MacPherson, H. 1916. "The Nature of Spiral Nebulae." *Observatory* 39 (March): 131–34.

———. 1919. "The Problem of Island Universes." *Observatory* 42 (September): 329–34.

"Mars." 1907. *Wall Street Journal,* December 28, p. 1.

Maunder, E. W. 1885. "The New Star in the Great Nebula in Andromeda." *Observatory* 8: 321–25.

Maxwell, J. C. 1983. *Maxwell on Saturn's Rings,* ed. S. G. Brush, C. W. F. Everitt, and E. Garber. Cambridge, Mass.: MIT Press.

Mayall, N. U. 1937. *"The Realm of the Nebulae,* by Edwin Hubble." *Publications of the Astronomical Society of the Pacific* 49: 42–47.

———. 1954. "Edwin Hubble: Observational Cosmologist." *Sky & Telescope* (January): 78–80, 85.

McCrea, W. 1990. "Personal Recollections." In *Modern Cosmology in Retrospect,* ed. B. Bertotti et al. Cambridge: Cambridge University Press.

McMath, R. R. 1942. "Heber Doust Curtis." *Publications of the Astronomical Society of the Pacific* 54 (April): 69–71.

———. 1944. "Heber Doust Curtis, 1872–1942." *Astrophysical Journal* 99 (May): 245–48.

McPhee, J. 1998. *Annals of the Former World.* New York: Farrar, Straus & Giroux.

McVittie, G. C. 1967. "Georges Lemaître." *Quarterly Journal of the Royal Astronomical Society* 8: 294–97.

Melotte, P. J. 1915. "A Catalogue of Star Clusters Shown on the Franklin-Adams Chart Plates." *Memoirs of the Royal Astronomical Society* 60: 168.

Messier, C. 1781. *Catalogue des Nébuleuses et Amas d'Étoiles Observées à Paris.* Paris: Imprimerie Royal.

Miller, H. S. 1970. *Dollars for Research.* Seattle: University of Washington Press.

Milne, E. A. 1932. "World Structure and the Expansion of the Universe." *Nature* 130 (July): 9–10.

"Mrs. Mizner Now Divorced." 1907. *New York Times,* August 25, p. 5.

"Mrs. Yerkes Marries Young San Franciscan." 1906. *New York Times,* February 1, p. 2.

"The New Director of Lick." 1898. *New York Tribune,* March 20, p. 7.

Newcomb, S. 1888. "The Place of Astronomy Among the Sciences." *Sidereal Messenger* 7: 69–70.

Newcomb, S., and E. S. Holden. 1889. *Astronomy.* New York: Henry Holt and Company.

Newton, I. 1717. *Opticks; or, A Treatise of the Reflections, Refractions, Inflections and Colours of Light,* 2nd ed. London: W. Bowyer.

Nichol, J. P. 1840. *Views of the Architecture of the Heavens in a Series of Letters to a Lady.* New York: H. A. Chapin.

———. 1846. *Thoughts on Some Important Points Relating to the System of the World.* Edinburgh: William Tait.

———. 1848. *The Stellar Universe.* Edinburgh: John Johnstone.

"Notables of World to Opening." 1931. *Los Angeles Times,* January 25, p. B14.

Nowell, C. E., ed. 1962. *Magellan's Voyage Around the World.* Evanston, Ill.: Northwestern University Press.

Noyes, A. 1922. *The Torch-Bearers—Watchers of the Sky.* New York: Stokes.

Olmsted, D. 1834. "Observations of the Meteors of November 13th, 1833." *American Journal of Science and Arts* 25 (January): 363–411.

———. 1866. *A Compendium of Astronomy.* New York: Collins & Brothers.

Öpik, E. 1922. "An Estimate of the Distance of the Andromeda Nebula." *Astrophysical Journal* 55: 406–10.

Osterbrock, D. 1976. "The California-Wisconsin Axis in American Astronomy, II." *Sky & Telescope* 51 1976: 91–97.

———. 1984. *James E. Keeler: Pioneer American Astrophysicist.* Cambridge: Cambridge University Press.

———. 1986. "Early Days at Lick Observatory." *Mercury* 15 (March–April): 53, 63.

———. 1993. *Pauper & Prince: Ritchey, Hale, & Big American Telescopes.* Tucson: University of Arizona Press.

———. 2001. "Astronomer for All Seasons: Heber D. Curtis." *Mercury* 30 (May–June): 25–31.

Osterbrock, D. E., R. S. Brashear, and J. A. Gwinn. 1990. "Self-Made Cosmologist: The Education of Edwin Hubble." In *Evolution of the Universe of Galaxies: Edwin Hubble Centennial Symposium,* ed. Richard G. Kron. San Francisco: Astronomical Society of the Pacific.

Osterbrock, D. E., and D. P. Cruikshank. 1983. "J. E. Keeler's Discovery of a Gap in the Outer Part of the A Ring." *Icarus* 53: 165–73.

Osterbrock, D. E., J. R. Gustafson, and W. J. S. Unruh. 1988. *Eye on the Sky: Lick Observatory's First Century.* Berkeley: University of California Press.

Paddock, G. F. 1916. "The Relation of the System of Stars to the Spiral Nebulae." *Publications of the Astronomical Society of the Pacific* 28: 109–15.

Pais, A. 1982. *"Subtle Is the Lord . . .": The Science and the Life of Albert Einstein.* Oxford: Oxford University Press.

Pang, A. S.-K. 1997. " 'Stars Should Henceforth Register Themselves': Astrophotography at the Early Lick Observatory." *British Journal of the History of Science* 30: 177–202.

Pannekoek, A. 1989. *A History of Astronomy.* New York: Dover.

Paul, E. 1993. *The Milky Way and Statistical Cosmology, 1890–1924.* Cambridge: Cambridge University Press.

Payne-Gaposchkin, C., with K. Haramundanis, ed. 1984. *Cecilia Payne-Gaposchkin: An Autobiography and Other Recollections.* Cambridge: Cambridge University Press.

Perrine, C. D. 1904. "A New Mounting for the Three-Foot Mirror of the Crossley Reflecting Telescope." *Lick Observatory Bulletin* 3: 124–28.

Pickering, E. C. 1898. *Harvard College Observatory Annual Report* 53: 1–14.

———. 1917. *Harvard College Observatory Bulletin,* no. 641 (28 July).

Plaskett, J. S. 1911. "Some Recent Interesting Developments in Astronomy." *Journal of the Royal Astronomical Society of Canada* 5 (July–August): 245–65.

"A Prize for Lemaître." 1934. *Literary Digest* 117 (March 31): 16.

Proctor, R. 1872. *The Orbs Around Us.* London: Longmans, Green.

Putnam, W. L. 1994. *The Explorers of Mars Hill.* West Kennebunk, Maine: Phoenix Publishing.

"Red Shift of Nebulae a Puzzle, Says Einstein." 1931. *New York Times,* February 12, p. 15.

"Relativity." 1930. *Los Angeles Times,* December 15, p. A4.

"Report of the Council to the Forty-Ninth General Meeting of the Society." 1869. *Monthly Notices of the Royal Astronomical Society* 29 (February): 109–91.

"Report of the RAS Meeting in January 1930." 1930 *Observatory* 53: 33–44.

"Report of the Seventeenth Meeting." 1914. *Popular Astronomy* 22: 551–70.

"Report of the Seventeenth Meeting (continued)." 1915. *Popular Astronomy* 23: 18–28.

Ritchey, G. W. 1897. "A Support System for Large Specula." *Astrophysical Journal* 5: 143–47.

———. 1901. "The Two-Foot Reflecting Telescope of the Yerkes Observatory." *Astrophysical Journal* 14: 217–33.

———. 1910a. "On Some Methods and Results in Direct Photography with the 60-Inch Reflecting Telescope of the Mount Wilson Solar Observatory." *Astrophysical Journal* 32: 26–35.

———. 1910b. "Notes on Photographs of Nebulae Made with the 60-Inch Reflector of the Mount Wilson Observatory." *Monthly Notices of the Royal Astronomical Society* 70 (June): 623–27.

———. 1910c. "Notes on Photographs of Nebulae Taken with the 60-Inch Reflector of the Mount Wilson Solar Observatory." *Monthly Notices of the Royal Astronomical Society* 70 (Suppl. 1910c): 647–49.

———. 1917. "Novae in Spiral Nebulae." *Publications of the Astronomical Society of the Pacific* 29: 210–12.

Rosse, the Earl of. 1850. "Observations on the Nebulae." *Philosophical Transactions of the Royal Society of London* 140: 499–514.

Rubin, V. 2005. "People, Stars, and Scopes." *Science* 309 (September 16): 1817–18.

Russell, H. N. 1913 "Notes on the Real Brightness of Variable Stars." *Science* 37: 651–52.

———. 1918. "Astronomy Notes." *Scientific American* 118: 412.

———. 1925. "Types of Variable Star Work." In *Reports and Recommendations, International Astronomical Union Meeting at Cambridge, July 14–22, 1925*, pp. 100–104.

Sandage, A. 1961. *The Hubble Atlas of Galaxies*. Washington, D.C.: Carnegie Institution of Washington.

———. 1989. "Edwin Hubble 1889–1953." *Journal of the Royal Astronomical Society of Canada* 83 (December): 351–62.

———. 2004. *Centennial History of the Carnegie Institution of Washington. Volume 1: The Mount Wilson Observatory*. Cambridge: Cambridge University Press.

Sanford, R. F. 1916–18. "On Some Relations of the Spiral Nebulae to the Milky Way." *Lick Observatory Bulletin* 9: 80–91.

Scheiner, J. 1899. "On the Spectrum of the Great Nebula in Andromeda." *Astrophysical Journal* 9: 149–50.

Schilpp, P. A., ed. 1949. *Albert Einstein: Philosopher-Scientist*. Evanston, Ill.: Library of Living Philosophers.

Schindler, K. S. 1998. *100 Years of Good Seeing: The History of the 24-Inch Clark Telescope*. Flagstaff, Ariz.: Lowell Observatory.

———. 2003. "The Slipher Spectrograph." *The Lowell Observer* (Spring): 5–6.

"Scientists Gather for 1920 Conclave." 1920. *Washington Post*, April 25, p. 38.

Seares, F. H. 1946. "Adriaan van Maanen, 1884–1946." *Publications of the Astronomical Society of the Pacific* 58: 89–103.

Seares, F. H., and E. P. Hubble. 1920. "The Color of the Nebulous Stars." *Astrophysical Journal* 52: 8–22.

Shapley, H. 1914. "On the Nature and Cause of Cepheid Variation." *Astrophysical Journal* 40: 448–65.

———. 1915a. "Studies Based on the Colors and Magnitudes in Stellar Clusters. First Paper: The General Problem of Clusters." *Contributions from the Mount Wilson Solar Observatory*, no. 115: 201–21.

———. 1915b. "Studies Based on the Colors and Magnitudes in Stellar Clusters. Second Paper: Thirteen Hundred Stars in the Hercules Cluster (Messier 13)." *Contributions from the Mount Wilson Solar Observatory* 116: 225–314.

——. 1917a. "Studies Based on the Colors and Magnitudes in Stellar Clusters. Fourth Paper: The Galactic Cluster Messier 11." *Contributions from the Mount Wilson Solar Observatory* 126: 29–46.

——. 1917b. "Note on the Magnitudes of Novae in Spiral Nebulae." *Publications of the Astronomical Society of the Pacific* 29: 213–17.

——. 1918a. "Studies Based on the Colors and Magnitudes in Stellar Clusters. Sixth Paper: On the Determination of the Distances of Globular Clusters." *Astrophysical Journal* 48: 89–124.

——. 1918b. "Studies Based on the Colors and Magnitudes in Stellar Clusters. Seventh Paper: The Distances, Distribution in Space, and Dimensions of 69 Globular Clusters." *Astrophysical Journal* 48: 154–81.

——. 1918c. "Studies Based on the Colors and Magnitudes in Stellar Clusters. Eighth Paper: The Luminosities and Distances of 139 Cepheid Variables." *Astrophysical Journal* 48: 279–94.

——. 1918d. "Globular Clusters and the Structure of the Galactic System." *Publications of the Astronomical Society of the Pacific* 30: 42–54.

——. 1919a. "Studies Based on the Colors and Magnitudes in Stellar Clusters. Ninth Paper: Three Notes on Cepheid Variation." *Astrophysical Journal* 49: 24–41.

——. 1919b. "Studies Based on the Colors and Magnitudes in Stellar Clusters. Tenth Paper: A Critical Magnitude in the Sequence of Stellar Luminosities." *Astrophysical Journal* 49: 96–107.

——. 1919c. "Studies Based on the Colors and Magnitudes in Stellar Clusters. Eleventh Paper: A Comparison of the Distances of Various Celestial Objects." *Astrophysical Journal* 49: 249–65.

——. 1919d. "Studies Based on the Colors and Magnitudes in Stellar Clusters. Twelfth Paper: Remarks on the Arrangement of the Sidereal Universe." *Astrophysical Journal* 49: 311–36.

——. 1919e. "On the Existence of External Galaxies." *Publications of the Astronomical Society of the Pacific* 31: 261–68.

——. 1920. "Star Clusters and the Structure of the Universe. Third Part." *Scientia* 27: 93–101.

——. 1923a. "The Galactic System." *Popular Astronomy* 31: 316–28.

——. 1923b. "Note on the Distance of N.G.C. 6822." *Harvard College Observatory Bulletin*, no. 796 (December): 1–2.

——. 1924. "Notes on the Thermokinetics of Dolichoderine Ants." *Proceedings of the National Academy of Sciences* 10 (October): 436–39.

——. 1929. "Note on the Velocities and Magnitudes of External Galaxies." *Proceedings of the National Academy of Sciences* 7 (July 15): 565–70.

——. 1930a. "The Super-Galaxy Hypothesis." *Harvard College Observatory Circular*, no. 350: 1–12.

——. 1930b. *Flights from Chaos*. New York: McGraw-Hill.

——. 1930c. *Star Clusters*. New York: McGraw-Hill.

——. 1969. *Through Rugged Ways to the Stars*. New York: Charles Scribner's Sons.

Shapley, H., and A. Ames. 1932. "A Survey of the External Galaxies Brighter Than the Thirteenth Magnitude." *Annals of the Astronomical Observatory of Harvard College* 88: 41–76.

Shapley, H., and H. D. Curtis. 1921. "The Scale of the Universe." *Bulletin of the National Research Council 2* (May): 171–217.

Sheehan, W., and D. E. Osterbrock. 2000. "Hale's 'Little Elf': The Mental Breakdowns of George Ellery Hale." *Journal for the History of Astronomy* 31: 93–114.

Shinn, C. H. c. 1890. "A Mountain Colony." *The Independent.*

Singh, S. 2005. *Big Bang.* London: Harper Perennial.

Slipher, V. M. 1913. "The Radial Velocity of the Andromeda Nebula." *Lowell Observatory Bulletin* 58, 2: 56–57.

———. 1915. "Spectrographic Observations of Nebulae." *Popular Astronomy* 23: 21–24.

———. 1917a. "The Spectrum and Velocity of the Nebula N. G. C. 1068 (M 77)." *Lowell Observatory Bulletin* 80, 3: 59–62.

———. 1917b. "Nebulae." *Proceedings of the American Philosophical Society* 56: 403–9.

———. 1921. "Dreyer Nebula No. 584 Inconceivably Distant." *New York Times,* January 19, p. 6.

Smart, W. M. 1924. "The Motions of Spiral Nebulae." *Monthly Notices of the Royal Astronomical Society* 84 (March): 333–53.

Smith, H. A. 2000. "Bailey, Shapley, and Variable Stars in Globular Clusters." *Journal for the History of Astronomy* 31: 185–201.

Smith, R. W. 1979. "The Origins of the Velocity-Distance Relation." *Journal for the History of Astronomy* 10: 133–65.

———. 1982. *The Expanding Universe.* Cambridge: Cambridge University Press.

———. 1983. "The Great Debate Revisited." *Sky & Telescope* 65 (January): 28–29.

———. 1990. "Edwin P. Hubble and the Transformation of Cosmology." *Physics Today* (April): 52–58.

———. 1994. "Red Shifts and Gold Medals." In *The Explorers of Mars Hill,* pp. 43–65. West Kennebunk, Maine: Phoenix Publishing.

———. 2006. "Beyond the Big Galaxy: The Structure of the Stellar System, 1900–1952." *Journal for the History of Astronomy* 37: 307–42.

Sponsel, A. 2002. "Constructing a 'Revolution in Science': The Campaign to Promote a Favourable Reception for the 1919 Solar Eclipse Experiments." *British Journal for the History of Science* 35 (December): 439–67.

Stebbins, J. 1950. Address at the Dedication of the Heber Doust Curst Memorial Telescope, University of Michigan, June 24, 1950.

"Stranger Than Fiction." 1929. *Los Angeles Times,* November 10, p. F4.

Stratton, F. J. M. 1929. *Transactions of the International Astronomical Union,* vol. 3. Cambridge: Cambridge University Press.

———. 1933. "President's Speech on Presenting Gold Medal." *Monthly Notices of the Royal Astronomical Society* 93 (1933): 476–77.

Strauss, D. 1994. "Percival Lowell, W. H. Pickering and the Founding of the Lowell Observatory." *Annals of Science* 51: 37–58.

———. 2001. *Percival Lowell: The Culture and Science of a Boston Brahmin.* Cambridge, Mass.: Harvard University Press.

Streissguth, T. 2001. *The Roaring Twenties.* New York: Facts on File.

Struve, O. 1960. "A Historic Debate About the Universe." *Sky & Telescope* 19 (May): 398–401.

Struve, O., and V. Zebergs. 1962. *Astronomy of the 20th Century.* New York: MacMillan.

Sutton, R. 1928. "The New Heavens." *Los Angeles Times,* September 12, p. A4.

———. 1930. "Caltech Scientists Plan Reception of Einstein." *Los Angeles Times,* December 28, p. A1.

———. 1933a. "Where Astronomy Is Taking Us." *Los Angeles Times,* September 24, p. G12.

——. 1933b. "Astronomy Stars That Are Human." *Los Angeles Times*, December 3, p. 14.

Swedenborg, E. 1845. *The Principia; or, The First Principles of Natural Things, Being New Attempts Toward a Philosophical Explanation of the Elementary World*, trans. A. Clissold. London: W. Newbery.

"Thirty-Third Meeting." 1925. *Publications of the American Astronomical Society* 5: 245–47.

"Thirty-Third Meeting of the American Astronomical Society." 1925. *Popular Astronomy* 33: 158–68, 246–55, 292–305.

Trimble, V. 1995. "The 1920 Shapley-Curtis Discussion: Background, Issues, and Aftermath." *Publications of the Astronomical Society of the Pacific* 107 (December): 1133–44.

Trollope, F. 1949. *Domestic Manners of the Americans*, ed. D. Smalley. New York: Alfred A. Knopf.

Tucker, R. H. 1900. "Obituary Notice." *Astronomische Nachrichten* 153: 399.

Turner, H. H. 1911. "From an Oxford Note-Book." *Observatory* 34: 350–54.

"The Universe, Inc." 1926. *The Nation* (February 10): 133.

"Universe Multiplied a Thousand Times by Harvard Astronomer's Calculations." 1921. *New York Times*, May 31, p. 1.

Van Maanen, A. 1916. "Preliminary Evidence of Internal Motion in the Spiral Nebula Messier 101." *Astrophysical Journal* 44: 210–28.

——. 1921. "Internal Motion in the Spiral Nebula Messier 33." *Proceedings of the National Academy of Sciences* 7 (January 15): 1–5.

——. 1923. "Investigations on Proper Motion. Tenth Paper: Internal Motion in the Spiral Nebula Messier 33, N.G.C. 598." *Astrophysical Journal* 57: 264–78.

——. 1925. "Investigations on Proper Motion. Eleventh Paper: The Proper Motion of Messier 13 and Its Internal Motion." *Astrophysical Journal* 61: 130.

——. 1930. "Investigations on Proper Motion. Sixteenth Paper: The Proper Motion of Messier 51, N.G.C. 5194." *Contributions from the Mount Wilson Observatory*, no. 408: 311–14.

——. 1935. "Internal Motions in Spiral Nebulae." *Astrophysical Journal* 81: 336–37.

——. 1944. "The Photographic Determination of Stellar Parallaxes with the 60- and 100-Inch Reflectors: Nineteenth Series." *Astrophysical Journal* 100: 55–56.

Very, F. W. 1911. "Are the White Nebulae Galaxies?" *Astronomische Nachrichten* 189: 441–54.

Webb, S. 1999. *Measuring the Universe*. London: Springer.

"Welfare of World Depends on Science, Coolidge Declares." 1925. *Washington Post*, January 1, pp. 1, 9.

White, C. H. 1995. "Natural Law and National Science: The 'Star of Empire' in Manifest Destiny and the American Observatory Movement." *Prospects* 20: 119–60.

Whiting, S. F. 1915. "Lady Huggins." *Astrophysical Journal* 42 (July): 1–3.

Whitney, C. 1971. *The Discovery of Our Galaxy*. New York: Alfred A. Knopf.

Wirtz, C. 1922. "Einiges zur Statistik der Radialberwegungen von Spiralnebeln und Kugelsternhaufen." *Astronomische Nachrichten* 215 (June): 349–54.

——. 1924. "De Sitters Kosmologie und die Radialbewegungen der Spiralnebel." *Astronomische Nachrichten* 222 (October): 21–26.

Wolf, M. 1912. "Die Entfernung der Spiralnebel." *Astronomische Nachrichten* 190: 229–32.

Wright, H. 1966. *Explorer of the Universe*. New York: E. P. Dutton.

———. 2003. *James Lick's Monument.* Cambridge: Cambridge University Press.

Wright, H., J. N. Warnow, and C. Weiner. 1972. *The Legacy of George Ellery Hale.* Cambridge, Mass.: MIT Press.

Wright, T. 1750. *An Original Theory; or, New Hypothesis of the Universe.* London: H. Chapelle.

Young, C. A. 1891. *A Textbook of General Astronomy for Colleges and Scientific Schools.* Boston: Ginn & Company.

Zwicky, F. 1929a. "On the Red Shift of Spectral Lines Through Interstellar Space." *Physical Review* 33: 1077.

———. 1929b. "On the Red Shift of Spectral Lines Through Interstellar Space." *Proceedings of the National Academy of Sciences* 15 (October 15): 773–79.

Index

Page numbers in *italics* refer to illustrations.

A NOTE ON THE TYPE

The text of this book was set in Electra, a typeface designed by W. A. Dwiggins (1880–1956). This face cannot be classified as either modern or old style. It is not based on any historical model, nor does it echo any particular period or style. It avoids the extreme contrasts between thick and thin elements that mark most modern faces, and it attempts to give a feeling of fluidity, power, and speed.

Composed by North Market Street Graphics,
Lancaster, Pennsylvania

Printed and bound by Berryville Graphics,
Berryville, Virginia

Designed by Virginia Tan